T0230687

# Nutritional Aspects of Aging

## Volume II

Editor

**Linda H. Chen, Ph.D.**
Professor and Chairman
Department of Nutrition
and Food Science
University of Kentucky
Lexington, Kentucky

**CRC Press**
Taylor & Francis Group
Boca Raton London New York

CRC Press is an imprint of the
Taylor & Francis Group, an informa business

First published 1986 by CRC Press
Taylor & Francis Group
6000 Broken Sound Parkway NW, Suite 300
Boca Raton, FL 33487-2742

Reissued 2018 by CRC Press

© 1986 by CRC Press, Inc.
CRC Press is an imprint of Taylor & Francis Group, an Informa business

No claim to original U.S. Government works

This book contains information obtained from authentic and highly regarded sources. Reasonable efforts have been made to publish reliable data and information, but the author and publisher cannot assume responsibility for the validity of all materials or the consequences of their use. The authors and publishers have attempted to trace the copyright holders of all material reproduced in this publication and apologize to copyright holders if permission to publish in this form has not been obtained. If any copyright material has not been acknowledged please write and let us know so we may rectify in any future reprint.

Except as permitted under U.S. Copyright Law, no part of this book may be reprinted, reproduced, transmitted, or utilized in any form by any electronic, mechanical, or other means, now known or hereafter invented, including photocopying, microfilming, and recording, or in any information storage or retrieval system, without written permission from the publishers.

For permission to photocopy or use material electronically from this work, please access www.copyright.com (http://www.copyright.com/) or contact the Copyright Clearance Center, Inc. (CCC), 222 Rosewood Drive, Danvers, MA 01923, 978-750-8400. CCC is a not-for-profit organization that provides licenses and registration for a variety of users. For organizations that have been granted a photocopy license by the CCC, a separate system of payment has been arranged.

**Trademark Notice:** Product or corporate names may be trademarks or registered trademarks, and are used only for identification and explanation without intent to infringe.

**Library of Congress Cataloging in Publication Data**

Main entry under title:

Nutritional aspects of aging.

  Includes index and bibliography.
  1. Aging--Nutritional aspects.  2. Aged--Nutrition.
3. Nutritionally induced diseases. I. Chen, Linda H.
QP86.N853    1986    613.2'0880565    85-9720
ISBN 0-8493-5737-3 (v. 1)
ISBN 0-8493-5738-1 (v. 2)

A Library of Congress record exists under LC control number: 85009720

Publisher's Note
The publisher has gone to great lengths to ensure the quality of this reprint but points out that some imperfections in the original copies may be apparent.

Disclaimer
The publisher has made every effort to trace copyright holders and welcomes correspondence from those they have been unable to contact.

ISBN 13: 978-1-315-89604-5 (hbk)
ISBN 13: 978-1-351-07514-5 (ebk)

Visit the Taylor & Francis Web site at http://www.taylorandfrancis.com and the
CRC Press Web site at http://www.crcpress.com

# PREFACE

The phenomena of aging are so obvious that most of us assume that we know what the term means. However, one will be surprised to find that people have a hard time agreeing on an objective scientific definition of aging. The aging process involves progressive changes with age after maturity in various organs leading to decreased functional ability. It reflects changes in many molecular, cellular, and systemic processes that take place with time. Often, aging is associated with chronic diseases although aging itself is not a disease.

Proper nutrition throughout the life span is of importance in maintaining mental and physical health. We believe that nutrition in early and middle life may have an important impact in determining the rate of aging. Therefore, the study of nutrition and aging, especially the study of nutrition and life span, cannot exclude nutrition at early life and middle years. In addition, some nutrition-related diseases which occur more frequently in later years may be prevented by adequate nutrition or can be managed by nutrition intervention. Nutrition is one of the environmental factors influencing biological aging which can be controlled.

Although we should deal with biological age rather than chronological age in gerontology, it is not easy to define biological age due to the lack of clear measures of biological age. The elderly, or the aged, are generally accepted as people who are at or after retirement age which is usually 65. As far as the Title IIIb National Nutrition Program for Older Americans is concerned, people at age 60 or above are qualified to participate. The number and the proportion of the elderly population in the U.S. are increasing rapidly. The Bureau of the Census reported that in 1977 there were 23 million people over 65 years of age which was about 10.6% of the total population. It is predicted that by the year 2050 there may be about 30 million people or 20% of the total U.S. population over the age of 65. The life expectancy at birth in 1977 was 76.5 years for women and 68.7 years for men, whereas the life expectancy at birth in the year 2050 is predicted to be 81 years for women and 71.8 years for men. With increasing life expectancy and increasing size of the elderly population, the quality of later life is of concern. The elderly have special health requirements but their socioeconomic status often produces difficulties for them to meet these requirements; thus they are at increasing risk of health problems. Nutritional inadequacy and the role of diet in pathogenesis of diseases are important components in health care of the elderly. The health care of the elderly is not limited to the treatment of diseases but also includes the prevention of diseases. The need for information on nutrition and aging is increasing.

The first section in Volume I describes the fundamentals of nutrition and aging which include research strategies for the study of nutrition and aging. The nutritional modulation of the aging process which has provided a major breakthrough in the field of nutrition and longevity is also discussed. In the second section, factors affecting the nutritional status of the elderly are discussed. These include biomedical influences, and social and psychological aspects. Section 3 includes dietary characteristics of the elderly population and methods for the assessment of nutritional status. The nutritional status of the elderly with respect to individual nutrients as determined by dietary survey and by biochemical methods is described in Section 4. Section 4 also includes discussion on nutrient metabolism, requirements, nutritional imbalances, and deficiencies of nutrients. Energy metabolism and obesity as a factor in pathogenesis of diseases are also discussed.

In the first section of Volume II, toxicological factors affecting nutritional status are discussed. Medications and alcohol may affect nutritional status. Section 2 provides a discussion of nutrition-related diseases which occur more frequently among the elderly. Cardiovascular diseases including coronary heart disease and cerebrovascular disease are the leading causes of death in the U.S. The relative importance of cardiovascular diseases, in terms of all deaths for the given age group, rises steadily with age. The death rate from these diseases is 28% for the middle age group (age 35 to 44), and is 69% for the old age group (age 75 and above). This reflects the continued progression of artherosclerosis with aging. Cancer is the second leading cause of death in the U.S. The death rate from cancer also rises steadily with age. The death rate from diabetes mellitus increases progressively with age and more rapidly after the age of 45. The incidence of diabetes mellitus is 0.23% under age 25 and 6.2% over age 45.

Osteoporosis is found in men and women over the age of 50. The incidence is especially high among women: about one third of women above the age of 60 have osteoporosis. Periodontal diseases are a major cause of tooth loss after the age of 35. The incidence in the U.S. is 24 to 69% among the age group of 19 to 26, and 100% above the age of 45.

Hypertension is found in persons of all ages. The incidence is 13 to 20% in the U.S. This disease is an important risk factor in the cardiovascular diseases. Diverticular disease is found in more than one third of the population over 60 years old in Western nations. Nutritional anemia including Fe deficiency anemia, megaloblastic anemia, and/or pernicious anemia due to deficiencies of Fe, folic acid, and/or vitamin $B_{12}$ are also found among the elderly. The etiology of anemia may also be secondary to diseases.

This book is designed to provide pertinent and useful information on the basic and applied aspects of nutrition and aging to health professionals. It may also be used as a reference book by researchers as well as professors and students for graduate courses at colleges and universities. It emphasizes recent advances and knowledge in the areas of basic, clinical, and community nutrition.

**Linda H. Chen, Ph.D.**

# THE EDITOR

**Linda Huang Chen, Ph.D.,** is currently Professor and Chairman of the Department of Nutrition and Food Science at the University of Kentucky.

Dr. Chen obtained her B.S. degree in Pharmacy in 1959 from the National Taiwan University in Taipei, Taiwan, Republic of China, and a Ph.D. in Biochemistry in 1964 from the University of Louisville. She received her postdoctoral training at the University of Louisville, School of Medicine, from 1964 to 1966. She served as an Assistant Professor from 1967 to 1972, an Associate Professor from 1972 to 1979, and has served as a Professor since 1979 at the University of Kentucky. It was in 1983 that she became the Chairman of the Department of Nutrition and Food Science.

Dr. Chen is a member of the American Institute of Nutrition, the American Society for Clinical Nutrition, the Gerontological Society of America, and the International Association of Vitamins and Nutritional Oncology. She served as a panelist for the National Program Standard in Gerontology for the Association for Gerontology in Higher Education, Washington, D.C. and the Gerontological Society of America. She also served a 4-year term as a member in the Cancer Research Manpower Review Committee of the National Cancer Institute. Her current research interests include vitamin interaction, nutrition and aging, nutrient-drug interaction, and nutrition and cancer.

# CONTRIBUTORS

**Thomas P. Almy, M.D.**
Distinguished Physician
Veterans Administration
Professor of Medicine and of Community
  and Family Medicine
Dartmouth Medical School
Hanover, New Hampshire

**Olav F. Alvares, B.D.S., Ph.D.**
Associate Professor
Department of Periodontics
The University of Texas Health Science
  Center
San Antonio, Texas

**Lynn B. Bailey, Ph.D.**
Associate Professor
Department of Food Science and Human
  Nutrition
University of Florida
Gainesville, Florida

**Charles H. Barrows, Sc.D.**
Chief
Laboratory of Nutritional Biochemistry
National Institute on Aging
Baltimore City Hospital
Baltimore, Maryland

**Rudy A. Bernard, Ph.D.**
Professor
Department of Physiology
Michigan State University
East Lansing, Michigan

**Judith Bond, Ph.D.**
Professor
Department of Biochemistry
Medical College of Virginia
Virginia Commonwealth University
Richmond, Virginia

**George A. Bray, M.D.**
Division of Diabetes and Clinical
  Nutrition
Department of Medicine
University of Southern California
School of Medicine
Los Angeles, California

**Toni M. Calasanti**
Instructor
Department of Sociology
University of Kentucky
Lexington, Kentucky

**Linda H. Chen, Ph.D.**
Professor and Chairman
Department of Nutrition and Food
  Science
University of Kentucky
Lexington, Kentucky

**Ching K. Chow, Ph.D.**
Professor
Department of Nutrition and Food
  Science
University of Kentucky
Lexington, Kentucky

**Harold H. Draper, Ph.D.**
Professor and Chairman
Department of Nutrition
College of Biological Science
University of Guelph
Guelph, Ontario, Canada

**Judy A. Driskell, Ph.D.**
Professor and Head
Department of Human Nutrition and
  Foods
Virginia Polytechnic Institute and State
  University
Blacksburg, Virginia

**Sandra E. Gibbs, Ph.D.**
Senior Associate
Human Factors Department
IBM Corporation
Lexington, Kentucky

**Janet L. Greger, Ph.D.**
Professor
Department of Nutritional Sciences
College of Agricultural and Life Sciences
University of Wisconsin
Madison, Wisconsin

Jon Hendricks, Ph.D.
Professor
Department of Sociology
University of Kentucky
Lexington, Kentucky

Kang-Jey Ho, M.D., Ph.D.
Department of Pathology
University of Alabama
Veterans Administration Medical Center
Birmingham, Alabama

Jeng M. Hsu, Ph.D., D.V.M.
Research Chemist
Medical Research
Veterans Administration Medical Center
Bay Pines, Florida

Glenville Jones, Ph.D.
Associate Professor
Departments of Biochemistry and
 Medicine
Queen's University
Kingston, Ontario, Canada

Nancy L. Keim, Ph.D.
Research Nutrition Scientist
USDA, ARS
Western Human Nutrition Center
Presidio of San Francisco, California

Gertrude C. Kokkonen, B.A.
Laboratory of Nutritional Biochemistry
Gerontology Research Center
National Institute on Aging
Baltimore City Hospital
Baltimore, Maryland

Calvin A. Lang, Sc.D.
Professor
Department of Biochemistry
University of Louisville
School of Medicine
Louisville, Kentucky

Loren G. Lipson, M.D.
Chief
Division of Geriatric Medicine
Associate Professor of Medicine and
 Gerontology
University of Southern California School
 of Medicine
Los Angeles, California

Fudeko Maruyama, Ph.D.
Extension Professor
Department of Nutrition and Food
 Science
University of Kentucky
Lexington, Kentucky

Malcolm J. McKay, Ph.D.
Department of Biochemistry
Medical College of Virginia
Virginia Commonwealth University
Richmond, Virginia

Jaime Miquel, Ph.D.
Guest Research Scientist
NASA Ames Research Center
Moffett Field, California

Betty Jane Mills, Ph.D.
Assistant Professor
Department of Biochemistry
University of Louisville
School of Medicine
Louisville, Kentucky

Michael J. Monzel, M.D.
Central Maine Medical Center
Lewiston, Maine

Eve Reaven, Ph.D.
Research Physiologist
Geriatric Research, Education, and
 Clinical Center
Veterans Administration Medical Center
Palo Alto, California

Gerald M. Reaven, M.D.
Nora Eccles Harrison Professor
Head, Division of Gerontology
Stanford University School of Medicine
Director, Geriatric Research, Education
 and Clinical Center
Veterans Administration Medical Center
Palo Alto, California

Lora Rikans, Ph.D.
Associate Professor
Department of Pharmacology
College of Medicine
University of Oklahoma Health Sciences
 Center
Oklahoma City, Oklahoma

**Howerde E. Sauberlich, Ph.D.**
Department of Nutrition Sciences
University of Alabama
Birmingham, Alabama

**Earl S. Shrago, M.D.**
Professor
Departments of Medicine and Nutritional
 Sciences
Clinical Nutrition Center
University of Wisconsin
Madison, Wisconsin

**Howard B. Turner, M.A.**
Research Associate
Department of Rural Sociology
University of Kentucky
Lexington, Kentucky

**Hans U. Weber, Ph.D.**
Manager
The Alexander Medical Foundation
San Carlos, California

**Aniece A. Yunice, Ph.D.**
Associate Professor
Department of Physiology and Biophysics
College of Medicine
Oklahoma University Health Sciences
 Center
Veterans Administration Medical Center
Oklahoma City, Oklahoma

# TABLE OF CONTENTS

## Volume I

Volume II

SECTION I: TOXICOLOGICAL FACTORS AFFECTING NUTRITIONAL STATUS

SECTION II: NUTRITIONAL BASIS FOR DISEASES IN AGING

*Section I*
*Toxicological Factors Affecting Nutritional Status*

Chapter 1

# DRUGS

**Lora E. Rikans**

## TABLE OF CONTENTS

## I. INTRODUCTION

Little information is available regarding the relationship between drugs and nutrition in the aged individual. Nevertheless, it is generally accepted that certain drugs affect the nutritional status of the elderly in an adverse fashion.[1,2] The likelihood of unwanted effects would seem greater in old people because they require more drugs and are more likely to have marginal nutrient intakes due to chronic illnesses and/or inadequate diets. Elderly patients on medications are generally the very ones who find it the most difficult to obtain and prepare adequate meals or to tolerate food prepared for them. The other side of the story is that nutritional status can have important effects on drug reactivity. Adverse drug reactions in the elderly may result, at least in part, from the effects of subclinical malnutrition on drug disposition.[3,4] Our information of drug/nutrient interactions in old age is based largely on clinical impressions and anecdotal observations; although it seems likely that the elderly are at greater risk for adverse effects from these interactions, there are no data from animal or human studies to support this impression.

## II. DRUG PATTERNS OF THE ELDERLY

### A. The Problem

Population statistics indicate that we now have more people aged 65 years and over than ever before, and that the aged form an increasingly larger proportion of our population.[5] For most people, health deteriorates with advancing years. Chronic conditions such as heart disease, arthritis, hypertension, and diabetes become prevalent and affect four of every five of the elderly. Of those affected, 35% have 3 or more such conditions.[5] It is not surprising that both the frequency of drug therapy and the number of drugs taken per person increase progressively with age.

Several studies agree that cardiovascular preparations (digitalis glycosides, diuretics, anti-arrhythmics, antihypertensives, and anticoagulants) and psychoactive drugs (antipsychotics, sedative-hypnotics) are the most commonly prescribed medications for elderly patients, while non-narcotic analgesics are the most commonly consumed nonprescription drugs. Laxatives, vitamins, and antacids are also widely used.[6-11] Some of these drug groups, namely the digitalis glycosides, diuretics, psychoactives, and laxatives, can profoundly influence the nutritional status of elderly patients. This aspect of drug toxicity will be discussed further in sections to follow.

Aging results in physiologic and pathologic changes which can profoundly affect the pharmacologic activity of drugs. The direction and magnitude of the changes in drug responsiveness are largely unpredictable, because a large number of factors influence drug activity and because there is a large degree of heterogeneity in the elderly population. Reports from surveillance systems which monitor drug toxicity during actual clinical use demonstrate a clear relation of age to untoward effects for certain sedative-hypnotic drugs and for the anticoagulant agent, heparin. A weak association of age to clinical toxicity has also been reported for the cardiac agents lidocaine and propranolol. For other drugs, age alone cannot be demonstrated to be the major determinant of toxicity.[7]

The elderly experience two to three times more adverse drug effects than middle-aged adults, partly because they take more drugs.[10] The widely used psychoactive and cardiovascular preparations are responsible for most of the serious drug reactions. In addition, many elderly patients take two to four different drugs daily and some take ten or more.[6,10-12] Studies have shown that the incidence of adverse drug reactions is directly related to the number of prescribed drugs.[13] Thus, the problem of increased drug toxicity in the elderly results from increased drug usage combined with age-associated changes in drug disposition.

resulting in a 40 to 50% reduction in total liver blood flow in elderly compared with young adult patients. Changes in hepatic blood flow are of major importance for drugs (e.g., propranolol) that are metabolized so rapidly that their rate of elimination is determined by the flow of blood to the liver.[23]

### d. Excretion

The kidney is the most important organ for the elimination of drugs and drug metabolites. Changes in renal function associated with aging have important implications for the clearance of many therapeutic agents. In fact, diminished renal function is probably the single factor most responsible for altered drug levels in the elderly. Renal excretion involves three processes: glomerular filtration, active tubular secretion, and passive tubular reabsorption; all these functions diminish with advancing years. Most important is the decline in glomerular filtration rate; a 35 to 50% reduction occurs between the ages of 20 and 70 years.

The earliest studies to demonstrate an effect of aging on drug elimination were concerned with antibiotics. The half-lives of penicillins, streptomycin, and tetracycline were substantially longer in elderly than in younger patients because of diminished renal function.[24] Reduced renal elimination in old age has also been found for digoxin, practotol, sulfamethiazole, phenobarbital, cimetidine, lithium, procainamide, and chlorpropamide. For drugs excreted by the kidney, total clearance declines in proportion to the reduction in glomerular filtration rate, so that appropriate adjustments in drug dosage regimens can be made on the basis of endogenous creatinine clearance.[25]

### 2. Tissue Sensitivity

Limited data indicate that age-related alterations in drug sensitivity due to changes in the drug receptor or the intrinsic properties of the responding tissue are an important consideration in geriatric pharmacology. For example, studies have shown that the elderly are more sensitive than the young to the central nervous system (CNS) depression produced by benzodiazepine sedative-hypnotics. One study found that a lower plasma level of diazepam was needed in older patients to achieve a given degree of CNS depression.[26] Another showed that CNS effects were greater in the elderly than in young patients at the same plasma level of nitrazepam.[27] Age-related changes in tissue sensitivity to the therapeutic effects of drugs have been demonstrated also for coumarin anticoagulants (warfarin) and for β-adrenergic agonists and antagonists (isoproteranol and propranolol). The mechanisms for age-related alterations in tissue responsiveness to drug actions could include changes in receptor number and/or affinity or in the translation of receptor binding into a response. Although a few studies have explored some of these possibilities (see reviews by Richey and Bender,[24] Crooks and Stevenson,[14] Ouslander,[6] Greenblatt et al.,[7] and Vestal[8]), the underlying reasons for age-associated changes in tissue sensitivity to drugs remain unclear.

The subject of altered tissue sensitivity to the toxic effects of drugs has received very little attention. In fact, the toxicology of old age has not been investigated systematically in either animals or humans. For example, evidence suggests that the incidence and severity of hepatic disease associated with isoniazid or dantrolene therapy increases with increasing age,[28,29] yet it is not known for certain if aging per se renders the elderly more susceptible to the hepatotoxic effects of these and other drugs.

Our group is presently involved in a study to define the influence of advancing age on the susceptibility of rats to liver injury from well-characterized hepatotoxicants. We found that aging had little or no effect on the damage produced by the administration of bromobenzene, carbon tetrachloride, or galactosamine. It was clear, however, that allyl alcohol-induced hepatotoxicity was more severe in older rats than in young ones. The extent of hepatocellular injury, as judged by light microscopy of liver sections, release of hepatic enzymes into the bloodstream, and destruction of the hepatic microsomal cytochrome $P_{450}$

## Table 1
## INFLUENCE OF AGING ON ALLYL ALCOHOL-INDUCED HEPATOTOXICITY IN RATS[a]

| | Age | | |
|---|---|---|---|
| | Young | Middle | Old |
| Extent of hepatocellular necrosis | 0 | +1 | +2 |
| Serum alanine aminotransferase (Rietman-Frankel U/m$\ell$) | 200 ± 50 | 2200 ± 500[b] | 5000 ± 1500[b] |
| Hepatic microsomal cytochrome P$_{450}$ | | | |
| nmol/g liver | 30.2 ± 1.4 | 23.4 ± 2.2[b] | 17.1 ± 1.5[b] |
| % of control | 92 | 70 | 62 |

[a]    Male Fischer rats 4 to 5, 14 to 15, and 24 to 25 months of age were give allyl alcohol (0.036 m$\ell$/kg, i.p.) 24 hr prior to sacrifice. Values are means ± SEM for 6 or more rats  Age-matched controls were given corn oil.[30]

[b]    Significantly different from young, $p < 0.05$

system, was much greater in the older rats (Table 1).[30] However, further work is needed to determine if this effect of aging on allyl alcohol hepatotoxicity represents a true increase in tissue sensitivity.

McMartin and Engel[31] recently investigated the effect of aging on gentamicin nephrotoxicity in rats. The study was designed to determine if the sensitivity of the kidney is increased as a function of age or if adverse effects arise from the higher drug levels that result from declining excretory function. By adjusting the doses of gentamicin to equalize renal exposure to the drug, they were able to demonstrate that renal tubular damage was dose related and was greater in old rats than in young ones at each blood concentration studied.[31] The implications of these results for geriatric medicine are obvious; increased tissue sensitivity probably contributes significantly to the drug toxicity problems of old age.

## III. DRUG EFFECTS ON NUTRITIONAL STATUS

### A. General Principles
There are several means by which pharmacological agents can produce nutritional deficiencies. Drug-induced nutrient depletion often occurs because of decreased absorption, increased elimination, or accelerated metabolism of nutrients. Well-known examples are the inhibition of GI absorption of fat-soluble vitamins by cholestyramine, the increased loss of K, Mg, and Zn due to inhibition of renal tubular function by thiazide diuretics, and the acceleration of vitamin D metabolism by the anticonvulsant drugs, phenytoin, phenobarbital, and primidone. Nutritional status can also be affected by drugs that depress appetite and food consumption, as when the nausea and vomiting associated with cancer chemotherapy cause an aversion to food. Finally, there are certain drugs that act as vitamin antagonists. The coumarin anticoagulants, for example, exert their pharmacological effects by antagonizing the actions of vitamin K; methotrexate is a folic acid antagonist.

Whether or not clinical malnutrition ensues as the result of drug administration depends on a number of factors, the most important of which are the adequacy of the individual's diet and the duration of medication. The nutritional hazards in geriatric patients are greatest in those already suffering from subclinical nutritional deficiencies and those for whom long-term drug regimens are necessitated by chronic disease. The problem may be exacerbated when more than one of the drugs being taken induces vitamin or mineral depletion. For example, chronic therapy with cholestyramine plus phenytoin would have additive effects

**Table 2**
## COMMONLY USED DRUGS WHICH CAN EXACERBATE NUTRITIONAL DEFICIENCIES IN THE ELDERLY[a]

| Drugs | Nutrients | Comment |
|---|---|---|
| Drugs affecting appetite and food intake | | |
| Antipsychotics and sedative-hypnotics | Protein/cal, other | Produce somnolence, disinterest in food |
| Digitalis glycosides | Protein/cal, other | Produce marked anorexia, nausea, and vomiting |
| Cancer chemotherapeutics | Protein/cal, other | Produce nausea and vomiting, aversion to food |
| Drugs affecting absorption or excretion of nutrients | | |
| Laxatives and cathartics[b] | Vitamins A and D, carotene, K | Problems stem from abuse |
| Diuretics | K, Mg, Zn | Produce more complications in the elderly |
| Corticosteroids | Vitamin D, Ca | |
| Cholestyramine | Vitamins A, D, K, and $B_{12}$ Folate | |
| Vitamin antagonists | | |
| Isoniazid, hydralazine, levodopa | Vitamin $B_6$ | Vitamin $B_6$ should be administered with isoniazid in elderly patients |
| Anticonvulsants | Vitamin D | Abnormal vitamin D metabolism |
| Anticoagulants | Vitamin K | Drug activity depends on vitamin antagonism; elderly are more sensitive |
| Methotrexate | Folate | Drug activity depends on vitamin antagonism |
| Other | | |
| Salicylates | Fe | Iron loss due to bleeding |

[a]  Information taken from References 1, 2, 32 to 35, and 93
[b]  Adverse effects of mineral oil have been questioned

on vitamin D and Ca status; osteomalacia would be much more likely to result from the combination of the drugs than from therapy with either drug alone. The use and misuse of nonprescription drugs such as laxatives, antacids, or analgesics can further complicate the picture. More complete and detailed discussions of interactions between drugs and nutrients can be found in the excellent reviews by Roe[32-34] and Matsui and Rozovski.[35]

Table 2 contains selected examples of commonly used drugs which, by virtue of their chronic use (or abuse), have the potential to precipitate nutritional deficiencies in the elderly. Some of these adverse drug/nutrient effects have special importance for geriatric medicine for reasons outlined in the following paragraphs.

## B. Specific Drug Groups
### 1. Digoxin

Digoxin is a drug that is very familiar to the geriatric practitioner. It is commonly prescribed for congestive heart failure and other cardiac disorders. However, the drug has a narrow margin of safety, and toxicity is commonly observed in the elderly.[13,36,37] Increased digoxin toxicity in elderly patients is due to a marked reduction in the rate of its renal clearance.[38-40] The most serious toxic effects are those that involve the heart, i.e., disturbances in cardiac rhythm. Cardiac toxicity is enhanced by hypokalemia, a condition which often occurs from the concurrent administration of diuretics that decrease body stores of K. This effect can be very dramatic in older patients who are predisposed to electrolyte imbalance from diuretic administration and possibly have inadequate intakes of dietary K. However, it is the noncardiac symptoms of digitalis toxicity that are harmful to the nutritional status of the elderly patient. Fatigue is very common, occurring in 95% of cases of digitalis intoxication, while anorexia, nausea, and vomiting occur in 50 to 80% of the cases.[41] These symptoms may produce marked reductions in food intake and cachexia.[42] Digoxin toxicity should be suspected in any digitalized patient who exhibits significant weight loss. Because of the well-known problems with digoxin toxicity in the aged, it probably is wise to monitor blood levels of digoxin and to reevaluate dosage schedules at frequent intervals.[43]

## 2. Diuretics

Diuretics are prescribed for the treatment of congestive heart failure and for hypertension. Unfortunately, geriatric patients appear to be more susceptible to the complications of diuretic therapy than young patients.[4] Reduced dietary potassium and a propensity to orthostatic hypotension may predispose the elderly to diuretic toxicity. A major complication is hypokalemia, with associated muscle weakness, electrocardiographic changes, and digitalis toxicity. Attention should be paid to the dietary intake of potassium in all elderly patients on diuretics or other potassium-losing drugs, and supplementation with K should be considered to correct or prevent the disturbances attributable to hypokalemia. Mg and Zn depletion may also result from the use of diuretics.

## 3. Psychoactive Drugs

Psychotropic drugs, especially sedative-hypnotics and antipsychotics, are the medications most frequently prescribed for aged patients in skilled nursing facilities.[13,36,44,45] As a result, the reactions to antipsychotic drugs account for about one third of the adverse drug responses in extended care settings.[36] An age-related increase in sensitivity to the neurologic side effects of phenothiazines, the most widely used antipsychotic agents, has been recognized for many years.[46,47] Ayd[46,47] reported in 1960 that the incidence of extrapyramidal reactions increased with age up to 70 years, and that one of the extrapyramidal syndromes, Parkinsonism, attained its peak incidence at 80 years of age. The mechanism responsible for the increased sensitivity is thought to be an age-related loss of CNS neurons.[48] More recently, information on benzodiazepine sedative-hypnotics has shown that these drugs also produce more toxicity in aged persons than in younger ones. Data from several surveillance programs now demonstrate that aging is responsible for an increase in untoward effects from chlordiazepoxide, diazepam, flurazepam, and nitrazepam.[7] Although some age-related changes in pharmacokinetics have been observed, data from pharmacodynamic studies demonstrate an increased sensitivity of the aging brain to the depressant effects of these drugs.[17,26,27] Thus, the elderly are more sensitive to the tranquilizing and sedative actions of both benzodiazepines and phenothiazines. Increased somnolence from these drugs may produce a disinterest in eating with ensuing malnutrition.[33]

## 4. Laxatives and Cathartics

Laxatives and cathartics pose a threat to the nutritional well being of the elderly, mainly because these popular medications are often misused.[1,9] Mineral oil and phenophthalein use can result in malabsorption of carotene and fat-soluble vitamins. With mineral oil, the nutrients presumably are dissolved in the oil and consequently become excreted. However, the importance of the malabsorption produced by mineral oil has been questioned.[49] In order for significant depletion to occur, the oil would need to be fed with meals and in quantities sufficient to dissolve most of the dietary supply of the fat-soluble nutrients. With phenophthalein and other cathartics, nutrients are lost because of intestinal hurry.[50-53] Anecdotal reports indicate that factitious diarrhea from long-term phenophthalein ingestion can lead to

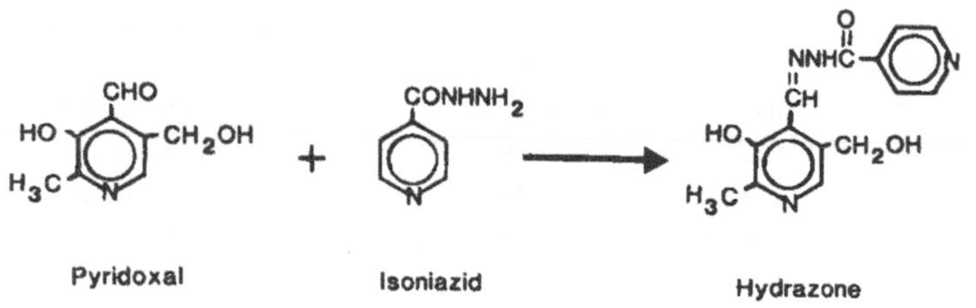

FIGURE 3.    Inactivation of pyridoxal by isoniazid.

osteomalacia or protein-losing gastroenteropathy.[53,54] Diarrhea produced by the ingestion of excessive amounts of cathartics can also result in K deficiency, dehydration, and the loss of other essential electrolytes.[55]

## 5. Isoniazid

One of the most interesting examples of interrelations between aging, nutritional status, and drug toxicity involves the drug isoniazid. This agent, considered to be the primary drug for the treatment of tuberculosis, impairs the actions of vitamin $B_6$. The mechanism involves the formation of a hydrazone of pyridoxal and pyridoxal phosphate, rendering them unavailable for enzymatic reactions (Figure 3). Since the hydrazone is readily excreted in the urine, isoniazid also increases the excretion of pyridoxine. As a consequence, the concentration of the vitamin in the plasma falls and the clinical picture resembles that of vitamin $B_6$ deficiency.[56] Other vitamin $B_6$ antagonists include the antihypertensive drug, hydralazine, and the anti-Parkinsonism drug, levodopa. The administration of pyridoxine prevents the neuropathologic changes associated with the toxicity of these drugs. However, the therapeutic efficacy of levodopa is reduced by supplementation with the vitamin.

Hepatic injury is another important untoward effect of isoniazid therapy. Although the mechanisms responsible for this toxicity are not known for certain, studies suggest that acetylhydrazine, a metabolite of isoniazid, is responsible.[56,57] Interestingly, the incidence of isoniazid-induced hepatic disease increases with increasing age.[28,58] The age-related increase in hepatotoxicity is probably not related to increased acetylhydrazine production because aging does not influence the rate of isoniazid acetylation.[59] On the other hand, it is well known that serum pyridoxal concentrations decrease with age throughout adult life.[60] Mitchell's group[61] has proposed, on the basis of studies in rats, that pyridoxine supplementation may provide protection against isoniazid-induced hepatitis. Thus, the available information suggests that isoniazid-induced hepatotoxicity is increased in elderly patients because they are deficient in vitamin $B_6$. Supplementation with vitamin $B_6$ may be especially important for geriatric patients who receive isoniazid.

## 6. Corticosteroids

The importance of dietary Ca in preventing age-related bone loss in man is controversial. Dietary protein, gonadal status, and plasma vitamin D levels also contribute to Ca status and are important in maintaining bone density.[62,63] The risk of osteoporosis is increased by any factor that inhibits the absorption of Ca, promotes the excretion of Ca, or interferes with the metabolism of vitamin D into its active forms. Such factors include the anticonvulsant drugs, phenytoin, phenobarbital, and primidone, and glucocorticoid preparations, such as cortisol, prednisone, or dexamethasone.

Prolonged therapy with corticosteroids produces disabling and potentially lethal effects in patients of all ages, but certain groups are especially vulnerable. Elderly, debilitated, or poorly nourished patients are particularly sensitive to the protein-wasting effects, while postmenopausal women are the most likely to develop osteoporosis and vertebral compression fractures.[64,65] Glucocorticoids inhibit bone formation by a direct effect on osteoblasts. They also inhibit Ca absorption by the intestine, which results in an increase in parathyroid hormone secretion and a stimulation of bone resorption by osteoclasts.[66] The mechanism of steroid hormone inhibition of Ca absorption involves effects on the metabolism of vitamin D as well as a direct effect on Ca transport.[67,68] Glucocorticoid-induced osteoporosis responds to large doses of vitamin D or small doses of vitamin D metabolite.[62]

## IV. NUTRITIONAL INFLUENCES ON DRUG REACTIVITY IN THE ELDERLY

Studies with adult animals and humans indicate that both pharmacokinetics and tissue sensitivity to drugs are affected by dietary factors. Although it seems reasonable to assume that nutritional influences may contribute substantially to the altered drug responsiveness of old age, experimental evidence in support of this hypothesis is scanty. In light of the large variation exhibited by the elderly in response to administered drugs, any information on the association between nutritional status and drug reactivity in the elderly would be of benefit.

### A. Pharmacokinetics
#### 1. Absorption
Most drugs are administered by the oral route. GI absorption of drugs is affected by the presence of food, but for most drugs the rate of entry into the systemic circulation is of little importance.[69] With regard to the influence of dietary factors on drug absorption in the elderly, age-related differences in nutritional status must be relatively unimportant since all the available information indicates that GI absorption of drugs is unaffected by old age.[70]

#### 2. Distribution
Age-related changes in drug distribution occur because of increased proportions of body fat and decreased concentrations of serum albumin. Both may be related to dietary intake, the former from overnutrition and the latter from undernutrition. The altered drug distribution probably has minimal effects on drug toxicity in the elderly. Although it has been suggested that decreased binding to plasma proteins could increase the number of adverse drug reactions that occur from the displacement of a bound drug during multiple drug therapy,[71] evidence for this effect has not been reported.

#### 3. Metabolism
Extensive animal research demonstrates that nutritional factors are important determinants of drug metabolism, especially biotransformation catalyzed by the hepatic cytochrome $P_{450}$-dependent system. Campbell and Hayes[72] have published an extensive review of the subject. Hepatic drug metabolism in animals is affected by macronutrient ratios (carbohydrates, lipid, lipotropes, and protein), vitamin deficiencies (vitamins A, C, E, niacin, riboflavin, and thiamin), and mineral deficiencies (Ca, Mg, Fe, I, Zn, Se, and Cu). Of these numerous influences, only the following have been found to have clinical significance in man: protein-calorie malnutrition, long-term ascorbic acid deficiency, changes in carbohydrate/protein ratio, and induction by indole-containing vegetables or charcoal-broiled meat.

The effect of malnutrition on drug metabolism has been demonstrated in studies of the disposition of drugs in infants and children with protein-energy malnutrition.[73,74] A recent study showed that the elimination of four drugs that undergo biotransformation was markedly decreased in malnourished children in relation to their age-matched controls. Nutritional rehabilitation restored the values to control levels.[73] These and other similar findings[74] indicate that humans suffering from undernutrition may experience adverse drug reactions because the rate of biotransformation of a drug is substantially diminished. It seems rea-

**Table 3**
## ASCORBIC ACID CONTENT IN HEPATIC
## MEMBRANES OF MATURE AND OLD RATS

|  | Mature | Old |
|---|---|---|
| Age (months) | 5—7 | 23—25 |
| Membrane-bound ascorbate[a] (mg/100 g wet wt liver) | 4.66 ± 0 32(4)[b] | 3 24 ± 0 24(3)[c] |
| Microsomal ascorbate[d] (mg/100 g wet wt liver) | 0.89 ± 0 06(9) | 0 53 ± 0.06(6)[c] |

[a] Assayed according to Summerwell and Sealock[94] in livers of female albino rats at 30 or 99 weeks of age. (From Patnaik[94] )

[b] Means ± SEM. Number of rats is given in parentheses

[c] Significantly different from value for mature rats, $p < 0.02$.

[d] Assayed according to Zannoni et al.[95] in washed microsomes from livers of male Fischer 344 rats at 5 or 25 months of age. (From Rikans, L. E. and Notley, B. A , Unpublished data.)

sonable that subclinical protein-calorie malnutrition in the aged may be a significant cause of increased drug toxicity.

Vitamin C deficiency has been shown to decrease the oxidative metabolism of a variety of drugs by guinea pig liver,[75,76] but data obtained in humans are contradictory. Human experiments which showed no effect of ascorbic acid deficiency were short-term studies,[77-79] and, although vitamin C concentrations in plasma or blood cells were decreased dramatically by the short-term deprivations, hepatic concentrations of the vitamin may not have been sufficiently depleted to affect the microsomal drug-metabolism system. Several lines of evidence suggest that ascorbic acid status in plasma does not necessarily reflect its status in other tissues.[80,81] Studies which did show an effect of vitamin C deficiency on hepatic drug metabolism were investigations in patients with preexisting, long-term deficiency and involved specific groups, the aged,[82] and patients with mild maturity-onset diabetes.[83] Supplementation with vitamin C restored drug elimination rates to normal levels.[82,83]

The precise role of ascorbic acid in the hepatic microsomal drug-metabolism system is poorly defined. One idea is that the vitamin is intimately associated with cytochrome $P_{450}$ and is needed for efficient interaction with NADPH-cytochrome $c$ ($P_{450}$) reductase.[84,85] Interestingly, both total membrane-bound and microsomal ascorbic acid concentrations are diminished in old rats (Table 3), rats in which hepatic cytochrome $P_{450}$ contents are also diminished.[21] One hesitates to extrapolate this data to the human situation because rats synthesize their own ascorbic acid while humans do not. Nevertheless, it is possible that aging liver has an increased requirement for the vitamin. Chronic marginal vitamin C deficiency may be more significant in the elderly than in the young and should be viewed, perhaps, as an additional risk factor which predisposes the elderly to toxic drug effects.

The protein/carbohydrate balance of calorically adequate diets produces consistent effects on hepatic drug biotransformation in experimental animals; microsomal drug metabolism activities are increased by feeding high-protein diets and decreased by feeding low-protein or high-carbohydrate diets. Pharmacokinetic studies in healthy male volunteers show that these relationships hold true in humans. The biotransformation of antipyrine or theophylline is increased by placing healthy, adult subjects on a high-protein low-carbohydrate diet and decreased by feeding them a low-protein high-carbohydrate diet.[86] It is not known, however, if macronutrient imbalance affects drug toxicity in the aged.

A study of the effects of dietary protein levels on hepatic microsomal drug metabolism during aging in Syrian hamsters yielded variable results depending on sex and generation.[87]

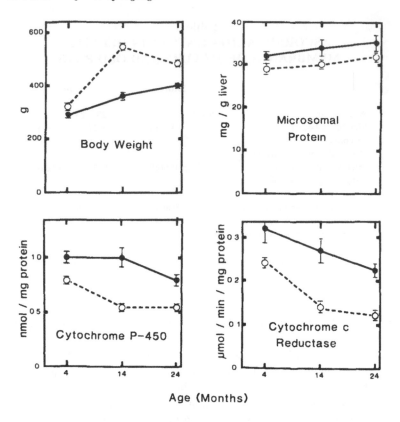

FIGURE 4.   Influence of diet on age-related changes in the hepatic microsomal cytochrome $P_{450}$ system. Male Fischer 344 rats were housed in our facilities under controlled environmental conditions for at least 1 month prior to the experiments. Values are means ± SEM for 7 or more rats fed commercial ration (-●—●-) vs. semipurified diet (-o—o-).[21,30]

Nevertheless, the investigators surmised that dietary protein levels modified both the magnitude and timing of age-related changes in microsomal drug-metabolism activities. Studies from this laboratory produced results consistent with the idea that dietary composition modifies the effects of aging on the hepatic microsomal enzyme system (Figure 4). Male Fischer 344 rats, obtained from the same source and maintained under identical conditions, were fed either a semipurified diet (AIN-76, ICN Nutritional Biochemicals) or a commercial ration (Purina). Feeding the commercial ration resulted in concentrations of hepatic microsomal enzymes that were 25% or more greater than those in the rats fed the semipurified diet. Furthermore, the age-related decreases in cytochrome $P_{450}$ content and cytochrome *c* reductase activity were smaller, and the decline occurred at a later age (Figure 4). These differences, due to a modulation by dietary factors of the magnitude and timing of aging effects, have important implications for future studies on the toxicology of aging.

Dietary constituents known to result in the induction of drug-metabolizing enzymes in humans include cabbage, brussel sprouts, and charcoal-broiled beef.[86] The consequences of induction for drug toxicity are complex; stimulation of biotransformation rates can increase production of toxic metabolites or enhance detoxification of the parent compound. The ability of the liver to respond to inducing drugs is modified in old age, both in animals[88,89] and in humans.[90] However, neither the effect of aging on inducibility by dietary constituents nor the effect of dietary influences on age-related changes in inducibility has been investigated.

## 4. Renal Excretion

Age-related studies of renal function in healthy human subjects indicate that glomerular filtration rates decline progressively after the 3rd decade, with values in the 8th decade that are one half to one third less than those measured in young adults. The loss of kidney function produces a marked decrease in the clearance of drugs eliminated primarily by renal excretion. In animals the age-dependent development of glomerular disease can be delayed by making food available only on alternate days or by limiting the amount eaten on an ad libitum schedule. In addition, the amount of dietary protein is important; glomerular sclerosis in rats subjected to renal ablation is enhanced by feeding high-protein diets and diminished by reducing dietary protein content.[91] On the basis of these and other findings, Brenner and colleagues[91] have proposed that modern eating habits (high-protein diets ad libitum) have imposed a functional burden on the human kidney and that the sustained hyperfiltration that results is ultimately detrimental to glomerular structure and function. It is intriguing to think that the progressive deterioration of kidney function might be minimized by changing our eating habits.

## B. Tissue Sensitivity

The issue of nutritional influences on age-related changes in tissue sensitivity to drugs is one that is largely unexplored. Isoniazid hepatotoxicity bears mentioning again; the age-associated increase in hepatic disease may be caused by a progressive worsening of pyridoxine status in elderly patients. Also, the effect of potassium deficiency on digitalis glycoside toxicity is due to an increase in the sensitivity of the heart to the arrhythmogenic properties of these drugs.

Foods yield protective substances, such as glutathione, that are capable of conjugating with toxic metabolites of drugs. Natural and synthetic antioxidants ($\alpha$-tocopherol, butylated hydroxytoluene, butylated hydroxyanisole) also play a protective role by scavenging free radicals produced by drugs.[92] It is remotely conceivable that age-associated changes in dietary patterns might affect the intake of these substances and influence drug toxicity.

## V. CONCLUSIONS

Aging, nutritional status and drug toxicity are related in an integral fashion. Although no conclusion is warranted yet concerning cause and effect relationships between malnutrition and the pathological changes that occur with advancing years, certain things are clear. Nutritional recommendations for the elderly should take into account the increased prevalence of chronic illnesses and the extensive use of therapeutic agents in older persons. Drug-induced nutritional deficiencies in geriatric patients should be avoided by supplementing their diets with the appropriate nutrients. Finally, there is a need to investigate the role of nutritional factors in modifying the age-associated alterations in drug effects.

## ACKNOWLEDGMENTS

The author's studies discussed in this chapter were supported in part by NIH grants AG01202 and AGO3642 from the National Institute on Aging. The author wishes to thank Ms. Annie Harjo for her expert secretarial assistance.

# REFERENCES

1. **Lamy, P. P.,** The food/drug connection in elderly patients, *Am Pharmacol* , NS18, 30, 1978.
2. **Roe, D. A.,** Nutrition and chronic drug administration. effects on the geriatric patient, *Am. Pharmacol.,* NS20, 273, 1980.
3. **Smithard, D. J. and Langman, M. J. S.,** Drug metabolism in the elderly, *Br. Med. J.,* 3, 520, 1977.
4. **Vestal, R. E.,** Drug use in the elderly: a review of problems and special considerations, *Drugs,* 16, 358, 1978
5. **Kovar, M. G.,** Health of the elderly and use of health services, *Public Health Rep.,* 92, 9, 1977.
6. **Ouslander, J. G.,** Drug therapy in the elderly, *Ann. Intern Med* , 95, 711, 1981.
7. **Greenblatt, D. J., Sellers, E. M., and Shader, R. I.,** Drug disposition in old age, *N. Engl. J. Med.,* 306, 1081, 1982.
8. **Vestal, R. F.,** Pharmacology and aging, *J Am. Geriatr. Soc* , 30, 191, 1982.
9. **Hale, W. E., Marks, R. G., and Stewart, R. B.,** Drug use in a geriatric population, *J. Am. Geriatr. Soc.,* 27, 374, 1979
10 **Basen, M. M.,** The elderly and drugs-problem overview and program strategy, *Public Health Rep* , 92, 43, 1977.
11 **Chen, L. H., Liu, S., Cook-Newell, M. E., and Barnes, K.,** Survey of drug use by the elderly and possible impact of drugs on nutritional status, *Drug-Nutr Interact* , 3, 73, 1985.
12 **Klein, L. E., German, P. S., and Levine, D. M.,** Adverse drug reactions among the elderly: a reassessment, *J. Am. Geriatr. Soc* , 29, 525, 1981.
13 **Williamson, J. and Chopin, J. M.,** Adverse reactions to prescribed drugs in the elderly: a multicentre investigation, *Age Ageing,* 9, 73, 1980.
14. **Crooks, J. and Stevenson, I. H.,** Eds., *Drugs and the Elderly,* University Park Press, Baltimore, 1979.
15. **Crooks, J. and Stevenson, I. H.,** Drug response in the elderly — sensitivity and pharmacokinetic considerations, *Age Ageing,* 10, 73, 1981
16. **Schmucker, D. L.,** Age-related changes in drug disposition, *Pharmacol Rev.,* 30, 445, 1979.
17. **Klotz, U., Avant, G. R., Hoyumpa, A., Schenker, S., and Wilkinson, G. R.,** The effects of age and liver disease on the disposition and elimination of diazepam in adult man, *J. Clin. Invest.,* 55, 347, 1975.
18. **Kato, R. and Takanaka, A.,** Metabolism of drugs in old rats I. Activities of NADPH-linked electron transport components and drug-metabolizing enzyme systems in liver microsomes of old rats, *Jpn. J. Pharmacol* , 18, 389, 1968.
19. **Kato, R. and Takanaka, A.,** Metabolism of drugs in old rats. II. Metabolism in vivo and effect of drugs in old rats, *Jpn. J. Pharmacol.,* 18, 389, 1968
20. **Rikans, L. E. and Notley, B. A.,** Decline in hepatic microsomal monooxygenase components in middle-aged Fischer 344 rats, *Exp. Gerontol.,* 16, 253, 1981.
21. **Rikans, L. E. and Notley, B. A.,** Age-related changes in hepatic microsomal drug metabolism are substrate selective, *J. Pharmacol Exp. Ther.,* 220, 574, 1982
22. **Gillette, J. R.,** Biotransformation of drugs during aging, *Fed Proc. Fed. Am. Soc. Exp. Biol.,* 38, 1900, 1979.
23 **Castleden, C. M. and George, C. F.,** The effect of ageing on the hepatic clearance of propranol, *Br. J. Clin Pharmacol.,* 7, 49, 1979
24. **Richey, D. P. and Bender, A. D.,** Pharmacokinetic consequences of aging, *Ann. Rev. Pharmacol. Toxicol.,* 17, 49, 1977.
25 **Kampmann, J. P. and Molholm-Hansen, J. E.,** Renal excretion of drugs, in *Drugs and the Elderly,* Crooks, J. and Stevenson, I. H., Eds., University Park Press, Baltimore, 1979, chap 8.
26 **Reidenberg, M. M., Levy, M., Warner, H., Coutinho, C. B., Schwartz, M. A., Yu, G., and Cheripko, J.,** Relationship between diazepam dose, plasma level, age, and central nervous system depression, *Clin. Pharmacol. Ther.,* 23, 371, 1978
27 **Castleden, C. M., George, C. F., Marcer, D., and Hallett, C.,** Increased sensitivity to nitrazepam in old age, *Br. Med. J.,* 1, 10, 1977.
28 **Black, M., Mitchell, J. R., Zimmerman, H. J., Ishak, K. G., and Epler, G. R.,** Isoniazid-associated hepatitis in 114 patients, *Gastroenterology,* 69, 289, 1975
29 **Utili, R., Boitnott, J. K., and Zimmerman, H. J.,** Dantrolene-associated hepatic injury: incidence and character, *Gastroenterology,* 72, 610, 1977.
30 **Rikans, L. E.,** The influence of aging on susceptibility of rats to hepatotoxic injury, *Toxicol. Appl Pharmacol* , 73, 243, 1984.
31. **McMartin, D. M. and Engel, S. G.,** Effect of aging on gentamicin nephrotoxicity and pharmacokinetics in rats, *Res Commun Chem. Pathol Pharmacol* , 38, 193, 1982
32. **Roe, D. A.,** Minireview: effects of drugs on nutrition, *Life Sci.,* 15, 1219, 1974.
33 **Roe, D. A.,** Interactions between drugs and nutrients, *Med. Clin. N. Am.,* 63, 985, 1979

34. **Roe, D. A.,** Dietary guidelines and drug therapies, *Comp Ther.,* 7, 62, 1981.

35. **Matsui, M. S. and Rozovski, S. J.,** Drug-nutrient interaction, *Clin. Ther.,* 4, 423, 1982.

36. **Cheung, A. and Kayne, R.,** An application of clinical pharmacy services in extended care facilities, *Calif Pharm.,* 23, 22, 1975.

37. **Baskin, S. I. and Kendrick, Z. V.,** Toxicity of digitalis in the aged, *Aging,* 6, 141, 1978

38. **Ewy, G. A., Kapadia, G. G., Yao, L., Lullin, M., and Marcus, F. I.,** Digoxin metabolism in the elderly, *Circulation,* 39, 449, 1969

39. **Chamberlain, D. A., White, R. J., Howard, M. R., and Smith, T. W.,** Plasma digoxin concentrations in patients with atrial fibrillation, *Br. Med. J ,* 3, 429, 1970.

40. **Falch, D.,** The influence of kidney function, body size and age on plasma concentration and urinary excretion of digoxin, *Acta Med. Scand.,* 194, 251, 1973.

41. **Lely, A. H. and van Enter, C. H. J.,** Non-cardiac symptoms of digitalis intoxication, *Am. Heart J.,* 83, 149, 1972.

42. **Banks, T. and Ali, N.,** Digitalis cachexia, *N. Engl. J. Med.,* 290, 746, 1974.

43. **Whiting, B., Wandless, I., Sumner, D. J., and Goldberg, A.,** Computer-assisted review of digoxin therapy in the elderly, *Br. Heart J.,* 40, 8, 1978

44. **Salzman, C., Shader, R. I., and van der Kolk, B. A.,** Clinical psychopharmacology and the elderly patient, *N.Y. State J. Med.,* 76, 71, 1976.

45. **Mitenko, P. A.,** Use of psychoactive drugs by the elderly, *Gerontology,* 28 (Suppl. 2), 5, 1982.

46. **Ayd, F. J.,** Trifluoperazine therapy for everyday psychiatric problems, *Curr. Ther. Res ,* 1, 17, 1959.

47. **Ayd, F. J.,** Tranquilizers and the ambulatory geriatric patient, *J. Am Geriatr. Soc.,* 8, 909, 1960.

48. **Hamilton, L. D.,** Aged brain and the phenothiazines, *Geriatrics,* 21, 131, 1966.

49. **Barrowman, J. A.,** Drug effects on fat-soluble vitamin absorption in, *Nutrition and Drug Interrelations,* Hathcock, J N. and Coon, J., Eds., Academic Press, New York, 1978, chap. 4.

50. **Smith, M. C. and Spector, H.,** Further evidence of the mode of action of vitamin D, *J. Nutr.,* 20, 197, 1940.

51. **Alexander, B., Lorenzen, E., Hoffmann, N. R., and Garfinkel, A.,** The effect of ingested mineral oil on plasma carotene and vitamin A, *Proc. Soc. Exp. Biol. Med.,* 65, 275, 1947

52. **Mahle, A. E. and Patton, H. M.,** Carotene and vitamin A metabolism in man: their excretion and plasma level as influenced by orally administered mineral oil and a hydrophilic mucilloid, *Gastroenterology,* 9, 44, 1947.

53. **Frame, B., Guiang, H. L., Frost, H. M., and Reynolds, W. A.,** Osteomalacia induced by laxative (phenophthalein) ingestion, *Arch. Intern Med ,* 128, 794, 1971.

54. **Heizer, W. D., Warshaw, A. L., Waldmann, T. A., and Laster, L.,** Protein-losing gastroenteropathy and malabsorption associated with factitious diarrhea, *Ann Intern. Med.,* 68, 839, 1968.

55. **Fingl, E.,** Laxative and cathartics, in *The Pharmacological Basis of Therapeutics,* 6th ed , Gilman, A G., Goodman, L S., and Gilman, A., Eds , Macmillan, New York, 1980, chap. 43.

56. **Mandell, G. L. and Sande, M. A.,** Antimicrobial agents: drugs used in the chemotherapy of tuberculosis and leprosy, in *The Pharmacological Basis of Therapeutics,* 6th ed., Gilman, A. G , Goodman, L S., and Gilman, A , Eds., Macmillan, New York, 1980, chap. 53.

57. **Whitehouse, L. W., Tryphonas, L., Paul, C. J., Solomonraj, G., Thomas, B. H., and Wong, L. T.,** Isoniazid-induced hepatic steatosis in rabbits: an explanation for susceptibility and its antagonism by pyridoxine hydrochloride, *Can J. Physiol. Pharmacol.,* 61, 478, 1983.

58. **Bailey, W. C., Weill, H., DeRouen, T. A., Ziskind, M. M., Jackson, H. A., and Greenberg, H. B.,** The effect of isoniazid on transaminase levels, *Ann. Inter. Med.,* 81, 200, 1974.

59. **Farah, F., Taylor, W., Dawlins, M. D., and James, O.,** Hepatic drug acetylation and oxidation: effects of aging in man, *Br. Med. J.,* 2, 155, 1977.

60. **Chen, L. H. and Fan-Chiang, W. L.,** Biochemical evaluation of riboflavin and vitamin $B_6$ status of institutionalized and non-institutionalized elderly in central Kentucky, *Int. J. Vitam. Nutr. Res ,* 51, 232, 1981.

61. **Lauterburg, B. H., Vaishnav, Y., and Mitchell, J. R.,** Rational approach to the prevention of isoniazid hepatitis, *Clin. Res.,* 27, 235A, 1979.

62. **Nordin, B. E. C.,** Calcium metabolism and bone, in *Metabolic and Nutritional Disorders in the Elderly,* Exton-Smith, A N. and Caird, F. I., Eds., John Wright & Sons, Bristol, 1980, chap. 11.

63. **Marcus, R.,** The relationship of dietary calcium to the maintenance of skeletal integrity in man — an interface of endocrinology and nutrition, *Metabolism,* 31, 93, 1982.

64. **Wylie, C. M.,** Hospitalization for fractures and bone loss in adults, *Public Health Rep.,* 92, 33, 1977.

65. **Windsor, A. C. M.,** Nutrition in the elderly, *Practitioner,* 222, 625, 1979.

66. **Haynes, R. C., Jr. and Murad, F.,** Adrenocorticotropic hormone; adrenocortical steroids and their synthetic analogs; inhibitors of adrenocortical steroid biosynthesis, in *The Pharmacological Basis of Therapeutics,* 6th ed., Gilman, A. G., Goodman, L S., and Gilman, A., Eds , Macmillan, New York, 1980, chap. 63

67. **Avioli, L. V., Berge, S. J., and Lee, S. W.,** Effects of prednisone on vitamin D metabolism in man, *J Clin. Endocrinol. Metab.,* 28, 1341, 1968.

68. **Kimberg, D. V.,** Effects of vitamin D and steroid hormone on the active transport of calcium by the intestine, *N. Engl. J Med.,* 280, 1396, 1969

69. **Melander, A.,** Food intake and drug bioavailability, *Prog. Clin. Biol. Res.,* 77, 747, 1981

70. **Stevenson, I. H., Salem, S. A. M., and Shepherd, A. M. M.,** Studies on drug absorption and metabolism in the elderly, in *Drugs and the Elderly,* Crooks, J. and Stevenson, I. H , Eds., University Park Press, Baltimore, 1979, chap 6.

71. **Wallace, S., Whiting, S., and Runcie, J.,** Factors affecting drug binding in plasma of elderly patients, *Br. J. Clin. Pharmacol.,* 3, 327, 1976.

72. **Campbell, T. C. and Haynes, J. R.,** Role of nutrition in the drug-metabolizing enzyme system, *Pharmacol. Rev.,* 26, 171, 1974.

73. **Mehta, S., Nain, C. K., Sharma, B., and Mathur, V. S.,** Drug metabolism in malnourished children, *Prog. Clin. Biol. Res.,* 77, 739, 1981.

74 **Monckeberg, F., Bravo, M., and Gonzalez, O.,** Drug metabolism and infantile undernutrition, in *Nutrition and Drug Interrelations,* Hathcock, J. N and Coon, J., Eds , Academic Press, New York, 1978, chap 14.

75. **Conney, A. H., Bray, G. A., Evans, C., and Burns, J. J.,** Metabolic interactions between L-ascorbic acid and drugs, *Ann. N.Y. Acad Sci.,* 92, 115, 1961.

76. **Zannoni, V. G., Sato, P. H., and Rikans, L. E.,** Ascorbic acid and drug metabolism, in *Nutrition and Drug Interrelations,* Hathcock, J. N. and Coon, J., Eds., Academic Press, New York, 1978, chap. 12.

77. **Trang, J. M., Blanchard, J., Conrad, K. A., and Harrison, G. G.,** The effect of vitamin C on the pharmacokinetics of caffeine in elderly men, *Am. J. Clin Nutr.,* 35, 487, 1982.

78 **Wilson, J. T., Van Boxtel, C. J., Alvan, G., and Sjoqvist, F.,** Failure of vitamin C to affect the pharmacokinetic profile of antipyrine in man, *J Clin Pharmacol.,* 16, 265, 1976.

79. **Holloway, D. E., Hutton, S. W., Peterson, F. J., and Duane, W. C.,** Lack of effect of subclinical ascorbic acid deficiency upon antipyrine clearance in man, *Am. J. Clin Nutr.,* 35, 917, 1982.

80 **Bates, C. J., Rutishauser, I. H. E., Black, A. E., Paul, A. A., Mandal, A. R., and Patnaik, B. K.,** Long-term vitamin status and dietary intake of healthy elderly subjects. II. Vitamin C, *Br. J. Nutr.,* 42, 43, 1977.

81. **Yew, M. S.,** Effect of streptozotocin diabetes on tissue ascorbic acid and dehydroascorbic acid, *Horm. Metab. Res.,* 15, 158, 1983

82. **Smithard, D. J. and Langman, M. J. S.,** The effect of vitamin supplementation upon antipyrine metabolism in the elderly, *Br. J. Clin. Pharmacol.,* 5, 181, 1978.

83. **Ginter, E. and Vejmolova, J.,** Vitamin C status and pharmacokinetic profile of antipyrine in man, *Br. J. Clin. Pharmacol ,* 12, 256, 1981.

84. **Sato, P. H. and Zannoni, V. G.,** Ascorbic acid and hepatic drug metabolism, *J. Pharmacol. Exp. Ther.,* 198, 295, 1976.

85. **Rikans, L. E.,** NADPH-dependent reduction of cytochrome P-450 in liver microsomes from vitamin C-deficient guinea pigs. effect of benzphetamine, *J. Nutr.,* 112, 1796, 1982.

86. **Alvarez, A. P., Pantuck, E. J., Anderson, K. E., Kappas, A., and Conney, A. H.,** Regulation of drug metabolism in man by environmental factors, *Drug Metab. Rev.,* 9, 185, 1979.

87. **Birt, D. F., Hruza, D. S., and Baker, P. Y.,** Effects of dietary protein level on hepatic microsomal mixed-function oxidase systems during aging in two generations of Syrian hamsters, *Toxicol. Appl Pharmacol.,* 68, 77, 1983.

88. **McMartin, D. N., O'Connor, J. A., Jr., Fasco, M. J., and Kaminsky, L. S.,** Influence of aging and induction on rat liver and kidney microsomal mixed function oxidase systems, *Toxicol. Appl. Pharmacol ,* 54, 411, 1980.

89. **Rikans, L. E. and Notley, B. A.,** Differential effects of aging on hepatic microsomal monooxygenase induction by phenobarbital and β-naphthoflavone, *Biochem. Pharmacol.,* 31, 2339, 1982.

90. **Salem, S. A. M., Rajjayabun, P., Shepherd, A. M. M., and Stevenson, I. H.,** Reduced induction of drug metabolism in the elderly, *Age Ageing,* 7, 68, 1978

91 **Brenner, B. M., Meyer, T. W., and Hostetter, T. H.,** Dietary protein intake and the progressive nature of kidney disease, *N. Engl. J. Med.,* 307, 652, 1982

92. **Carr, C. J.,** Food and drug interactions, *Ann. Rev. Pharmacol. Toxicol.,* 22, 19, 1982.

93. **Gilman, A. G., Goodman, L. S., and Gilman, A., Eds.,** *The Pharmacological Basis of Therapeutics,* 6th ed., Macmillan, New York, 1980.

94. **Patnaik, B. K.,** Change in the bound ascorbic acid content of muscle and liver of rat in relation to age, *Nature (London),* 218, 393, 1968.

95. **Zannoni, V., Lynch, M., Goldstein, S., and Sato, P.,** A rapid micromethod for the determination of ascorbic acid in plasma and tissues, *Biochem. Med.,* 11, 41, 1974.

96. **Rikans, L. E. and Notley, B. A.,** Unpublished data.

Chapter 2

ALCOHOL

**Aniece A. Yunice and Jeng M. Hsu**

TABLE OF CONTENTS

## I. INTRODUCTION

The word "alcohol", which is used interchangeably with ethanol in this review, is derived from the Arabic word *alkuhul*, meaning the eye cosmetics made of powdered antimony; in time, it came to mean "essential spirit of wine".[1] It is a small molecule that has been described extensively by a wide variety of people ranging from romantic poets to the exploring scientists. It provides calories and hence is defined as an aliment which is a nonspecific compound[2] used by the organism for the production of energy. By this definition then, alcohol is not a nutrient since it does not have any of the properties used for growth, repair, and reproduction, albeit its sparing effect of major nutrients. Its use throughout history was based primarily on its tranquilizing and euphoric effects and this property alone has paradoxically caused much human misery and degradation by only a few who abused its use to the point of intoxication, leading them to a host of organ and cellular dysfunctions. Although the molecular structure of ethanol is very simple, it is well established now that aside from depriving us of certain vital nutrients, it is a toxic substance,[3] which in excess can lead to the death of the cell. Perhaps the most serious consequence of this phenomenon is the hepatic dysfunction, or what is commonly known as post-alcoholic cirrhosis.

The historical genesis of the ethanol molecule is unknown. It is known to have existed in many of the ancient civilizations of Egypt, Persia, the Near East, Rome, and Greece. Fermentation products of honey (mead), cereal (beer), and grape (wine) were well known among the Greek historians. However, archaeological findings of these civilizations allude only to the art of fermentation, but the chemistry of distillation did not come to the fore until the dawn of the Arab civilization at about the 8th century A.D. The Arabian alchemist, Jabir ibn Hayyam, known to the West as Geber, was probably the first to use distillation as a means of obtaining alcohol. The practical value of this distillation, however, was envisioned by Arnauld de Villeneuve at the University of Montpellier around the 13th century.[1] The European alchemists believed that alcohol was an antidote to senility and hence called it the "water of life".[4]

Nonetheless, it should be mentioned that the presence of the basic ingredients required for alcohol production, namely sugar, water, yeast, and moderate temperature have been with us since time immemorial, and hence its use by man began before recorded history. Ancient man considered alcoholic beverages as sacred liquid used during religious ceremonies, sometimes as useful medicine, and at other times as nutritious food.[5] Although the production of alcohol by the action of yeast on sugars appeared to have occurred long before man, it should be noted that small amounts are produced endogenously throughout man's life and in other mammalian organisms as well.[6,7]

Innovative approaches to the preparation of a variety of alcoholic beverages, now in existence all over the world, continued to be discovered and appraised up through the 19th century. Since then, the little molecule has been gaining high esteem as it has occupied the attention of notable scientists and physicians in trying to unravel the mystery of its physiological and biochemical action in man and animals. Although the pathways of ethanol oxidation and the manner in which it is disposed of in the body is well known, the mechanism of action is still highly debatable and far from clear.

The long history of the medical use of alcohol for the treatment and prevention of disease is still accepted today in the medical community. Beneficial effects of alcoholic beverages in the treatment of disease is, however, not well documented by scientific evidence in spite of the fact that some physicians still prescribe it for chronic diseases such as diabetes, arthritis, and coronary diseases, particularly as a sedative for geriatric patients. As a tranquilizer, alcohol in moderation has enjoyed a great reputation and still has a significant role in religious ceremonies. In excess, however, alcohol throughout history has been condemned as evil.[8]

Perhaps the oldest description on record of its evil was found 3000 years ago in Egypt: "Take not upon thyself to drink a jug of beer, thou speakest, and an unintelligible utterance issueth from thy mouth. If thou fallest down and thy limits break, there is none to hold out a hand to thee. Thy companions in drink stand up and say: 'away with this sot' and thou art like a little child".[1] Similar admonitions are expressed in other cultures, including the cultures of the Old (Leviticus 10:9)[9] and New Testaments (Ephesians 5:18).[10]

The historical events that led to the introduction of alcoholic beverages and the public sentiment accompanying its use and abuse in the U.S. is interesting. After it was brought to America in 1607, its use in excess was alarming. This was coupled by increased brewing and wine production in Virginia and Massachusetts Bay, which led some religious groups to advocate moderation in drinking, particularly among the Puritans, a voice that culminated in 1830 in the temperance movement,[11] followed by the intemperance movement, and finally to total abstinence. This led the country to total prohibition in 1920, thus giving birth to the 18th Amendment to the U.S. Constitution which prohibited the sale of all alcoholic beverages. During this period of prohibition (1920 to 1933), bootlegging of illicit liquor caused a variety of public and social problems. The emergence of the right to use alcoholic beverages after the Prohibition era ended brought about the dilemma of alcoholism and excessive drinking and all the public health problems it poses in the 20th century.[5]

Prior to 1860, physiologists believed that alcohol was completely oxidized in the body. Following this period, experimental research has shown that alcohol could be recovered in urine as well as expired air.[12] The modern study of alcohol metabolism did not start, however, until 1870, when a group of investigators came to the conclusion that the body can utilize alcohol as a foodstuff. Widmark's[4] micromethod for alcohol determination in 1922 was an important step toward the study of alcohol. In the following 20 years, the general facts about alcohol absorption, distribution, and elimination in the animal and the human body were well established.[13] Sequelae of alcohol-induced malnutrition are perhaps the most widely reviewed subject which encompasses a host of manifestations supported both by clinical and experimental evidence in both man and animals. We would be doing an injustice to the field if we attempted to cover the gamut of malnutrition in alcoholism, since it would take more than one review to do a good job. Therefore, we have limited this review to the interaction of three nutrients; Zn, ascorbic acid, and Mg in alcoholism and their relationship to the aging processes. Readers interested in exhaustive reviews of the malnutrition in alcoholism are referred to the excellent articles in References 14 to 18.

## II. CURRENT STATUS OF ETHANOL METABOLISM

Although the production and use of alcohol, henceforth referred to in this review as "ethanol", goes back to ancient history, much of our current scientific knowledge of the physiological, biochemical, pharmacological, and nutritional aspects of ethanol intoxication has its foundation in scientific advances made only in the last century. Control of fermentation and later, the discovery of distillation, led to the production of a wide variety of distilled spirits from different sources.[19]

### A. Factors Affecting Absorption, Removal, Degradation, and Measurement

Ethanol is absorbed into the bloodstream unchanged and undigested all along the GI tract. About 20% is rapidly absorbed from the stomach and 80% from the small intestine. Several factors have been ascribed to affect the rate of absorption: concentration, nature of alcoholic drink, permeability of the GI tract, rate of gastric emptying, presence of food in the stomach, and individual variations. About 90% of the absorbed ethanol is metabolized in the body, primarily in the liver, yielding 7.0 kcal/g on complete oxidation to carbon dioxide and water. The remainder is excreted unchanged in the urine, sweat, and expired air.[20,21]

It has previously been assumed that elimination of ethanol from the blood of animals and humans is independent of the blood concentration (zero-order kinetics), but recently this assumption has been questioned.[22]

Although the most commonly used method for the rate of ethanol metabolism is the rate of ethanol disappearance from the blood (β-60), or what is sometimes referred to as "Widmark gradient", it has been found to be linear only above 10 to 20 mg/dℓ and exponential below this level.[23] Other methods commonly used are total body ethanol concentrations and the assessment of complete oxidation of ethanol to $CO_2$ by measuring the rate of its production from an administered dose of $^{14}C$-labeled ethanol.[24]

It is generally recognized that the most important of the three enzyme systems responsible for ethanol oxidation is hepatic alcohol dehydrogenase (ADH), the major enzyme for ethanol metabolism in vivo. Other ethanol oxidative systems include the microsomal-ethanol oxidizing system (MEOS), the reduced nicotinamide adenine dinucleotide phosphate (NADPH)-cytochrome *c* reductase, and the peroxidase-xanthine oxidase-catalase systems.

## B. Pathways of Ethanol Metabolism
### 1. Alcohol Dehydrogenase (ADH)

This system is dependent and to a large extent influenced by the availability of $NAD^+$. The steps involved in ethanol oxidation are binding of ethanol and NAD to the enzyme, transfer of a hydrogen ion, and dissociation of the complex to give enzyme plus the reaction products,[25] thus:

$$Ethanol + NAD \rightarrow Acetaldehyde + NADH + H^+$$

The activity of this system is inhibited by several compounds, the best known of these being pyrazole, which is used to distinguish the ADH from the non-ADH-dependent systems.[26,27]

Human hepatic ADH has two functional atoms of Zn.[28] The active enzyme consists of similar polypeptide chains, each containing a reaction-SH group of cysteine residue.[29] Dissociation of the dimer and its recombination has led to the discovery of an atypical human hepatic ADH which consists of several enzymes (polymorphism).[30] This enzyme is nonspecific and can be found primarily in hepatic cytosol, although lesser amounts can be found in the gastric mucosa, kidney, and brain.[31]

Another distinguishing characteristic of this system is the ability of ethanol to alter the reduction-oxidation state of NAD-coupled substrates as reflected by the increased ratio of lactate to pyruvate.[32] Since the ratio of the oxidized NAD to the reduced form plays a major role in this system, it has been used to assess the rate of several other metabolite couples such as pyruvate-lactate and acetoacetaldehyde-β-hydroxybutyrate.[33] However, it is generally accepted now that the availability of NAD appears to be the most important factor involved in the regulation of ADH activity in vivo,[34] rather than the amount of ADH, as was once thought.

Factors other than ADH activity in vivo have been described by Larsen[35] such as extrahepatic metabolism, availability of NAD, altered physiological activity, human individuality, animal species, and different stages of growth. Recently, it has been found by Videla and Israel[36] that when the ADH system becomes limited, increased mitochondrial reoxidation of NADH and its subsequent removal can accelerate ethanol metabolism.

### 2. Microsomal-Ethanol Oxidizing System (MEOS) and its Relation to Ethanol Metabolism

Perhaps the most significant advance in our understanding of metabolic tolerance was the discovery by Lieber and DeCarli[37] of the microsomal ethanol oxidizing system, which plays a significant role in increasing ethanol metabolism. This system was identified first in the rat[38] and later in man.[39] It has been differentiated from ADH by its subcellular localization (microsomal vs. cytosolic), co-factor requirement (NADPH vs. NAD), pH optimum (7.4 vs. 9.0), and its insensitivity to pyrazole inhibition.[40,41] The steps involved in ethanol oxidation through this system are

$$\text{Ethanol} + \text{NADPH} + \text{H}^+ + \text{O}_2 \rightarrow \text{Acetaldehyde} + \text{NADP} + 2\text{H}_2\text{O}$$

Although MEOS is unique in its versatility and its ability to be induced by a variety of agents, its contribution to accelerated ethanol metabolism after chronic consumption has not been completely understood.

Recent evidence suggests that under normal conditions, and when ethanol concentration is low, ADH appears to be the only active ethanol-oxidizing enzyme. However, when ethanol concentration is high (above 10 m$M/\ell$), ADH accounts for over 60% of ethanol oxidation.[42] This has been demonstrated in both rats and baboons. In rats administered ethanol chronically and whose ADH and catalase activities had been inhibited by pyrazole and sodium azide, respectively, the remaining ethanol-oxidizing activity was found to be higher in ethanol-fed rats than control animals.[40] More recently, rates of ethanol elimination were measured in baboons fed alcohol for 2 to 7 years and their controls, using constant ethanol infusion. Results suggest that a non-ADH system contributes significantly to ethanol elimination in alcohol-fed baboons.[43]

### 3. The Catalase System in Alcohol Metabolism

The steps involved in the reaction of this system depend largely on the production of hydrogen peroxide, which is generated by either xanthine oxidase or NADPH oxidase systems, thus:

$$\text{Ethanol} + \text{H}_2\text{O}_2 \rightarrow \text{Acetaldehyde} + 2\text{H}_2\text{O}$$

Inhibition of this system by sodium azide distinguishes it from the other two systems. Although the contribution of this system may be very small under normal conditions, its activation can be triggered during chronic ethanol intoxication and under a variety of pathophysiological conditions, particularly under conditions conducive to the production of hydrogen peroxide.[37] It should be mentioned, however, that species differences in the activity of these enzymes in various organs are not clear.

Other factors have also been described to contribute to the enhancement of ethanol disposition during chronic ethanol intoxication. Physiological, pharmacological, or dietary conditions may be influenced to a great extent either by regeneration of NAD (such as that of fructose action), by direct oxidation of the ADH-NADH system, or by metabolic adaptation.[44,45]

Adaptational responses in the development of compensatory mechanisms have also been described in chronic administration of ethanol vs. acute ethanol effects. For example, where acute ethanol administration causes inhibition of ATPase activity (Na$^+$-K$^+$), chronic mode of action results in compensatory overactivity.[46] Also, whereas acute administration of ethanol in doses up to 0.5 g/kg produces depression of norepinephrine uptake, chronic intoxication causes significant increases.[47] On the other hand, Ca$^{2+}$ uptake, which is increased only slightly with acute ethanol toxicity, is increased markedly with chronic dosages.[48]

All the reactions described above lead to the formation of acetaldehyde after ethanol is oxidized. Acetaldehyde then is oxidized to acetate via a different enzyme, aldehyde dehydrogenase, using NAD as a cofactor:

$$\text{Acetaldehyde} + \text{NAD} + \text{H}_2\text{O} \rightarrow \text{Acetate} + \text{NAD} + \text{H}^+$$

The final disposition of the acetate molecule into $CO_2$ and $H_2O$ is via the tricarboxylic acid cycle. It should be mentioned here that several compounds have been described to inhibit aldehyde dehydrogenase, thus leading to the accumulation of acetaldehyde in the blood with subsequent aversion to ethanol. The most widely used compound for this purpose is Antabuse® (disulfiram).

## III. ALCOHOL EFFECTS AND THE AGE FACTOR

It has been known for some time that a moderate intake of alcohol by the elderly is beneficial and is oftentimes recommended not only for its tranquilizing characteristics, but for therapeutic modalities as well.[49] Among the ancient records, history reveals the use of wine as an antidote and as a balm to boredom in advancing years;[50] even in Biblical records Paul advised his Christian brethren to take a little wine for their stomachs (New Testament, I Timothy 5:23).[10] Advocates of the idea that a moderate intake of alcohol can relieve the aged of the stresses and strains of life have emphasized the use of alcohol in certain patients to alleviate their anxiety or boredom. Among them are such authors as Leake and Silverman,[51] Chafetz,[52] and Lolli.[53]

Well-controlled studies, however, were lacking until several investigators from a variety of disciplines set out to prove on a scientifically controlled basis that alcohol in moderate amounts exerts a favorable physical as well as psychological response. Physical response included a variety of parameters such as obesity, decreased incontinence, and blood pressure. Psychological parameters included sleeplessness, behavior modification, outlook on life, and outright relaxation. Studies conducted by Burrill et al.[54] in two groups of subjects ranging in age between 36 and 60 years with chronic brain syndrome did not demonstrate any social responsiveness in the experimental group that received one bottle of beer (equivalent to 12.5 g ethanol) vs. the control group that received soft drinks. Although the subjects in this study were confined to a wheelchair, they can hardly be considered elderly. However, work by Mishara et al.[55] demonstrated a beneficial effect of alcohol in two different elderly populations with two different environmental settings. Beverages containing alcohol in an amount of 0.4 fl oz were given to volunteer subjects 5 days a week for 18 weeks. The control group received nonalcoholic drinks for the first 9 weeks and was then switched to alcoholic drinks for another 9 weeks. Improvements in cognitive performance, morale, and sleep pattern were more frequent in the experimental group than control subjects.

Studies of obese patients conducted by Lolli[56] after control periods in which subjects were given 10 g of alcohol (as wine) before, during, and after dinner for 5 weeks showed a loss of weight; some others had experienced less fatigue following wine drinking.

In another study that was conducted in a double-blind cross-over fashion by Kastenbaum and Slater,[57] increased social interaction was observed in the group receiving wine after 1 month of observation. The same group[57] demonstrated a significant change of attitude and improvement in urinary incontinence after 2 months of drinking 12.6 g ethanol as beer each day. Damrau et al.[58] reported an improvement of sleep patterns among 28 men and 22 women aged 35 to 85 years after receiving 18.6 g of ethanol in the form of a bottle of stout before retiring to bed. Thus, these studies, as well as others,[59-61] support the contention that the moderate intake of alcohol in the elderly can make their lives more pleasant, their personality more stable and alert, and their outlook on life more optimistic. It must be borne in mind, however, that these studies were conducted with very moderate amounts of alcohol. But how about the consequences of large amounts of alcohol in the elderly? It is reasonable to assume, considering the slow metabolic status of the aged, that aging, coupled with excess ethanol, can be devastating to the nutritional status of the elderly as it will act synergistically to deprive the individual of essential nutrients as we shall see in the next section.

## IV. ROLE OF NUTRIENTS IN ETHANOL INTOXICATION

Although there has been considerable debate regarding the role of nutritional deficiencies in the pathogenesis of alcoholism, many workers in this field believe that inadequate diet or imbalance between supply and need for nutrients influences the sequelae of ethanol intoxication. Early studies by Best et al.[62] suggest that the malnutrition associated with alcoholism, rather than ethanol per se, was the cause of liver disease. However, recent studies in the baboon fed ethanol indicate that normal nutrition along with ethanol may not spare the alcoholic from developing cirrhosis,[63] suggesting that ethanol has a direct toxic effect. This, however, has been questioned by Thomson,[15] who postulated that the diet may not contain adequate levels of all the nutrients or, alternatively, that nutrients are not made available for utilization at the subcellular level. Although the biochemical mechanisms of ethanol toxicity have not been adequately defined, most observers believe that an imbalance between supply and demand induces nutritional deficiencies which will eventually lead to impairment of gluconeogenesis, reduced albumin and transferrin synthesis, increased lipoprotein synthesis, and decreased fatty acid oxidation.[64-66]

Nutritional deficiencies have been well documented by a large number of studies. Dietary inadequacies are often exaggerated by malabsorption of thiamin, folic acid, vitamin $B_{12}$, methionine, Ca, and Mg.[67-70] Since injury to most cells of the body reduces their vitamin storage capacities to such an extent as to impair their ability to alter these vitamins into metabolically useful forms, the increased need for DNA/RNA synthesis falls short of protecting the remaining cells to produce new hepatocytes. Failure to properly utilize the absorbed nutrients includes diminished ability to synthesize albumin and to regulate carbohydrate and lipid metabolism.

It is well established now that excessive intake of alcohol can lead to deficiencies of vitamins and minerals.[14,15,71,72] Alcoholism is now recognized as the major cause of malnutrition in the world.[73,74] Several workers in the field have demonstrated in clinical as well as experimental settings deficiencies of vitamins[75-77] and minerals[78-82] in chronic alcoholism.

Clinically, malnutrition in the alcoholic can be of significant importance when combined with the toxic properties of ethanol and this fact may be responsible for many of the ailments seen in alcoholics. Deficiencies of vitamins and minerals have been implicated in the etiology of Wernicke-Korsakoff syndrome,[83] cerebral syndrome,[84] myopathy,[85] polyneuritis,[86,87] and post-alcoholic cirrhosis.[78]

Alcohol may influence nutrient interaction in a variety of ways. Its toxic properties in pharmacological doses and the direct result of its metabolism, particularly acetaldehyde, can have an additive effect. Additionally, the interplay of directly displacing essential nutrients, or slowing or preventing absorption of nutrients from the GI tract can contribute considerably to body depletion of nutrients. It has previously been assumed that deficiency of essential nutrients was the major culprit for the pathological sequelae of alcoholism.[88] However, recent evidence suggests that ethanol has a direct toxic effect of its own on tissues.[89-91] Chronic administration of alcohol in humans and primates resulted in hepatic steatosis in spite of a nutritionally adequate diet.[91] The clinical implications of these findings are, of course, obvious. A patient cannot guard himself against ethanol toxicity by simply supplementing his diet with vitamins and minerals,[14] although this may mitigate the consequences of the deficiency.

### A. Vitamins in General

Vitamins (particularly Group B), sugars, amino acids, and triiodothyronine have been observed to accelerate the removal of ethanol from the blood.[92-95] The effects of ascorbic acid in large doses and/or trace metals have been only sparsely alluded to in the literature. Pawan,[95] studying the effects of vitamins and sugars on the rate of ethanol metabolism in humans, has demonstrated that only fructose produces a significant increase in ethanol removal. None of the vitamins he tested, including 600 mg ascorbic acid, changed the slope of ethanol levels.

Most of the vitamins are affected to some extent by alcohol. This subject has recently been reviewed in a series of articles by Bonjour.[96-99] First, it is believed that large intake can produce a nutritional imbalance, particularly if the amount of alcohol consumed exceeds two thirds of the caloric intake.[90] Second, inadequate diet can be exasperated by existing malabsorption from the GI tract, such as is the case with vitamin $B_1$ (thiamin), $B_{12}$ (cobalamine), or folic acid,[15] Third, in the presence of toxic levels of ethanol, pathological alteration of hepatocyte and its injury can lead either to the release of essential vitamins[100] or the failure of the hepatocyte to convert the vitamins to their useful forms.

Relatively little is known about vitamin A deficiency in the alcoholic. It is known, however, that ethanol inhibits retinol formation in vitro,[101] but whether chronic ingestion of alcohol leads to chronic deficiency of activated vitamin A is not known.[102] Some have suggested an altered synthesis or release of retinol-binding protein in which the role of Zn deficiency has been implicated.[103] However, the observation of night blindness[104] and dysfunction of the germinal cells in both man and animals support this contention.[105] In a recent study,[106] 129 patients with chronic alcoholism were assessed for their nutritional status. Vitamin A and β-carotene levels in alcoholics and patients with Korsakoff's psychosis were lower than control, but the difference was not significant.

The possibility exists that ethanol per se could inhibit absorption of vitamins or, alternatively, other nutrient deficiencies caused by ethanol intake could trigger malabsorption syndromes such as that found in decreased vitamin $B_1$ absorption during vitamin $B_6$ deficiency.[107] There is evidence that alcohol reduces the active form of vitamin $B_6$.[108] Recent reports suggest that deficiency of vitamin $B_1$ (thiamin) or $B_6$ (pyridoxine), or both, is implicated in neurological disorders which may be reversible.[109] Supplementing alcoholic beverages with the vitamin prevented this disorder according to Centerwall and Crigui.[110]

Clinical evidence in humans given ethanol (1.5 g/kg) either parenterally or orally led to a 50% reduction of thiamin absorption in 4 of 12 healthy subjects.[111] However, other studies in which ethanol was administered intravenously had no effect on $B_1$ absorption.[112] Vitamin $B_2$ (riboflavin) status in the alcoholic is associated with other vitamin deficiencies.[75] Alcoholism causes deficiencies of other vitamins of the B family: folic acid and $B_{12}$ (cobalamine). This can be precipitated by malabsorption, decreased utilization, increased requirement, or increased destruction. Alcoholics with malnutrition have been found to have lower levels of serum folate than those adequately nourished.[113] Tissue uptake of the vitamin is also found to be reduced during ethanol ingestion. With regards to vitamin $B_{12}$, the reports are conflicting.[114 115] Most observers, however, estimate that 5% of alcoholics have a low level of vitamin $B_{12}$. A recent review by Thomson et al.[15] on the interrelationship of folic acid and vitamin $B_{12}$ discusses in greater detail the status of these two vitamins in the alcoholic. Deficiency of vitamin C in alcoholics will be discussed in a separate section of this review.

With regards to vitamin D, it is known that cholecalciferol (vitamin $D_3$) reaches the liver from the diet or from precursors activated at the skin, which is then hydroxylated in the liver to the 25-OH compound. Further hydroxylation to 1,25-dihydrocholecalciferol occurs in the kidney[116] in which it becomes biologically active for Ca absorption from the gut. This may explain the reason why alcoholics with liver cirrhosis have lowered Ca absorption, as demonstrated by the work of Dechavanne et al.[117]

Perhaps susceptibility of chronic alcoholics to vitamin E (α-tocopherol) deficiency is due to reduced intake as well as intestinal malabsorption of a variety of nutrients. Whether the relationship between vitamin E deficiency and testicular lesion on one hand and the hypogonadism described in alcoholics and vitamin E status on the other are interrelated is not revealed in the literature. It is known that vitamin E prevents peroxidation and that its deficiency reduces testosterone levels, but whether alcohol intake has any effect on this vitamin is unknown.[105]

Although vitamin E has been known for some time to play a vital role in lipid oxidation processes, its function in ethanol-induced hepatic injury remains controversial.[118-124] A protective effect of vitamin E against acute ethanol toxicity has not been demonstrated, nor has it been shown whether it has any influence on blood ethanol clearance in animals. If indeed vitamin E is a potent antioxidant, then the possibility that it might have a protective effect against ethanol toxicity should be considered.

## B. Minerals in General

The two most notable cations that are depleted in chronic alcoholism are Zn and Mg, which will be discussed in separate sections due to their clinical implications in a variety of alcoholic diseases.[5] All the bulk metals (Na, K, Ca, and Mg) and the known essential trace metals (Zn, Cu, Fe, and P) are affected to some degree by excessive intake of alcohol. This could result from malabsorption, malnutrition, or dysfunction of the renal tubules.[81,82] Malabsorption is caused by syndromes secondary to disorders of the liver, pancreas, or the small intestine. Malnutrition can be provoked by increased requirements for nutrients during alcoholism or alternatively decreased utilization such as occurs in delirium tremens, where cerebral oxidation is diminished due to deficiencies of thiamin pyrophosphate, or of Mg, which is needed for the action of thiamin.[125,126] Increased excretion via the renal tubule is another mechanism of nutrient loss, particularly in the presence of impaired liver function superimposed on kidney dysfunction. In both cases nutrient storage or utilization is impaired.

It may be mentioned in this section that plasma Ca, P, Fe, and K are reduced during alcoholism. Kalbfleisch et al.[68] have demonstrated more than twofold urinary increases in Ca excretion with a concomitant decrease in serum. The interrelation between Ca and Mg during ethanol intoxication and withdrawal has been discussed recently by Harris.[29]

Mg deficiency as well as alcoholic hypocalcemia may contribute significantly to the hypophosphatemia seen in a large number of hospitalized alcoholic patients.[127] Other preexisting conditions may also contribute to this deficiency, such as vomiting and diarrhea. Potassium depletion in the alcoholic is affected in a similar manner as Ca and Mg.

The relationship between Fe and folic acid in alcoholics, particularly in post-alcoholic cirrhosis, has been established by McDonald et al.[128] Rats fed diets high in Fe deposited more Fe in the liver in the presence of folic acid and/or choline deficiency which could be reversed by supplementation. It is also known that some wines contain large amounts of Fe which could contribute to its accumulation in the liver[129] and that alcohol increases Fe absorption from the GI tract.[130] In support of these findings, Miralles-Garcia et al.[131] found that alcoholics who drank 2 $\ell$ of wine daily for 15 years had higher hepatic Fe levels and a higher grade of siderosis than controls. Similar findings have been reported by Cosnett.[132]

## V. MAGNESIUM AND CALCIUM AS AFFECTED BY ETHANOL

Mg and Ca, which are known to be depleted in chronic alcoholics and adversely affected during acute alcohol intoxication, are two cations that play a major role in the body due to their ability to alter electrical properties of tissues and their regulation of many enzyme activities. It is known that smooth muscle contraction is stimulated by cellular calcium influx, which may involve enhancement of intracellular enzymes as well as altered membrane properties. Both cations have been reported to alter fluidity of cellular membrane,[133] which is also adversely affected by ethanol.[47] A number of vital enzyme activities are controlled by Ca and Mg. The adenosine triphosphatases are responsible for the release of chemical energy stored as ATP as well as the activating mechanism of the transport system across the cellular membrane. In addition to that, they are involved in the liberation of neurotransmitters and are necessary for the synthesis of protein RNA and DNA.[134] Additional reviews on the subject can be found in these excellent articles.[135-137]

The first clinical evidence of magnesium deficiency in alcoholics undergoing withdrawal symptoms was reported by Flink et al.[138] Subsequent work by the same authors provided further evidence of Mg as well as Ca[69] depletion in human alcoholics. Several workers have confirmed these findings and have further identified the clinical manifestations displayed by alcoholic patients during withdrawal from alcohol such as tetany, tremor, convulsion, and, in extreme cases, delirium tremens. In both alcoholic and nonalcoholic subjects, administration of 0.34 g/kg of ethanol was found by Kalbfleish et al.[68] to increase urinary Ca significantly. However, other investigators, using higher doses in humans, did not observe this increase in Ca.[70,139] In chronic studies of alcohol administration to humans given 4.4 g/kg/day for 17 days, Ca decreased in plasma with a concomitant decrease in urinary excretion.[70] In hospitalized alcoholics, several investigators observed diminished serum Ca levels and in several other studies this was coupled with hypomagnesemia.[140,141]

Hypocalcemia may indicate manifestation of intestinal malabsorption, pancreatitis, or malnutrition with hypoalbuminemia.[17] In order to explain the hypocalcemia seen in alcoholics, some have suggested that under conditions of acute and/or chronic pancreatitis as well as post-alcoholic cirrhosis, there is an increased secretion of corticosteroids, which in turn leads to body depletion of Ca.[142,143] A possible mechanism for the hypocalcemia in hypomagnesemia noted in most hospitalized alcoholics may be related to decreased secretion of parathormone. Mg has been suggested to be necessary for the secretion of parathormone in response to hypocalcemia.[141] Depletion of Ca and Mg during withdrawal symptoms has been observed in serum, in erythrocytes, and in skeletal muscles of alcoholics undergoing treatment as well as in experimental subjects.[144] The significance of these deficiencies in the etiology of the alcohol withdrawal syndrome is not clear. Whether reduced binding and/or transport of Ca in skeletal and cardiac muscles is directly related to cardiomyopathies seen in chronic alcoholics is not well established.

Although many studies have repeatedly demonstrated diminished circulating Mg concentrations, few studies have focused on altered tissue concentration, particularly in the heart, liver, and kidney. In Mg-deficient animals and in biopsies from the hearts of cirrhotic patients with cardiomyopathies, swelling of the mitochondria have been noted.[145] Delirium tremens, which occurs in severe cases of alcohol withdrawal, shares the same clinical manifestation as that of Mg deficiency. This led some to speculate that ethanol intoxication leads to the failure of the Na-K ATPase pump in which Mg deficiency is involved.[146] However, so far, Mg supplementation failed to reverse delirium tremens in alcoholics, suggesting that it is not caused by Mg deficiency[147] alone.

In dogs[148] and rats[149] given alcohol in doses of 3 and 6 g/kg orally, plasma Ca concentration decreased, but not to the same extent seen in humans. Daily administration of ethanol in doses of 3 g/kg to dogs for several days had no effect on plasma Ca levels in spite of increased tubular reabsorption in the kidney.[149] This is probably a compensatory response for the decreased Ca absorption from the GI tract as seen in rats ingesting alcohol chronically.[150] It is therefore concluded from these studies that acute ethanol administration in humans leads to increased excretion of both ions without significant change in plasma levels. In animals, ethanol intake increased urinary Mg excretion with a concomitantly slight increase in serum levels.

In order to compare the effect of ethanol and ascorbic acid on tissue cations (Na, K, Ca, Mg, Zn, Cu), we carried out a chronic study in male Sprague-Dawley rats (Charles River), 200 to 250 g body weight. They were randomized into four groups, seven rats per group. Animals were housed 2 to a cage in a temperature-controlled room and were left to acclimatize for 2 weeks before starting the treatment. All animals received rat chow ad libitum. The first control group received distilled water; the second group received 20% ethanol (v/v) in drinking water; the third group received ascorbic acid in a concentration of 100 mg/d$\ell$ of 20% ethanol; the fourth group received 100 mg/d$\ell$ ascorbic acid in distilled water. Fresh stock solutions were changed daily and the volume consumed was recorded to determine the total daily intake. After 10 weeks of treatment, each animal was transferred to a metabolic cage and a 24-hr urine was collected and measured.[151]

Results are summarized in Tables 1 to 5. With the exception of the brain, all tissues examined in the experiment seemed to be affected to some degree by this treatment. Following these findings, we raised the question as to whether the influence of ethanol on kidney metabolism was due to hemodynamic changes, such as altered glomerular filtration rate, or to a reduction in the reabsorptive capacity of the kidney.

To answer these questions we utilized our recently developed dog model to test the effect of intravenous infusion of various nutrients on renal handling of Zn.[152] Along with our saline control and ascorbic acid groups, we also used a third ethanol group.

Three groups of five or ten experiments each were performed as follows:

1.  Ten dogs were infused with 33% ethanol (v/v) in physiological saline at the rate of 3.3 m$\ell$/min per dog for 60 min (total = 51.0 g of absolute alcohol)
2.  Ten dogs, infused with an ascorbic acid solution of 50 mg/m$\ell$ physiological saline at the rate of 1.1 m$\ell$/min per dog for 60 min (total = 3300 mg)
3.  Five control dogs, infused with physiological saline throughout the experiment and subjected to the same surgical procedure as the test animals

The experiment was designed such that each dog served as its own control and was therefore divided into three periods: (1) a control period, before infusion of test material, (2) an infusion period, and (3) a post-infusion period, during which time infusion of test material was replaced with saline. These three periods were each divided into three 20-min subperiods during which samples of blood and urine were collected. [125]I Iodothalamate was used for the determination of glomerular filtration rate (GFR) and $p$-aminohippurate for the determination of effective renal blood flow (ERPF). A first blood sample (Bo) was obtained prior to the injection of primary doses of 0.1 m$\ell$ [125]I-iodothalamate (Abbott) and 20% $p$-aminohippurate solution (8 mg/kg, Merck, Sharp and Dohme) followed by a sustaining dose of 0.10 m$\ell$ [125]I-iodothalamate and 0.25 mg $p$-aminohippurate per kilogram per minute in isotonic saline infused at the rate of 0.05 m$\ell$/kg/min throughout the course of the experiment. Blood samples were drawn at the midpoint of each urine collection.

Serum and urinary Na, K, Ca, Mg, Zn, and Cu were determined on each sample by conventional methods.[125]I-Iodothalamate concentrations were quantified on a Nuclear-Chicago gamma counter and $p$-aminohippurate was measured by a colorimetric method adapted for a Technicon Auto Analyzer.

Each of the three periods of collections was averaged. The mean and standard errors were determined for each period for each treatment. Student's paired $t$-tests were used to compare data collected in the pre-infusion periods to determine the significance of differences observed. The control group was utilized to exclude any variable other than the treatment.

Data presented in Tables 6 and 7 and Figures 1 and 2 show that the mechanisms of action of ethanol and ascorbic acid on the renal handling of the cations examined are diametrically opposed to each other. While both showed increases in urinary flow, it was ascorbic acid that consistently displayed significant urinary excretions of Na, K, Ca, Mg, Zn, and Cu with concomitant increases in urinary osmolality at a time when plasma osmolality remained unchanged. Ethanol, on the other hand, did not affect urine excretion of cations, except for Zn and Cu, in spite of a steady increase in plasma osmolality. These divergent effects suggest that dissimilar transport mechanisms of the two agents across the renal tubular membrane are present, although both were similar with respect to their effect on GFR and ERPF.[153,154]

## Table 1
## ADRENALS[a]

|  | Saline control | ETOH | ASA[b] | ETOH & ASA |
|---|---|---|---|---|
| No. of rats | 7 | 6 | 6 | 7 |
| Dry weight (g) | 0.040 ± 0.006 | 0.025 ± 0 002 | 0.025 ± 0.003 | 0.024 ± 0.004 |
| Wet weight (g) | 0.051 ± 0.007 | 0.029 ± 0 003 | 0.031 ± 0.004 | 0.027 ± 0.004 |
| Dry/wet × 100 | 85.4 ± 4.3 | 84 2 ± 1.1 | 78.8 ± 4 5 | 91.8 ± 6.5 |
| (mg/g dry weight) |  |  |  |  |
| Sodium | 1.6 ± 0.2 | 2.1 ± 0.2 | 2.5 ± 0.4 | 2.1 ± 0.2 |
| Potassium | 5.1 ± 0.5 | 6.6 ± 0.6 | 8.5 ± 1 2[d] | 6 8 ± 0.9 |
| Calcium | 0.356 ± 0.050 | 0 543 ± 0.054[d] | 0.392 ± 0 060 | 0.367 ± 0.036 |
| Magnesium | 0.237 ± 0 024 | 0.327 ± 0.024[d] | 0.344 ± 0.036[d] | 0 320 ± 0 047 |
| (µg/g dry weight) |  |  |  |  |
| Zinc | 27.1 ± 3 0 | 37.8 ± 13.5 | 35.3 ± 3 9 | 30.9 ± 4 6 |
| Copper | 5 3 ± 0 6 | 6.2 ± 0.4 | 14.2 ± 2.8[d] | 6.1 ± 0 9 |

[a]   Means ± SEM.
[b]   ASA = Ascorbic acid.
[c]   $p < 0.01$.
[d]   $p < 0.05$.

## Table 2
## BRAIN[a]

|  | Saline control | ETOH | ASA[b] | ETOH & ASA |
|---|---|---|---|---|
| No. of rats | 7 | 6 | 6 | 7 |
| Dry weight (g) | 0.316 ± 0.022 | 0.318 ± 0 038 | 0.376 ± 0.007 | 0.374 ± 0.008 |
| Wet weight (g) | 1.249 ± 0.09 | 1 142 ± 0 19 | 1.496 ± 0.04 | 1.426 ± 0.06 |
| Dry/wet × 100 | 25.4 ± 0.8 | 30 1 ± 3 2 | 25.2 ± 0.7 | 26.4 ± 0.8 |
| (mg/g dry weight) |  |  |  |  |
| Sodium | 5.2 ± 0.1 | 5.0 ± 0.1 | 5.2 ± 0 05 | 5.1 ± 0 1 |
| Potassium | 17 3 ± 0.1 | 17 4 ± 0 2 | 18.5 ± 0.1 | 17.9 ± 0.2 |
| Calcium | 2.196 ± 1.023 | 2.172 ± 0.590 | 0.924 ± 0.243 | 4.480 ± 3.000 |
| Magnesium | 0.634 ± 0.050 | 0.668 ± 0.024 | 0 679 ± 0.010 | 0.723 ± 0.050 |
| (µg/g dry weight) |  |  |  |  |
| Zinc | 54.1 ± 2.3 | 54.5 ± 2.6 | 54 3 ± 3.0 | 59.7 ± 2.4 |
| Copper | 11 0 ± 0.2 | 12.2 ± 0.3 | 12.6 ± 0.2 | 11.4 ± 0.3 |

[a]   Means ± SEM.
[b]   ASA = Ascorbic acid.
[c]   $p < 0.01$.
[d]   $p < 0.05$.

## Table 3
## HEART[a]

|  | Saline control | ETOH | ASA[b] | ETOH & ASA |
|---|---|---|---|---|
| No. of rats | 7 | 6 | 6 | 7 |
| Dry weight | 0.41 ± 0.02 | 0.31 ± 0.02 | 0.37 ± 0.02 | 0.27 ± 0.02 |
| Wet weight | 1.42 ± 0.05 | 1.04 ± 0 10[d] | 1.31 ± 0.05 | 0.91 ± 0.08[d] |
| Dry/wet × 100 | 28.8 ± 0.7 | 30.0 ± 1.0 | 28.0 ± 0.6 | 30.8 ± 1.5 |
| (mg/g dry weight) |  |  |  |  |

## Table 3 (continued)
## HEART[a]

| | Saline control | ETOH | ASA[b] | ETOH & ASA |
|---|---|---|---|---|
| No. of rats | 7 | 6 | 6 | 7 |
| Sodium | 4 1 ± 0.2 | 4 2 ± 0 1 | 3 9 ± 0 1 | 4 1 ± 0.1 |
| Potassium | 12 8 ± 0.4 | 13 8 ± 0 2 | 12 5 ± 0.3 | 13 7 ± 0.2 |
| Calcium | 0.222 ± 0.007 | 0.269 ± 0.016[d] | 0 250 ± 0 025 | 0 217 ± 0.008 |
| Magnesium (μg/g dry weight) | 0.780 ± 0.026 | 0.848 ± 0 016[d] | 0 804 ± 0 022 | 0 814 ± 0 018 |
| Zinc | 63.6 ± 2.8 | 60 7 ± 1.0 | 59.8 ± 1.4 | 61.4 ± 2.0 |
| Copper | 18.9 ± 1 3 | 22.3 ± 0.3[d] | 21.2 ± 0.6 | 21.7 ± 0.6 |

[a]  Means ± SEM
[b]  ASA = Ascorbic acid.
[c]  $p < 0.01$.
[d]  $p < 0.05$.

## Table 4
## KIDNEYS[a]

| | Saline control | ETOH | ASA[b] | ETOH & ASA |
|---|---|---|---|---|
| No. of rats | 7 | 6 | 6 | 7 |
| Dry weight (g) | 0 84 ± 0.01 | 0.66 ± 0.03[c] | 0.87 ± 0.02 | 0.60 ± 0.03[c] |
| Wet weight (g) | 2.88 ± 0.05 | 2.22 ± 0 15 | 3.14 ± 0.06 | 2.08 ± 0 14 |
| Dry/wet × 100 (mg/g dry weight) | 29.0 ± 0.2 | 29.8 ± 1 2 | 27.7 ± 0.5[d] | 29.3 ± 1.1 |
| Sodium | 4.9 ± 0.2 | 4.9 ± 0.2 | 5.0 ± 0.1 | 5 0 ± 0 2 |
| Potassium | 10.9 ± 0 1 | 11.6 ± 0.2[d] | 10 7 ± 0.2 | 11 6 ± 0 1[c] |
| Calcium | 0.241 ± 0.007 | 0.235 ± 0.006 | 0 304 ± 0 022[d] | 0.213 ± 0.004 |
| Magnesium (μg/g dry weight) | 0 688 ± 0.013 | 0.789 ± 0.025 | 0.738 ± 0.012 | 0.710 ± 0.013 |
| Zinc | 87.1 ± 1.4 | 87 5 ± 2.3 | 23.7 ± 0.5[c] | 89.0 ± 2.0 |
| Copper | 33.9 ± 3 1 | 22.0 ± 1.2[c] | 32.9 ± 1.9 | 22.4 ± 1.0[c] |

[a]  Means ± SEM.
[b]  ASA = Ascorbic acid.
[c]  $p < 0.01$.
[d]  $p < 0 05$.

## Table 5
## LIVER[a]

| | Saline control | ETOH | ASA[b] | ETOH & ASA |
|---|---|---|---|---|
| No. of rats | 7 | 6 | 6 | 7 |
| Dry weight (g) | 0.447 ± 0.042 | 0.298 ± 0 008 | 0 436 ± 0.046 | 0.230 ± 0.022 |
| Wet weight (g) | 1.595 ± 0.218 | 0 942 ± 0.037 | 1 413 ± 0.155 | 0.754 ± 0.074 |
| Dry/wet × 100 (mg/g dry weight) | 29.2 ± 2.3 | 32 0 ± 0.8 | 31.0 ± 0.9 | 30.5 ± 0.3 |
| Sodium | 2.027 ± 0.058 | 2.155 ± 0.093 | 2.181 ± 0.125 | 2.319 ± 0 160 |
| Potassium | 10.9 ± 0.1 | 11.0 ± 0.2 | 11.1 ± 0.3 | 10.6 ± 0.3 |
| Calcium | 0.206 ± 0.011 | 0.241 ± 0.007 | 0 172 ± 0.007[d] | 0.236 ± 0.018 |

## Table 5 (continued)
## LIVER[a]

| | Saline control | ETOH | ASA[b] | ETOH & ASA |
|---|---|---|---|---|
| No. of rats | 7 | 6 | 6 | 7 |
| Magnesium (μg/g dry weight) | 0.336 ± 0.028 | 0 432 ± 0 017 | 0 382 ± 0.029[d] | 0.237 ± 0 024[c] |
| Zinc | 61.2 ± 5.4 | 65.0 ± 2 2 | 72.1 ± 4.9 | 36.3 ± 3 2[c] |
| Copper | 9.0 ± 0 4 | 9.2 ± 0.3 | 12 1 ± 1 4 | 6 1 ± 0.4[c] |

[a]   Means ± SEM
[b]   ASA = Ascorbic acid.
[c]   $p < 0.01$
[d]   $p < 0 05$.

## Table 6
## EFFECT OF INTRAVENOUS INFUSION OF ASCORBIC ACID (2.2 mg/kg/min) ON RENAL FUNCTION IN DOGS[a]

| | Pre-infusion | During infusion | t | Post-infusion | t |
|---|---|---|---|---|---|
| $U_{flow}$ (mℓ/min) | 0.17 ± 0.02 | 0.31 ± 0 03 | 8 5[b] | 0 33 ± 0.035 | 5.35[b] |
| MAP (mmHg) | 129 ± 4 | 129 ± 4 | 0.0 | 127 ± 5.0 | 1.20 |
| Hct (%) | 38 ± 2.7 | 38 ± 2 6 | 0.0 | 38 ± 2.7 | 0.14 |
| $C_1$ 125 (mℓ/min) | 69 ± 9 | 77 ± 10 | 2.14 | 67 ± 8 | 0.82 |
| $C_{PAH}$ (mℓ/min) | 175 ± 24 | 185 ± 19 | 0.4 | 172 ± 18 | 0.03 |
| $U_{Osm}$ (mOsm/kg $H_2O$) | 1680 ± 104 | 1825 ± 102 | 1.6 | 1945 ± 23 | 2.93[c] |
| $P_{Osm}$ (mOsm/kg $H_2O$) | 303 ± 3 | 305 ± 2 | 0 04 | 304 ± 3 | 0 05 |
| $U_{Na}$ V (μEq/min) | 16.5 ± 4.7 | 48 5 ± 11 2 | 4.51[b] | 60.9 ± 11.4 | 3 24[c] |
| $U_K$ V (μEq/min) | 27.6 ± 2.0 | 46.6 ± 6.2 | 3 40[c] | 54.9 ± 8.7 | 3.00[c] |
| $U_{Ca}$ V (μEq/min) | 1.25 ± 0.40 | 2 43 ± 0 74 | 3.25[c] | 2.09 ± 0.45 | 3.58[c] |
| $U_{Mg}$ V (μEq/min) | 3.03 ± 0 45 | 4 65 ± 0 60 | 5.40[b] | 4.64 ± 0.80 | 3 56[c] |
| $U_{Zn}$ V (nEq/min) | 2.03 ± 0.26 | 2.73 ± 0 26 | 5.85[b] | 2.55 ± 0 29 | 4.29[b] |
| $U_{Cu}$ V (nEq/min) | 1.43 ± 0.15 | 1.92 ± 0.14 | 4.52[b] | 1.84 ± 0 19 | 2 77[c] |

[a]   Values represent the means ± SEM of 3 20-min clearance periods, pre-infusion, during infusion, and post-infusion.
[b]   $p < 0.01$.
[c]   $p < 0 05$.

## Table 7
## EFFECT OF INTRAVENOUS INFUSION OF ETHANOL (34 mg/kg/min) ON RENAL FUNCTION IN DOGS[a]

| | Pre-infusion | During infusion | t | Post-infusion | t |
|---|---|---|---|---|---|
| $U_{flow}$ (mℓ/min) | 0.18 ± 0 01 | 0.21 ± 0 02 | 1.85 | 0 30 ± 0.05 | 2.23[b] |
| MAP (mmHg) | 135 ± 5 | 136 ± 6 | 0 64 | 135 ± 8 | 0.02 |
| Hct (%) | 41 ± 1 6 | 41 ± 1.6 | 0 0 | 41 ± 1.9 | 0.2 |
| $C_1$125 (mℓ/min) | 66 ± 7 | 65 ± 6 | 0.45 | 62 ± 7 | 1.0 |
| $C_{PAH}$ (mℓ/min) | 216 ± 24 | 225 ± 31 | 0.3 | 201 ± 38 | 0.3 |
| $U_{Osm}$ (mOsm/kg $H_2O$) | 1560 ± 285 | 1443 ± 262 | 1 20 | 1419 ± 254 | 1 20 |
| $P_{Osm}$ (mOsm/kg $H_2O$) | 313 ± 5 | 365 ± 12 | 4.0[c] | 378 ± 12 | 5.0[b] |
| $U_{Na}$ V (μEq/min) | 70.1 ± 2 60 | 58.3 ± 18 4 | 1.09 | 52.1 ± 13 6 | 0.72 |
| $U_K$ V (μEq/min) | 40.7 ± 3 1 | 40.0 ± 5 0 | 0.30 | 38.2 ± 5 2 | 0.50 |
| $U_{Ca}$ V (μEq/min) | 2.2 ± 0.6 | 2 4 ± 0 7 | 0.80 | 1 6 ± 0.4 | 2.13 |

**Table 7 (continued)**
## EFFECT OF INTRAVENOUS INFUSION OF ETHANOL (34 mg/kg/min) ON RENAL FUNCTION IN DOGS[a]

|  | Pre-infusion | During infusion | t | Post-infusion | t |
|---|---|---|---|---|---|
| $U_{Mg}$ V ($\mu$Eq/min) | 3.2 ± 0.8 | 3.7 ± 0 6 | 0.92 | 3.2 ± 0.6 | 0.06 |
| $U_{Zn}$ V (nEq/min) | 2.41 ± 0.37 | 3 54 ± 0.72 | 1 85 | 4.3 ± 0.92 | 2.34[c] |
| $U_{Cu}$ V (nEq/min) | 3 30 ± 0 70 | 3.61 ± 0 80 | 0.65 | 4 2 ± 0.90 | 3.70[b] |

[a]  Values represent the means ± SEM of 3 20-min clearance periods, pre-infusion, during infusion, and post-infusion.
[b]  $p < 0.05$.
[c]  $p < 0.01$.

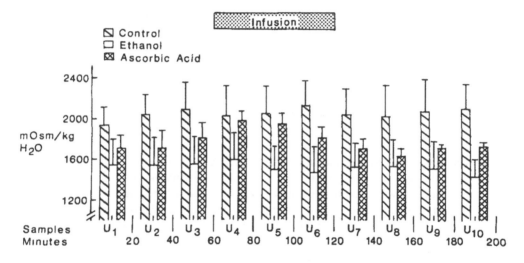

FIGURE 1.   Effect of ethanol (34 mg/kg/min) and ascorbic acid (2.2 mg/kg/min) vs. saline control group on urine osmolality (mOsm/kg H₂O) of urine samples collected at the times indicated. From time 0 to 60 min, from 60 to 120 min, and from 120 to 180 min are pre-infusion, infusion, and post-infusion periods, respectively.

Several investigators have reported on the effect of acute ethanol administration on cation excretions and renal hemodynamics in animals, but unfortunately, their experimental designs are so widely divergent that it becomes difficult if not impossible to make any reasonable comparisons. For example, acute vs. chronic experiments, oral vs. direct infusion, sampling time, alcohol dosages, and different animal species all may account for these discrepancies. Hemodynamically, the reports are also conflicting, with some investigators finding no changes, others finding increases, and still others finding decreases. We have chosen this experimental model of acute infusion because it lends itself to minimum variability and better control.

## VI. EFFECT OF ZINC AND OTHER TRACE METALS ON ETHANOL METABOLISM

There is sufficient evidence in the literature to indicate that selected inorganic ions and certain organic nutrients can be used to modulate and/or enhance ethanol metabolism in animals and man. Of the inorganic ions tested, Li, Cs, Rb, and V[155-158] are known to influence ethanol oxidation.

Several investigators have demonstrated that supplemental Zn accelerates healing of surgically inflicted wounds in Zn-deficient animals.[159-161] In addition, healing of wounds in

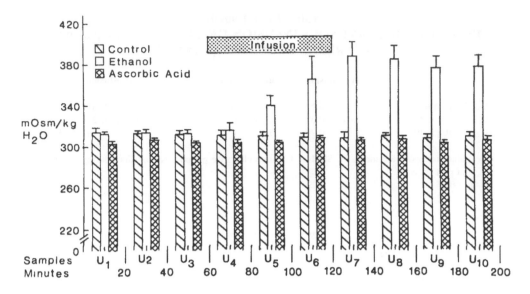

FIGURE 2    Effect of ethanol (34 mg/kg/min) and ascorbic acid (2.2 mg/kg/min) vs. saline control group on serum osmolality (mOsm/kg H₂0) of serum samples collected at the times indicated. From time 0 to 60 min, from 60 to 120 min, and from 120 to 180 min are pre-infusion, infusion, and post-infusion periods, respectively

young males treated similarly was accelerated upon Zn supplementation.[162] Zn also has been recognized to be a rate-limiting factor in DNA and RNA synthesis.[163,164] We have previously reported on the apparent increase in survival rates of ethanol-intoxicated mice and rats following treatment with ascorbic acid and/or Zn. Profiles of temporal blood ethanol concentrations in rats have also shown an enhanced ethanol metabolism in treated groups compared to saline control or to pyrazole-treated animals.[165] Confirmation of these results in animals has subsequently been made by others.[166]

One of the consequences of post-alcoholic cirrhosis (originally described by Vallee and colleagues[167]) was the increased urinary output of Zn with concomitant decreases of hepatic and serum Zn levels. Subsequently, several workers were able to describe the same phenomenon in animal models. These and other observations, which have been repeatedly confirmed, seem to indicate that alcohol produces a *conditioned* Zn deficiency. Further work provides evidence that hepatic alcohol dehydrogenase activity is significantly reduced in Zn-deficient animals and that deficiency may inhibit alcohol oxidation.[168] Furthermore, Vallee's discovery that alcohol dehydrogenase of horse liver is a Zn-containing metalloenzyme seems to suggest that the observed zincuria may be related to ethanol ingestion and enzyme inhibition.

As for the interaction of Zn with ethanol, firstly, evidence has demonstrated abnormal Zn metabolism in post-alcoholic cirrhosis in man and animals. This is characterized by hypozincemia, hyperzincuria, and diminished levels of hepatic Zn.[78,169-171] Secondly, in Zn-deficient animals, hepatic alcohol dehydrogenase activity is diminished up to 60% of normal.[172] Thirdly, plasma Zn levels were significantly increased in Zn-deficient dams receiving alcohol during pregnancy, suggesting a greater demand for Zn release into circulation in the presence of ethanol.[173] This is analogous to the increased release of ascorbic acid from the adrenal gland during ethanol intoxication.[174] Fourthly, Zn has been reported to play an important role in exerting a stabilizing effect on biomembranes in liver injury induced by toxic agents.[175] Basic adaptational response to chronic ethanol administration occurs in the membrane. Recent studies seem to indicate that the physical and chemical properties of cell membrane are altered in the presence of ethanol from its highly organized structure to a disorganized state for which the term "fluidity" has been applied.[176] If indeed a stabilizing

Table 8
SERUM OR PLASMA ZINC CONCENTRATIONS AND
URINARY ZINC EXCRETIONS (MEAN ± SD) IN
PATIENTS WITH CIRRHOSIS OF THE LIVER
COMPARED TO NORMAL CONTROL SUBJECTS

| Cirrhosis | | Normal subjects | | |
|---|---|---|---|---|
| Serum ($\mu$g/100 cm³) | Urine ($\mu$g/24 hr) | Serum ($\mu$g/100 cm³) | Urine ($\mu$g/24 hr) | Ref. |
| 67 ± 19 | 1016 ± 196 | 121 ± 19 | 457 ± 120 | 167 |
| 64 ± 13 | | 124 ± 16 | | 177 |
| 63 | Increased | 93 ± 11 | 375 ± 150 | 178 |
| 64 ± 19 | 1431 ± 490 | 104 ± 14 | 643 ± 198 | 179 |
| 79 ± 13 | 1116 ± 1026 | 91 ± 17 | 525 ± 254 | 180 |
| 63 ± 14 | | 96 ± 12 | | 181 |
| 103 ± 29 | | 123 ± 23 | | 182 |
| 49 ± 12 | 1358 ± 1032 | 90 ± 13 | 641 ± 241 | 183 |

effect is offered by either Zn and/or ascorbic acid in membrane fluidity, then it is possible that the responses to ethanol, such as altered configuration of structural proteins, alteration of Ca binding, or changes in membrane-associated enzymes, could be modulated or delineated by these two agents.

Low serum or plasma concentrations and high urinary Zn excretions in patients with Laennec's cirrhosis have been reported by numerous investigators,[167,177-183] as shown in Table 8. Earlier studies from our laboratory confirmed these findings. The mean serum Zn concentration in 12 cirrhotic patients was 49 ± 12 (mean ± SD) $\mu$g/100 m$\ell$, with all but one below the lower limits of normal. The mean urinary Zn excretion was significantly increased to 1358 ± 1032 $\mu$g ($p$ <0.05), when compared to a control group of 16 subjects with an average urinary output of 641 $\mu$g/24 hr.[183]

We have previously reported on the effect of vitamin C, with or without Zn, on the toxicity and metabolism of ethanol in rodents.[165] In this study we demonstrated a significant protection in rats and mice following ethanol intoxication, which was produced so that the effect on survival and on ethanol metabolism could be determined in animals pretreated with ascorbic acid and/or zinc sulfate (Table 9). Only 13 of 40 mice injected i.p. with a fixed amount of ethanol on 2 consecutive days survived. In contrast, the survival rates in matched animals pretreated with 25 mg ascorbic acid or 1 $\mu$M Zn were 100 and 90%, respectively. Smaller amounts of Zn (0.2 and 0.4 $\mu$M) improved survival after the first injection of ethanol (24 hr), but not after the second (48 hr). Similar observations were made in rats given repeated injections of ethanol over a 4-week period. We also determined blood ethanol concentrations in rats pretreated with ascorbic acid or Zn, a combination of ascorbic acid and Zn, pyrazole (an alcohol dehydrogenase inhibitor), or saline (Figure 3). Blood ethanol concentrations were significantly lower ($p$ <0.01) 1 hr after i.p. ethanol injection in animals pretreated with ascorbic acid and/or Zn when compared to saline control animals. Pyrazole, in contrast, maintained increased blood ethanol levels. It was concluded from these studies that both ascorbic acid and Zn exert a protective effect in ethanol-intoxicated animals.

## VII. ROLE OF ASCORBIC ACID AND ITS EFFECT ON ETHANOL METABOLISM

First, it has been shown that excessive drinking (in chronic alcoholism) is frequently accompanied by a subclinical deficiency of this vitamin, which may be due to either inadequate intake, malabsorption, or increased requirement.[75,184,185] Second, a correlation has

## Table 9
### EFFECT OF PRETREATMENT WITH ASCORBIC ACID, ZINC SULFATE (Zn++) AND/OR MANGANESE CHLORIDE (Mn++) ON SURVIVAL RATES OF MICE INJECTED INTRAPERITONEALLY WITH 4.1 m$M$ ETHANOL ON DIFFERENT SCHEDULES (1, 24, AND/OR 168 HR)

| | No. of mice | Treatment (0 hr) | Ethanol (4.1 m$M$/mouse) Times (hr) | No. alive 24 hr | 48 hr | 10 day | Survival |
|---|---|---|---|---|---|---|---|
| **Group 2a** | | | | | | | |
| 1 | 10 | 0.2 μ$M$ Zn++ | 1, 24, 168 hr | | | 3 | 30 |
| 2 | 9 | 1 2 μ$M$ Zn++ | 1, 24, 168 hr | | | 6 | 67[a] |
| 3 | 10 | 0.2 μ$M$ Mn++ | 1, 24, 168 hr | | | 4 | 40 |
| 4 | 10 | 1 2 μ$M$ Zn++ | 1, 24, 168 hr | | | 3 | 30 |
| 5 | 10 | Saline (control) | 1, 24, 168 hr | | | 3 | 30 |
| 6 | 10 | Saline (control) | 1, 24, 168 hr | | | 10 | 100 |
| **Group 2b** | | | | | | | |
| 7 | 30 | 1 μ$M$ Zn++ | 1 hr | 23 | | | 77[a] |
| 8 | 35 | Saline (control) | 1 hr | 5 | | | 14 |
| **Group 2c** | | | | | | | |
| 9 | 20 | Saline (control) | 1, 24 hr | 20 | 20 | | 100 |
| 10 | 20 | Saline (control) | 1, 24 hr | 6 | 6 | | 30 |
| 11 | 20 | 0.2 μ$M$ Zn++ | 1, 24 hr | 20 | 4 | | 20 |
| 12 | 20 | 0.4 μ$M$ Zn++ | 1, 24 hr | 20 | 2 | | 10 |
| 13 | 20 | 1.0 μ$M$ Zn++ | 1, 24 hr | 20 | 18 | | 90[a] |
| 14 | 20 | Saline (control) | 1, 24 hr | 12 | 7 | | 35 |
| 15 | 20 | 25 mg ascorbic acid | 1, 24 hr | 20 | 20 | | 100[a] |
| 16 | 20 | 25 mg ascorbic acid + 1 0 μ$M$ Zn++ | 1, 24 hr | 16 | 16 | | 80[a] |

[a]   Significant difference from controls $p < 0.05$.

been noted between low activity of alcohol dehydrogenase (ADH) and low tissue levels of ascorbic acid.[186] Third, ascorbic acid in various concentrations exerted a powerful protection against the intoxicating effect of alcohol in mice as measured by the swimming test.[187] Fourth, in the rat it has been found that large doses of ascorbic acid offer a high degree of protection against the lethal dose of acetaldehyde, the first intermediary formed in ethanol metabolism.[188,189]

Ascorbic acid is known to participate in several metabolic reactions, particularly the oxidative-reductive processes. The most agreed upon function is that of a cofactor of proline hydroxylase in the hydroxylation reactions concerned with collagen formation,[190,191] which is necessary in wound healing.[192] It is unlikely that it participates directly in the ethanol-acetaldehyde-acetate reaction in spite of its well-known property as a strong reducing agent. The possibility does exist, however, that ascorbic acid can function as an electron donor similar to nicotinamide adenine dinucleotide (NAD) in ethanol metabolism. This would explain its sparing effect upon NAD/NADH available for ethanol removal and the acceleration of this reaction in its final disposition of ethanol to $H_2O$ and $CO_2$.

There is now increasing evidence, documented by several workers, that ascorbic acid may influence ethanol clearance from the blood: a gradient known as β-60 and expressed as milligrams of ethanol cleared per 100 m$\ell$ of blood per hour.[96,187,188,193-195] Krasner's[196] work in human volunteers receiving 1.0 g ascorbic acid per day for 2 weeks showed that their β-60 highly correlated with leukocyte ascorbic acid, suggesting an association between ascorbic acid and ethanol. In subsequent work,[186] the same authors demonstrated a high

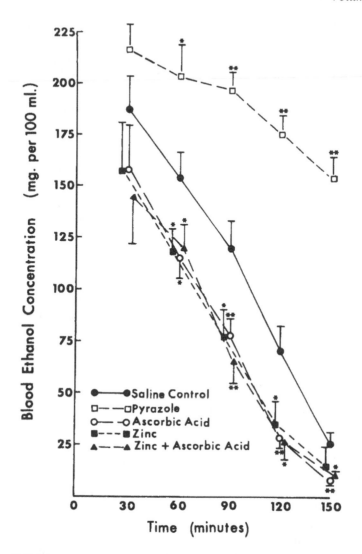

FIGURE 3. Blood ethanol concentrations in milligrams per 100 mℓ of treated groups (pyrazole, ascorbic acid, Zn, Zn and ascorbic acid) and saline control at 30, 60, 90, 120, and 150 min post-injection. Each point represents the mean ± SEM of 7 rats Differences from control did not become statistically significant (p <0.05) until 60 min after injection *p <0 05, **p <0 01 when treated groups were compared to control.

correlation between low ADH activity and low tissue levels of ascorbic acid. Also, in patients with nonalcoholic liver disease, they found a high correlation between ADH activity and leukocyte ascorbic acid, although the same phenomenon could not be demonstrated in vitro.[197]

Sprince and colleagues[189] documented that ascorbic acid exerts a dose-related protection against acetaldehyde, an intermediary of ethanol oxidation. Acetaldehyde is implicated in several disorders such as cerebral degeneration, alcohol cardiomyopathies, and disulfiram-ethanol reaction. It is hypothesized by these authors that the high doses of ascorbic acid alone or in combination with other agents may modulate the excessive release of catecholamines provoked by acetaldehyde.[188]

In spite of these positive findings, other workers were unable to find a similar relationship between ascorbic acid and ethanol metabolism. For example, in human volunteers, Pawan[95] was unable to find any change in the rate of ethanol metabolism following the administration of 600 mg ascorbic acid for 5 days. In guinea pigs, Dow and Goldberg[198] found no difference in the rate of ethanol disappearance from the blood between scorbutic and nonscorbutic animals dosed with 2.5 g of ethanol per kilogram. It is very likely that the discrepancy between the positive and negative findings is related to the amount of ascorbic acid administered. Unless a high level of tissue saturation is achieved by an adequate amount of ascorbic acid, effective protection cannot be demonstrated.

More recently, however, Busnel and Lehmann[187] examined the effect of ascorbic acid on impaired swimming behavior in mice treated with ethanol. The results demonstrated that ascorbic acid in various concentrations, particularly in doses of 500 mg/kg, exerted powerful protection against the intoxicating effects of ethanol. The effect was extended even after elimination of ethanol from the blood, suggesting that the effect is maintained by one or more ethanol metabolites. At doses of 250 mg/kg or lower, protection against ethanol toxicity was not as significant.

On the other hand, several workers have demonstrated depressed serum ascorbic acid levels in patients with post-alcoholic cirrhosis as well as lower leukocyte ascorbic acid in patients with a variety of alcoholic liver diseases.[184,198-200] However, no differences were observed in patients who manifested liver disease of other etiologies.[115]

Ascorbic acid seems to influence a variety of drug-metabolizing enzymes in man and animals. In guinea pigs, vitamin C deficiency resulted in decreased metabolism of various pharmacological agents.[201] The salutary effect of ascorbic acid supplementation in alcoholics and in ethanol-drinking rats has also been reported.[200]

We have recently published a study[380] in which we attempted to design an acute and chronic experimental model for alcoholism in the guinea pig so that alcohol administration and blood sampling could be performed under physiological conditions free of anesthetic agents by permanent cannulation of the carotid artery in order to study the effect of marginal vs. excess ascorbic acid on ethanol metabolism.

An experiment using 24 guinea pigs cannulated via the carotid artery were randomized 2 weeks after post-operative recovery and divided into three groups:

- Group 1. Received equal volume of saline as experimental groups. The diet was similar to experimentals except that it contained a high level of ascorbate (2000 mg/kg diet).
- Group 2. Received ethanol by infusion; the diet was similar to Group 1, except that it contained marginal amounts of ascorbate (200 mg/kg diet).
- Group 3. Received ethanol by infusion and the diet was similar to Group 1, i.e., high ascorbate.

Groups 1 and 3 were pair-fed the average food intake that the second group had consumed the day before.

Ascorbic acid-deficient diet was purchased from Bioserve (Frenchtown, N.J.). The basal diet contained casein, DL-methionine, corn oil, dextrose, corn starch, and sucrose. To maintain the isocaloric design of the diet, 0.889 g of sucrose per gram of food was added to the Group 1 diet and 0.636 g/g of food was added to Groups 2 and 3. In addition, salt mix, vitamin mix, choline chloride, and fiber (as cellulose) were added at 5.0, 2.0, 1.3, and 14%, respectively. For every gram of food consumed by Groups 2 and 3, 0.14 g of ethanol as 22% (v/v) was infused to maintain isocaloric design of the experiment so that 30% of caloric intake was derived from ethanol in Groups 2 and 3. Animals were maintained on this regimen for 8 weeks. Body weight was taken and infusion was made daily, 5 days a week. Random blood samples were obtained weekly for the analysis of ascorbic acid and ethanol.

At the end of 8 weeks each animal was placed in a plastic metabolic cage for 24-hr urine collection, after which the animal was sacrificed by exsanguination, and blood, liver, heart, kidneys, adrenals, and brain were obtained. Plasma was separated for the analysis of ascorbic acid, total protein, and malonylaldehyde. One aliquot of liver was placed in formalin for pathological examination and another was frozen immediately, with the adrenals, for ascorbic acid determination. All other tissues were weighed and then frozen for trace metal analysis.

Results are illustrated in Figures 4 to 8 (for Figures 7 and 8, see glossy insert) and Tables 10 to 15. Figure 4 shows the composition of the diet in terms of percentage caloric values of both control group and ethanol group, and Table 10 lists composition of the basal diet and the actual calories derived from protein, fat, and carbohydrates per kilogram of diet. Amounts of fat and protein were the same in all groups and isocaloric amounts of ethanol substituted for 50% of the carbohydrate portion of the diet and 30% of total caloric intake.

Figure 5 demonstrates the fluctuations in body weight over the 8-week experimental period. There was a slight drop in body weight in Group 1, but the weight loss in Groups 2 and 3 was significant at the 2nd week of dieting. Although their body weight improved in subsequent weeks, it stayed below the control group throughout the experimental period.

Figure 6 depicts blood ethanol levels 3 hr post-infusion in both groups that received marginal and high levels of ascorbic acid in their diets. It can be seen that animals that received high levels of ascorbic acid had significantly ($p < 0.01$) lower blood ethanol levels, suggesting that ascorbic acid enhances the rate of ethanol clearance from the blood.

Absolute organ weights and organ to body weight ratios are shown in Table 11. No significant difference was noted. Ascorbic acid concentrations of liver, kidneys, and adrenals are shown in Table 12. All tissues were significantly lower in the groups that received ethanol (Groups 2 and 3). Since the adrenal gland is considered to be a good index of ascorbic acid status in animals, it is interesting to note that its concentration in the ethanol-ascorbic acid-supplemented group was not much different from that of the ethanol-ascorbic acid-deficient group, suggesting that ethanol induces a severe ascorbic acid deficiency in spite of adequate intake.

Table 13 shows metal and ascorbic acid concentrations of urine. Cationic concentrations expressed in terms of 24-hr urinary volume are shown. As can be seen from the table, ascorbic acid concentration in the urine was higher in the ethanol-treated groups, particularly in the group that received marginal ascorbic acid, although the difference did not reach statistical significance. Nevertheless, this suggests that ethanol increases, by an as yet unknown mechanism, the urinary output of this vitamin.

Table 14 depicts various parameters in plasma, the most significant of which are the higher malonylaldehyde levels in the ethanol group, suggesting an increased peroxidation and the increased levels of α-tocopherol in the ethanol-treated ascorbic acid-deficient group. No changes are observed in the cationic (Ca, Mg, Zn) concentration of plasma.

Histological examination of the liver showed moderate steatosis (Grade 2) (Figure 7) in five and minimal changes in two (Grade 1) (Figure 8) ethanol-treated ascorbic acid-deficient animals and three out of four with minimal fatty infiltration (Grade 1) in ethanol-treated high ascorbic acid animals. There was no evidence of fibrosis or cirrhosis (8-week experiment). Analysis of tissues for Ca, Mg, Zn, and Cu are shown in Table 15. No significant differences were observed except for a diminished liver and skeletal muscle Mg in the ethanol-treated group that received marginal ascorbic acid in their diet.

Similar studies by Hsu and Hsieh[202] were recently carried out in the rat to determine the influence of ethanol ingestion on ascorbic acid metabolism. Three groups of rats were used. A control group that received tap water, an ethanol group that received 20% ethanol in their drinking water, and an isocaloric group that received sucrose were pair-fed with the ethanol-treated group. The experiment was conducted for 24 days or for 1 year. Ascorbic acid was determined in urine, plasma, liver, kidney, spleen, and adrenals. In addition, activities of

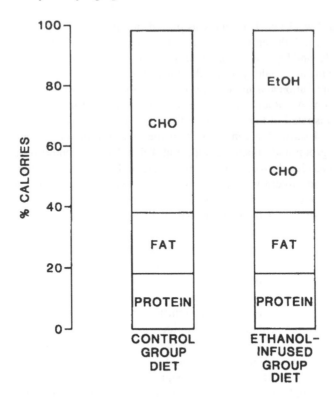

FIGURE 4.    Composition of the guinea pig diet in terms of percentage caloric values of both control group and ethanol group  Amounts of fat and protein were the same in all groups and isocaloric amounts of ethanol substituted for 50% of the carbohydrate portion of the diet and 30% of the total caloric intake.

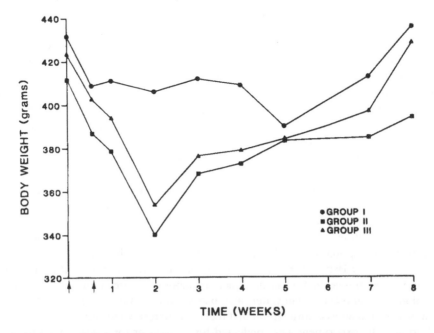

FIGURE 5.    Body weights of the three chronic ethanol-treated groups during the 8-week experimental period. Saline control, ●-●-● (6 guinea pigs), ascorbic acid deficient group, ▲-▲-▲(5 guinea pigs), and ascorbic acid-supplemented group ■-■-■ (4 guinea pigs).

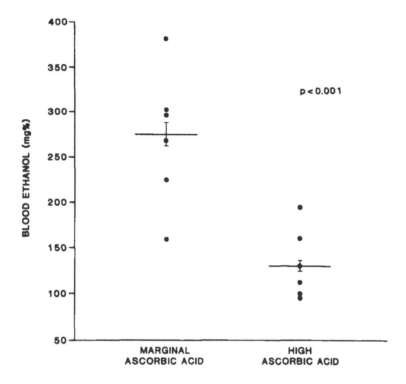

FIGURE 6. Ethanol clearance at 3 hr post-ethanol infusion of marginal ascorbic acid and high ascorbic acid groups Each point represents the average of seven blood samples taken randomly (—I—) Refers to mean ± SEM

## Table 10
## GUINEA PIG DIET[a]

| | Control (kcal/kg)[b] | % Cal | Ethanol (kcal/kg)[c] | % Cal |
|---|---|---|---|---|
| Protein | 720 | 18 | 720 | 18 |
| Fats | 800 | 20 | 800 | 20 |
| Carbohydrates | 2400 | 60 | 1200 | 30 |
| Ethanol | 0 | 0 | 1200 | 30 |
| Total | 3920 | 98 | 3920 | 98 |

*Note:* Ethanol as 22% (v/v) at 0 148 g/g of diet consumed was used.

[a] Basal diet contained the following ingredients in g%: casein, 21.96: DL-methionine, 0.34; corn oil, 10.09; dextrose, 22.53; corn starch, 22.53; sucrose, 22.53.

[b] To maintain isocaloric design of the diet in control group, 0.889 g of sucrose was added per gram of food.

[c] To maintain isocaloric design of the diet in the ethanol-infused group, 0.636 g of sucrose was added per gram of food.

## Table 11
## TISSUE WEIGHTS AND ORGAN TO BODY WEIGHT RATIOS[a]

| Group | Ascorbic acid | Ethanol | Tissue | Tissue weights (g) | Organ to body weight ratio |
|---|---|---|---|---|---|
| 1 (6) | 2000 mg/kg | — | Liver | 17.05 ± 2.82 | 0 0396 ± 0.0025 |
| | | | Kidney | 3.28 ± 0.12 | 0.0078 ± 0.00010 |

## Table 11 (continued)
### TISSUE WEIGHTS AND ORGAN TO BODY WEIGHT RATIOS[a]

| Group | Ascorbic acid | Ethanol | Tissue | Tissue weights (g) | Organ to body weight ratio |
|-------|---------------|---------|--------|--------------------|----------------------------|
|       |               |         | Heart   | 1.33 ± 0.61   | 0 0030 ± 0.00020   |
|       |               |         | Adrenal | 0 31 ± 0.09   | 0 00162 ± 0 00030  |
| 2. (5) | 200 mg/kg    | +       | Liver   | 14 60 ± 2 11  | 0 0400 ± 0.0024    |
|       |               |         | Kidney  | 3 29 ± 0.43   | 0 0095 ± 0 00080   |
|       |               |         | Heart   | 1.36 ± 0 21   | 0 0040 ± 0.0004    |
|       |               |         | Adrenal | 0.38 ± 0 06   | 0.0010 ± 0.00008   |
| 3. (3) | 2000 mg/kg   | +       | Liver   | 14.78 ± 1.98  | 0.0350 ± 0.0005    |
|       |               |         | Kidney  | 3 29 ± 0 39   | 0.0077 ± 0.00016   |
|       |               |         | Heart   | 1 58 ± 0 35   | 0.0037 ± 0.00016   |
|       |               |         | Adrenal | 0.48 ± 0 13   | 0.0087 ± 0 00003   |

*Note.* ( ) — Number of guinea pigs in each group

ᵃ   Mean ± SEM.

## Table 12
### CONCENTRATION OF ASCORBIC ACID IN TISSUES

| Group | Ascorbic acid in diet (mg/kg) | Ethanol | Liver | t | Kidney | t | Adrenal | t |
|-------|-------------------------------|---------|-------|---|--------|---|---------|---|
| 1. (6) | 2000 | − | 211 ± 30.2 |       | 241 ± 17 96 |       | 1077 ± 11.5 |       |
| 2. (5) | 200  | + | 102 ± 17.9 | 3 12ᶜ | 123 ± 6.70  | 5.81ᵇ | 174 ± 56.3  | 8 41ᵇ |
| 3. (4) | 200  | + | 116 ± 14 5 | 2.88ᶜ | 190 ± 7.50  | 2.62ᶜ | 150 ± 28.0  | 7 86ᵇ |

Ascorbic acid (μg/g of wet tissue)[a]

*Note:* ( ) — Number of guinea pigs

ᵃ   Means ± SEM.
ᵇ   $p < 0.01$.
ᶜ   $p < 0.05$.

## Table 13
### URINE

| Group | Dietary ascorbic acid (mg/kg) | Ethanol | Ascorbic acid[a] (μg/24 hr) | Ca[a] (μg/24 hr) | Mg[a] (μg/24 hr) |
|-------|-------------------------------|---------|------------------------------|-------------------|-------------------|
| 1. (5) | 2000 | − | 121 ± 52  | 569 ± 420 | 414 ± 320 |
| 2. (5) | 200  | + | 340 ± 121 | 225 ± 139 | 15 ± 10   |
| 3 (4)  | 200  | + | 195 ± 160 | 358 ± 48  | 95 ± 80   |

*Note.* ( ) — Number of guinea pigs.

ᵃ   Means ± SEM
ᵇ   $p < 0.01$.
ᶜ   $p < 0.05$.

the enzyme systems responsible for the synthesis of ascorbic acid from glucuronolactone and gulonolactone were determined in liver tissue. Results of the experiments are illustrated in Tables 16 to 18. Table 16 shows concentration of 24-hr urine ascorbic acid in the three

**Table 14**
**PLASMA**[a]

| | Group I | Group II | t | Group III | t |
|---|---|---|---|---|---|
| Hematocrit | 42 4 ± 6 69 | 33 25 ± 13.15 | NS | 39.00 ± 1.73 | NS |
| Total protein (g/100 m$\ell$) | 4.38 ± 0 82 | 4.18 ± 0.55 | NS | 4.14 ± 0.58 | NS |
| Malonylaldehyde ($\mu M$) | 5 28 ± 3 19 | 8.82 ± 2.62 | $p < 0.01$ | 10 33 ± 3.83 | $p < 0.01$ |
| $\alpha$-Tocopherol ($\mu$g/m$\ell$) | 8.57 ± 2.54 | 13.62 ± 2.90 | $p < 0.05$ | 9.14 ± 0.53 | NS |
| Ca (mEq/$\ell$) | 7.02 ± 1.08 | 7 13 ± 0.29 | NS | 7.29 ± 0.20 | NS |
| Zn ($\mu$g%) | 95.00 ± 13.40 | 90.00 ± 25.07 | NS | 90.30 ± 25 50 | NS |
| Mg (mEq/$\ell$) | 1.07 ± 0 19 | 1 14 ± 0.17 | NS | 1.14 ± 0.08 | NS |

[a]  Mean ± SEM

groups after 24 days. It was significantly higher in the group that received ethanol when compared to either control or the pair-fed group. Table 17 shows increases of ascorbic acid in all organs examined in the group that received ethanol when compared to control, but the difference was significant only in kidney tissue when compared to pair-fed group. Table 18 shows changes in glucuronolactone and gulonolactone enzymes. Both increased significantly in the ethanol group whether the results are expressed in terms of per gram liver or per milligram protein, suggesting an enhancement of ascorbic acid synthesis during ethanol intoxication.

## VIII. PHYSIOLOGICAL AND PHARMACOLOGICAL EFFECTS OF ETHANOL TOXICITY

A few definitions by highly authoritative workers in their field seem to be necessary at the outset of this section in order to distinguish the true character of alcoholism from the misconception that is held by the general public. Jellinek,[203] a former consultant on alcoholism to the World Health Organization, defines alcoholism as "a progressive disease characterized by an uncontrollable drinking". "Progressive" and "uncontrollable" seem to be the key words in the definition. Clinebell[204] defines an alcoholic as "anyone whose drinking interferes frequently or continuously with any other important life adjustments or interpersonal relationships". While these definitions are true statements and portray excellent characterizations of the disease, they are, nevertheless, descriptive in the nature of the manifestation of the illness and hence do not display the mechanism of the disease itself.

The relationship between well-defined disease entities and the quantity and/or frequency of ethanol intake in humans is difficult to establish in a precise manner. Although toxic levels of ethanol in most animal species used for alcohol research are well defined, such levels are difficult to measure rigorously in humans because of the multitude of variables that exist. Some recommended that blood ethanol levels be taken as the sole criteria for determining toxicity,[205] but that in itself is subject to a variety of factors that may or may not be controlled properly. Such variables as body weight, body water, duration of drinking, alcohol content of the drink, number of drinks per occasion, or whether drinks have been taken with or without food[206] have to be carefully evaluated if any reasonable correlation is to be made between ethanol toxicity and its physiological manifestations.

In this section the assumption is made that ethanol intake at a level sufficient to provoke altered cellular metabolism and organ damage will be made since it is well established now that altered pathology of any organ affected by ethanol is directly related to the dose.[207,208] It is likely that an intake of 160 g of ethanol or more, practiced on a regular basis, is potentially cirrhogenic. It is necessary to mention here that the rate of ethanol metabolism differs a great deal between species of animals and between animals and man. The average

**Table 15**

**TISSUE METAL CONCENTRATIONS (PER GRAM DRY WEIGHT)[a]**

| Group | Ascorbic acid (mg/kg) | Ethanol | Tissue | Mg (mg) | Ca (mg) | Zn (µg) | Cu (µg) |
|---|---|---|---|---|---|---|---|
| 1.(6) | 2000 | − | Liver | 0.476 ± 0.024 | 0.275 ± 0.022 | 107 ± 12 | 35 ± 21 |
| | | | Kidney | 0.502 ± 0.010 | 0.986 ± 0.131 | 95 ± 5 | 14 ± 0.08 |
| | | | Heart | 0.496 ± 0.039 | 0.489 ± 0.080 | 80 ± 6 | 15 ± 1.3 |
| | | | Brain | 0.448 ± 0.084 | 4.261 ± 1.513 | 70 ± 3 | 10 ± 3 |
| | | | Muscle | 0.665 ± 0.019 | 0.314 ± 0.014 | 15,199 ± 2,740 | 679 ± 112 |
| 2. (5) | 200 | + | Liver | 0.385 ± 0.026[b] | 0.288 ± 0.021 | 79 ± 7 | 79 ± 19 |
| | | | Kidney | 0.527 ± 0.024 | 0.728 ± 0.064 | 118 ± 9 | 14 ± 3.7 |
| | | | Heart | 0.527 ± 0.031 | 0.680 ± 0.254 | 88 ± 3 | 14 ± 3 |
| | | | Brain | 0.436 ± 0.019 | 4.321 ± 1.535 | 63 ± 1 | 7 ± 1.87 |
| | | | Muscle | 0.606 ± 0.036[b] | 0.329 ± 0.015 | 17,232 ± 357 | 673 ± 31 |
| 3 (3) | 2000 | + | Liver | 0.404 ± 0.040 | 0.322 ± 0.101 | 97 ± 16 | 26 ± 10 |
| | | | Kidney | 0.499 ± 0.030 | 0.830 ± 0.210 | 101 ± 7 | 16 ± 0.8 |
| | | | Heart | 0.462 ± 0.046 | 0.525 ± 0.070 | 80 ± 3 | 14 ± 0.7 |
| | | | Brain | 0.436 ± 0.008 | 5.609 ± 1.946 | 70 ± 1 | 10 ± 1.02 |
| | | | Muscle | 0.642 ± 0.087 | 0.342 ± 0.003 | 16,751 ± 4,459 | 658 ± 119 |

*Note:* ( ) — Number of animals in each group

[a]   Mean ± SEM.
[b]   $p < 0.05$.

## Table 16
## URINARY EXCRETION OF ASCORBIC ACID IN
## YOUNG ADULT RATS[a]

| Treatment | Final body wt in 24 days (g) | Daily feed intake (g) | Ascorbic acid (μg/24 hr) |
|---|---|---|---|
| None | 219 ± 5[b] | 19.8 ± 1 1 | 1079 ± 159 |
| Ethanol | 184 ± 14[c] | 12.2 ± 0.7[d] | 1433 ± 313[c] |
| Sucrose | 212 ± 11 | 12.2 ± 0.3[d] | 698 ± 294[e] |

[a] Weanling rats were kept on experiment for 24 days The 24-hr urine specimens were collected daily in the last 4 consecutive days of the experiment. Values in ascorbic acid were the means of four 24-hr urinary excretions.
[b] Mean of 6 rats ± S D.
[c] Ethanol vs. none ($p < 0.05$)
[d] Ethanol or sucrose vs. none ($p < 0.01$).
[e] Sucrose vs ethanol or none ($p < 0.05$).

## Table 17
## ASCORBIC ACID CONTENTS IN TISSUES

| Treatment | No. of rats | Liver (μg/g) | Kidney (μg/g) | Spleen (μg/g) | Adrenal (μg/100 mg) |
|---|---|---|---|---|---|
| None | 6 | 311 ± 13[a] | 166 ± 16 | 683 ± 33 | 410 ± 36 |
| Ethanol | 8 | 361 ± 3[b,d] | 244 ± 19[b,d] | 744 ± 28[c,c] | 473 ± 25[c,e] |
| Sucrose | 7 | 319 ± 16 | 207 ± 36[f] | 691 ± 35 | 391 ± 35 |

[a] Mean ± S D.
[b] Ethanol vs none ($p < 0.01$)
[c] Ethanol vs. none ($p < 0.05$).
[d] Ethanol vs. sucrose ($p < 0.01$).
[e] Ethanol vs. sucrose ($p < 0.05$).
[f] Sucrose vs none ($p < 0.05$).

## Table 18
## ACTIVITY OF ENZYME SYSTEMS FORMING
## ASCORBIC ACID FROM GLUCURONOLACTONE
## AND GULONOLACTONE[a]

| Treatment | Substrate used | Ascorbic acid synthesized in 2 hr | |
|---|---|---|---|
| | | Per g liver | Per mg protein |
| Sucrose (pair-fed) | D-Glucuronolactone | 285 ± 17[b] | 1.95 ± 0.08 |
| Ethanol | D-Glucuronolactone | 370 ± 25[c] | 2.68 ± 0.04[c] |
| Sucrose (pair-fed) | L-Gulonolactone | 271 ± 18 | 1 73 ± 0.07 |
| Ethanol | L-Gulonolactone | 320 ± 19 | 2 25 ± 0.08[c] |

[a] Weanling rats were kept on experiment for 28 days, after which their livers were isolated for enzyme studies.
[b] Mean of 6 rats ± SD
[c] $p < 0 05$.

rate of metabolism in man is about 0.15 g/kg/hr or 252 g/24 hr.[209] An intake of 22 g during the fasting state can bring the blood ethanol level to 45 mg% in about 30 min.[210] When the same amount of ethanol is taken with food, however, blood concentration is lower, probably due to diminished absorption from the GI tract. In this section, physiological effects of ethanol toxicity on liver, heart, kidney, central nervous system, pancreas, fetal physiology, and endocrine function will be described briefly. Readers interested in more details on the subject are referred to several reviews.[23,211-213]

## A. Effects of Ethanol on the Liver

Excessive chronic ethanol intake has been shown to induce hepatic injury, which includes fatty liver, alcoholic hepatitis, and liver cirrhosis.[214] According to Jellinek,[203] the clinical manifestations resembling the condition which is generally designated today as cirrhosis were described in the early 2nd century B.C. In the Greek literature there are certain indications that damage to the liver by alcohol was alluded to by the medical writings of that era. The first significant description of this condition was made by Mathew Baulie in the 18th century, although it did not exclude other etiological factors in producing the disease.

In 1826, Laennec (after whom the syndrome was named) described the nodular characteristics of the atrophic liver which he specifically identified as the alcoholic form of cirrhosis, although this condition was later produced by agents other than alcohol. The term is still used today whether the condition is alcoholic or nonalcoholic in its origin.

Perhaps morphological changes of the hepatocyte in alcohol toxicity have received more attention by investigators than any other cell in the body in view of the fact that the hepatocyte of the liver is the prime target of ethanol toxicity. Recently, Popper[212] reviewed the subject extensively. Morphological alterations at the subcellular level, including lysosomes, rough endoplasmic reticulum, smooth endoplasmic reticulum, mitochondria, and microsomes, have been examined to assess the anabolic organelle responses to ethanol injury. Development of hypoactive hypertrophic smooth endoplasmic reticulum is emphasized as a result of anabolic as well as catabolic adaptive responses.[212]

Evolutionary changes of the hepatocyte are described step by step as the disease progresses from alcoholic hepatitis to steatosis to fibrosis. Moon[215] suggests three essential features of the cirrhotic process:

1.  Degeneration and destruction of liver cells
2.  Regeneration of liver cells from those which escaped destruction
3.  Proliferation of connective tissue

Probably the most concise and accurate classification of the cirrhotic process was given by Boyd.[216] He states that after the original use of the term by Laennec, who employed the Greek work *kirros* (towny) to describe the yellowish nodules on the surface of the liver, it was realized that fibrosis was a striking feature of the lesions, and hence cirrhosis came to be synonymous with fibrosis. The unfortunate thing, he indicates, is that the name suggests a disease entity, but according to him, the essense of cirrhosis is not fibrosis or scarring, but destruction of liver parenchyma, followed by reconstruction of a new lobular pattern with the formation of regenerative nodules.

Boyd[216] contends that cirrhosis may be regarded as a chronic hepatitis with changes involving both the parenchyma and mesenchyma, and hence fibrosis is secondary in nature rather than primary. Boyd[216] classifies cirrhosis into biliary, which can be either intrahepatic or extrahepatic and portal or Laennec, which can be either nutritional (such as in alcoholism), post-hepatic, or post-necrotic in character. Furthermore, a miscellaneous variety can include other descriptive terms such as pigmentation, cardiac, toxic, and infectious.

The sequence of events that culminates in full-fledged post-alcoholic cirrhosis has been described by many authors.[212,217-223] During Laennec's cirrhosis, formation of fibrous septa connecting central and portal canals results in post-alcoholic cirrhosis, which is often associated with steatosis.[217] Although in the early stages of the disease (alcoholic hepatitis), mitochondria is not affected except for slight enlargement; the centrolobular areas are quite conspicuous in the peripheral zone. Most often, central localization leads to necrosis[218,219] around the central vein followed by fibrosis[220] and then destruction of the veins themselves.[212]

It is agreed among most investigators that the first manifestations of excessive intake of alcohol is an enlargement of the liver (hepatomegaly), followed by simple steatosis in which portal and periportal fibrosis is noted. Massive steatosis associated with acute hepatic failure is the next stage, in which deep jaundice is often seen. Lesions of the entire lobular parenchyma may set in during the next stage, which is identified as alcoholic hepatitis. The final stage of post-necrotic cirrhosis is far more common among alcoholics.[223] This may represent the end stage of the disease. Cholestasis and latent scar formation may ensue as a result of impaired hepatic circulation.[212]

It should be mentioned here that not all alcoholics (only a small proportion) develop cirrhosis. This suggests that other factors may enter into play such as genetic determinant, malnutrition, or environmental influences. Recently, a mechanism of ethanol-induced liver damage has been proposed.[209] It is hypothesized that a combination of an increased need for oxygen coupled with a decreased availability results in a rapid fall in oxygen tension. This results in centrolobular anoxia and necrosis. The hypothesis is supported by animal experimentation in which chronic feeding of ethanol to animals was found to increase oxygen consumption by liver slices in vitro. Centrolobular necrosis can be induced in these animals if placed in reduced oxygen tension for a short time.

With regards to the increased size of the hepatocyte, it is believed to be due to the rapid increase in fat and protein content coupled with increased water in the liver. The direct effect of ethanol on hepatic collagen deposition may be the cause of the fibrotic disease since it is known that ethanol inhibits hepatic degeneration by depressing DNA and protein synthesis.[224]

Other hypotheses have also been advanced to explain the development of ethanol-induced liver cell necrosis, i.e., lipid peroxidation. DiLuzio and Hartman,[119] and Tappel and Dillard[122] presented evidence to support the hypothesis that ethanol ingestion results in increased hepatic lipid peroxidation in animals. More recently, Suematsu et al.[225] conducted studies in humans to determine the relation of liver lipoperoxides and serum lipoperoxides to the development of liver damage and recovery from it in alcoholic patients. On admission, chronic alcoholics showed a very high level of serum lipoperoxide as demonstrated by the thiobarbituric acid test. In addition, a significant positive correlation was observed between serum lipoperoxides levels and SGOT activity which was reversed during 2-week abstinence. The findings that heavy drinkers with high levels of liver lipoperoxides had a higher incidence of liver cell necrosis than those with low levels of liver peroxidation adds credence to this hypothesis.[22]

## B. Effects of Ethanol on the Heart

The effect of alcohol on the heart and circulation has been surrounded by many misconceptions. A drink of whiskey makes the heart beat faster, thus giving a sensation of warmth to the skin. This perhaps is due to a transient irritation of the nerve endings in the mouth and stomach. This in turn increases blood flow, thus dilating the peripheral vessels, which in turn produces the feeling of warmth.[12] Gillespie,[227] on the other hand, concluded that the vasodilator effect of alcohol is not produced solely by peripheral sympathetic blockade or central inhibition, but rather as a vasodilator only when partly metabolized in the liver. Intraarterially infused alcohol has a direct vasoconstricting effect.

Eliaser and Giensircusa,[228] however, believe that the therapeutic effects of alcohol in coronary artery disease are attributable to cerebral responses rather than a demonstrable increase in coronary blood flow. This is supported by the data of Regan et al.,[229] who found no decrease of blood flow in chronic alcoholics. There is no evidence that moderate amounts of alcohol cause heart disease, high blood pressure, or atherosclerosis. Getting drunk, however, is believed to tax the heart and would be dangerous for a man with heart disease.[228]

Perkoff et al.[230] described a clinical syndrome associated with biochemical abnormalities in chronic alcoholism that resembles hereditary phosphorylase deficiency. The main features of this syndrome are muscle tenderness, increased serum creatine phosphokinase, poor lactic acid response to ischemic exercise, and a variable muscle phosphorylase activity. According to Wendt et al.,[231] alcoholism and cardiovascular changes were recognized as early as 1873 by Walshe. Recently, Eliaser and Giensircusa[228] identified three cardiovascular syndromes and designated them as alcoholic myocardosis, nutritional heart disease, and beriberi heart disease. Evans[232] suggests an early diagnosis of alcoholic cardiomyopathy with EKG, which can be manifested in a depressed T wave, the presence of extrasystole, a bundle-branch block, and a depression of S-T segment.

Brigden[233] recognized several etiologic groups of cardiomyopathies; among them is a condition he calls "alcoholic cardiomyopathies", which does not manifest itself with the development of beriberi. Gallop rhythm was common, and left ventricular failure was frequently encountered with progressively diminishing response to therapy, including large doses of thiamin. Burch and Walsh[234] indicated that alcohol per se may have a direct toxic effect on the heart. They reviewed eight cases of alcoholics in which they found a classical picture of beriberi heart disease as well as a typical nonspecific heart failure. Wendt et al.[231] tend to support the concept that chronic alcoholism results in altered myocardial metabolism even in those patients who do not display any clinical evidence of heart or liver disease. This led them to classify chronic alcoholic patients into four groups on the basis of the hemodynamic as well as the metabolic changes noted: (1) those with clinical evidence of heart or liver disease, (2) those with alcoholic cardiomyopathy, (3) those with Laennec's cirrhosis, and (4) those with both Laennec's cirrhosis and heart disease. They postulated, however, that with repeated bouts of alcoholism, the cellular and mitochondrial permeability as well as the metabolic pathways are altered. In support of this hypothesis, Ferrans and co-workers[235] made a recent study of the cardiac muscle with alcoholic cardiomyopathy by electron microscopy and found the presence of mitochondrial swelling, a decrease in intramitochondrial enzymes, and the deposition of liquid droplets in the myocardial cells.

In a review article, Rubin[236] recently emphasized the danger of chronic ethanol abuse which may lead to alcoholic cardiomyopathy, one of the more common congestive cardiomyopathies in the Western world. Acute intoxication exerts a negative inotropic effect (decreased cardiac contractility), increases blood pressure, is arrythmogenic, and destroys the function of subcellular constituents of the cardiac cell.

The question as to whether ethanol is toxic or tonic to the myocardium is still unresolved and at what level of intake the cutoff point between good and bad can be established is unknown. There are a number of reports in the literature that indicate no association between myocardial disease and ethanol.[232,237-239] On the other hand, there are several other recent reports that demonstrated negative correlation. The data of Stason et al. from the Framingham study,[240] Yano et al. from the Honolulu Heart Study,[241] and Barboriak et al.[242] established a definite and significant trend of negative association between alcohol intake and coronary heart disease. The amount of ethanol consumed to produce this beneficial effect did not exceed 56 g/day. The Kaiser Permanente group, in which an association between blood pressure and ethanol intake was sought, found a significant increase in blood pressure among drinkers (36 g/day) than nondrinkers.[243] These data are opposite to those of Barboriak.[242] The beneficial effect of ethanol is believed to be due to increases in high-density lipid concentration, which is protective against formation of atherosclerotic plaques.[244] Similar data were obtained by Hennekens et al.[245,246] in a large group of white males with appropriately selected controls. A protective effect was demonstrated whether wine, beer, or liquor was consumed.

## C. Effects of Ethanol on the Kidney

Ethanol is increasingly recognized as having a toxic effect on a variety of extrahepatic tissues. The kidney is no exception to the rule. It has frequently been observed that renal dysfunction is associated with alcoholic hepatic impairment, a disease commonly known as hepatorenal syndrome.[247] It is characterized by azotemia, diminished glomerular filtration rate, oliguria, and dilutional hyponatremia despite increased Na retention. Patients with this disease usually die within a short period of time after manifestations of renal failure set in. This is in spite of the fact that histological examination of the kidney shows no abnormalities.[248] This led some investigators to speculate that a shunt in renal flow brought about by a release of vasoactive substances from the liver could accelerate the changes seen in renal hemodynamics.[247]

It must be recognized, however, that the majority of patients with liver disease and associated renal dysfunction are alcoholics.[247] In addition, renal enlargement and lipid accumulation have been observed in ethanol-fed rats.[249] Van Thiel et al.[250] have recently demonstrated a reduced renal function and interstitial edema in ethanol-fed rats when compared with their isocaloric controls. The renal hypertrophy observed was due to increased absolute amounts of protein, fat, and water, again suggesting the deleterious effect of ethanol on kidney tissue.

Other physiological effects on the kidney occur through the pituitary gland, where a reduction in its activity, and hence a reduction of the antidiuretic hormone, produces copious urine. Other glands effected are the adrenals,[251] which increase their secretion in the presence of large doses of ethanol, hence the disturbance of mineral metabolism in the body.

## D. Effects of Ethanol on the Central Nervous System

The most pronounced physiological effect of alcohol is on the nervous system.[252] A blood-alcohol concentration of about 0.05% depresses the uppermost level of the brain — the center of inhibition, restraint, and judgment. At a concentration of 100 mg% in the blood, which can result from drinking 5 to 6 oz of whiskey, alcohol depresses the lower motor area of the brain; hence, it depresses sensory and motor functions. A blood-alcohol concentration of 200 mg%, which can result from drinking 10 oz of whiskey, disturbs the midbrain, where emotional behavior is largely controlled. At 300 mg%, from about 1 pint of whiskey, the drinker becomes stuporous. When the alcohol reaches the level of 400 or 500 mg%, the whole perception area of the brain is suppressed and the drinker falls into a coma.

The physiological effect of alcohol on the nervous system is probably the most pronounced effect to be described in detail by many workers in the field. Evidence is accumulating that alcohol or its metabolites produce at least some of their effects through the interaction of various brain amines. Davis et al.[253,254] provide evidence supporting the fact that there is a "shift" in norepinephrine (NE) metabolism from a normal oxidative route to a reductive pathway. The mechanism involved is the increased production of NADH resulting from ethanol oxidation, hence a secondary "shift" in the metabolism of 3,4-dihydroxymandelic aldehyde and 3-methoxy-4-hydroxymandelic aldehyde to the corresponding glycols. Furthermore, the acetaldehyde form competes with the aromatic aldehyde for alcohol dehydrogenase, thus inhibiting their oxidation to the acids.

Duritz and Truitt[255] did not find any significant increases either in norepinephrine or in 5-hydroxytryptamine (5HT) after ethanol administration in rats. However, increased acetaldehyde blood levels produced by disulfiram pretreatment before the ethanol doses or by administration of acetaldehyde itself caused statistically significant decreases in brain norepinephrine, but had no effect on 5HT.

It has been suggested[255] that the direct myocardial release of NE is due to acetaldehyde, which has a stimulating action similar to that of tyramine. Towne[256] found amine levels of nerve cell bodies in the catecholamine cell groups to be significantly affected by ethanol.

Thus, the metabolism of acetaldehyde could contribute to a number of pathologic effects through the mechanism of $NADH_2$ generation as well as through increased catecholamine release.[257] Some ethanol effects on the nervous system have been attributed directly to the accumulation of acetaldehyde.[255,256,258,259]

Drugs are now available which, in very small doses, can cause dramatic changes in the behavioral patterns of the alcoholic. For example, chlorpromazine, a commonly used tranquilizer, delays the disappearance of alcohol from the blood in man.[260] It also has the capacity to inhibit alcohol dehydrogenase although this has been challenged by Edwards and Price,[261] who observed inhibition in vitro but not in vivo. Metronidazole has also been reported to inhibit ADH activity and hence decreases alcohol consumption in man.[261] The drug requirement for inhibitor is much higher in vitro than in vivo. Antabuse®, as well as metrondiazole, inhibit xanthine oxidase, a molybdenum-activated enzyme, which is implicated in the metabolism of acetaldehyde.

### E. Effects of Ethanol on the Pancreas and the Hemopoietic System

Pancreatitis, as a sequelae of chronic ethanol ingestion, has been known for a long time and its severity can be more pronounced when ethanol ingestion is accompanied with smoking.[262] Clinically it is characterized by episodes of acute abdominal pain with pancreatic changes, including inflammation, fibrosis, and calcification.[263] However, it is believed that a protein-deficient diet and hyperlipidemia may be contributing factors.[264-266]

The so-called Zieve's syndrome has been described by Zieve[267] and subsequently by others.[268-270] It is characterized by hyperlipemia, jaundice, fatty liver, and hemolytic anemia. Sobel and Waye[270] found that the incidence of pancreatic lesions was twice as high in patients with portal vs. post-necrotic cirrhosis. This suggested to them that pancreatic lesions are more correlated with alcoholism than with liver disease.

Although the mechanism of hemolytic anemia is unknown, work by Beard and Knott[271] indicate that the chronic administration of ethanol results in the development of a normochromic normocytic anemia accompanied by a concomitant reticulocytopenia. There was also a generalized hypercellularity of the bone marrow with no change in the myeloid/erythroid ratio. These findings led them to conclude that the pathophysiological sequelae of chronic alcohol ingestion may result from a direct depressive and toxic action of ethyl alcohol on the hematopoietic system. These data have subsequently been confirmed by several investigators.[113,272]

Sullivan and Herbert[273] were able to correct this suppression of the hematopoietic response with large doses of folic acid or by cessation of alcohol; hence, the improvement seen in anemic alcoholics after hospitalization was due to both ingestion of folate-containing food and to the cessation of alcohol ingestion. The mechanism involved here may be partly due to an effect on folate metabolism by slowing down its movement from liver cells.[274] However, Lindenbaum and Lieber[275] indicated that the dose-related vacuole formation in RBC precursors due to alcohol could not be corrected with vitamin supplementation, including large doses of folic acid.

Herbert et al.[276] studied 70 patients with alcoholism and varying degrees of hepatic dysfunction. They found a significant correlation between serum folate deficiency and alcoholism, less significance with either macrocytosis or anemia, and borderline with cirrhosis and fatty hepatitis.

### F. Fetal Alcohol Syndrome

It has been known for many years that an association exists between heavy drinking in pregnant women and the underdeveloped fetus, resulting in mental and growth retardation and frequently, congenital malformation. However, definite characterization of the syndrome has only recently been made by Lemoine et al.[277] in France, and Jones and Smith.[278] In the

U.S., Jones and colleagues[278] thus coined the term "fetal alcohol syndrome" after they observed the clinical manifestations in eight unrelated infants born of chronic alcoholic mothers. Characteristic features of the syndrome are microcephaly, joint abnormalities, growth deficiency (both pre- and post-natal), impaired mental development, and fine motor dysfunction.[279] At the time of birth, impairment of body length is often greater than body weight. Prenatal mortality among these infants is high and those that survive usually display neurological manifestations.

Severity of syndrome manifestations depends on several factors: age of the mother, amount and frequency of alcohol consumed during gestation, and nutritional status of the mother. There are some reports, however, that indicate a high incidence of some anomaly even among rare drinkers.[280] There was a significant difference between babies born of rare and moderate drinkers, but the difference was highly significant, with the highest anomalies among mothers who are heavy drinkers. It was concluded by Turner et al.[206] that an amount of ethanol in excess of 150 g/day consumed by a pregnant mother is likely to lead to birth of a baby with fetal alcohol syndrome. The mechanism of action for these manifestations is presently unknown, but is likely to be the direct effect of ethanol toxicity on the fetus, malnutrition caused by a lack of certain essential nutrients, or both.

## G. Ethanol and the Endocrine System

It has been a common observation that alcoholics become sexually impotent and infertile, but scientific evidence for the development of these manifestations of altered gonadal function as a consequence of alcohol abuse did not come to the fore until methods of assessing endocrine functions became available to us, in particular the development of radioimmunoassay.[281] It should be mentioned here that cirrhotic patients have been known to have low concentrations of testosterone, although reduction can occur in the absence of cirrhosis or nutritional deficiencies.[282] The clinical manifestations of this disorder are well defined: impotency, scanty pubic hair, hypogonadism (small testes), and reduced libido.

Data on plasma testosterone levels in alcoholics are not consistent, however, since the secretion occurs in episodic bursts and the fluctuation can be over several-fold range.[283-286] According to Van Thiel,[105] even moderate intermittent ethanol ingestion can transiently depress plasma testosterone levels. Depression of spermatogenesis as a result of gonadal dysfunction has been explained by the authors[281] to be related to a direct metabolic effect of ethanol on the interstitial germinal epithelial cell.

## H. Effects of Ethanol on Bone Formation

We have previously indicated that the effect of ethanol on the growth of the fetus in alcoholic mothers is more significant in its height than its weight,[278] suggesting that ethanol retards growth of the skeletal system. The data of Saville and Lieber[287] demonstrated that isocaloric administration of ethanol in experimental animals does not support growth as much as a matched isocaloric sucrose group. Based on data in animals, they concluded that restriction of growth due to ethanol is of a different nature than that of total diet restriction.[288]

When bone density in nonalcoholic subjects was compared with those of alcoholics, it was found to be diminished after 50 years in nonalcoholic women and after 70 years of age in nonalcoholic men, whereas all alcoholics (age 45) had a bone density similar to that of 70 years of age, suggesting retarded bone formation in the alcoholic.[289] Nilsson and Westlin[290] also demonstrated that bone mass in alcoholics under age 70 was significantly decreased as compared to age-matched controls, and the difference increased with increasing age.

In view of the adverse effects ethanol has on mineral homeostasis, which is important in bone formation, it is not surprising that bone formation in the alcoholic can be retarded depending on the degree and duration of intoxication. We have seen how alcoholics lose in their urine an appreciable amount of cations such as Ca, Mg, P, and Zn which are needed in bone formation. In chronic alcoholics, the urinary losses of these cations, the impaired fractional absorption of dietary Ca, and the malabsorption of vitamin D due to steatorrhea, will eventually lead to diminished bone mass. Thus, alcoholics are known to suffer from osteoporosis, osteomalacia, and an increased risk of fractures of the hip, wrist, spine, and upper humorus. In aged alcoholics, these contributing factors may aggravate the situation further and set the stage for serious deformity and irreversible disability.

Other organs affected by ethanol toxicity in the aged are the oropharynx, larynx, and esophagus. Epidemiological evidence suggests that epidermoid carcinoma is not uncommon among aged alcoholics[291] as demonstrated by the increased mortality from esophageal cancer in France due to increased ethanol consumption.[292]

## IX. EFFECT OF ETHANOL ON LIPIDS, CARBOHYDRATE, AND PROTEIN SYNTHESIS

The subject of ethanol toxicity and its effect on lipid, carbohydrate, and protein synthesis has recently been reviewed by Lieber and DeCarli,[213] Badawy,[18] and Marks.[293] The intent of this section is to provide further evidence from the literature, which abounds with material in this regard, on how ethanol abuse can contribute to the general ill health in the young and the old by simply altering the metabolism of major nutrients. It has been shown to alter the micronutrients as well. A factor that should be considered in the metabolism of ethanol is the fatty acids, which are increased in starvation due to the increased production of acetyl CoA and possibly by inhibition of glycolysis by citrate. It is not known, however, whether fatty acids are produced *de novo* or are mobilized from adipose tissue into the liver.

Bouchier and Dawson[294] infused 1.0 g ethanol per kilogram body weight over 30 min and noted a decrease of free fatty acids (FFA) during infusion and an increase after the infusion had been stopped. They concluded that FFAs are mobilized from adipose tissue via adrenaline. Brown[295] administered ethanol to rats orally and found a lower serum FFA and lower liver triglycerides. DiLuzio and Poggi[296] provided further support for the concept that depressed intrahepatic triglyceride is a factor in the development of steatosis following acute ethanol intoxication. Contrary to that, Elko et al.[297] presented data indicating a significant increase in hepatic triglycerides 4 to 16 hr after alcohol intubation into rats. They do not subscribe to the idea that ethanol intoxication enhances lipid mobilization. Lieber et al.[298] observed a fall in FFA levels during ethanol administration, but were unable to find any evidence that ethanol induces the mobilization of FFA from the extremities. Thorpe and Shorey[299] found decreased serum FFA levels after 4 months of alcohol administration, but found no change in serum triglyceride and albumin levels.

To assess peripheral lipid mobilization, Feinman and Lieber[300] investigated the effect of ethanol on plasma glycerol and FFA in alcoholic volunteers. They found the drop in plasma FFA levels and glycerol to be 46 and 36%, respectively, from the baseline values. They did not detect any change in glucose levels. Hence, their data support the findings of a reduction in peripheral lipid mobilization after ethanol consumption. French[301] found an increased level of phospholipid in livers of both male and female rats fed ethanol. It would seem, therefore, that the lipid that accumulates in liver has three obvious sources: dietary, whose magnitude depends on the amount of fat taken in, endogenous synthesis, and mobilization of peripheral fat depots, which may be important only when large quantities of alcohol are consumed.

As to the mechanism and cause of fatty liver development in the alcoholic, Salaspuro and Maenpaefe[302] and Lieber and DeCarli[303] presented data to indicate that the oxidation of fatty acids and the citrate cycle are strongly inhibited during ethanol oxidation as demonstrated by lowered carbon dioxide production. Hence, the ''shift'' toward fatty acid synthesis due to its replacement by ethanol in the energy-yielding process and the similarity of the NADH

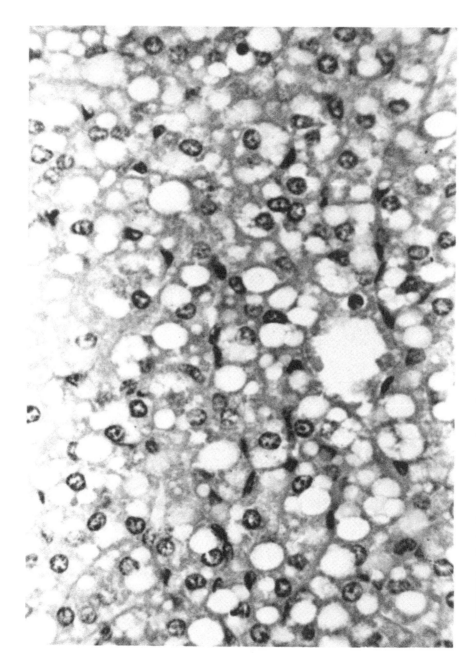

FIGURE 7    Grade 2 hepatic steatosis of ethanol-treated ascorbic acid-deficient guinea pigs.  (H & E, magnification × 40 )

FIGURE 8. Grade 1 hepatic steatosis of ethanol-treated ascorbic acid-supplemented guinea pigs (H & E; magnification × 40.)

generated from the oxidation of ethanol to that generated from the oxidation of fatty acids is the main cause for the development of a fatty liver in the alcoholic.[304-306]

It would seem, therefore, from the foregoing evidence that ethanol interacts with NAD, which causes a fall of the $NAD:NADH_2$ ratio in the liver. Consequently a rise of the lactate to pyruvate ratio follows. The decrease in the pyruvate causes an inhibition of gluconeogenesis accompanied by a rise in $\alpha$-glycerophosphate concentration in the liver. All of these changes lead to the accumulation of fat in the liver, hypoglycemia, hyperlactemia, and hyperuricemia. Krebs[307] contends that the $NAD:NADH_2$ ratio of the cytoplasm and mitochondria are maintained and regulated within relatively narrow limits and any condition that upsets their ratio leads to pathogenesis of the liver. In support to this contention, Dajani and Kouyoumjian[308] conducted in vitro studies and were able to find a correlation between the concentrations of hepatic FFA as well as triglyceride with the levels of liver ADH and $NADH_2$.

As to the effect of ethanol on ATP, French[309] presented evidence indicating an increase in acute treatment with ethanol and a decrease after chronic ingestion. This was attributed to degenerative changes in liver mitochondria as reported by Kiessling and Tilander[310] and Kiessling and Pilstrom.[311,312]

With regards to the effect of ethanol on glucose metabolism it is well known that hypoglycemia is the major consequence of ethanol ingestion, which often impairs brain function to such an extent to cause cerebral damage.[313] Ethanol interferes with hepatic gluconeogenesis and counteracts the mechanisms of maintaining carbohydrate homeostasis as shown by an insulin tolerance test.[314] Hypoglycemia is believed to be due to inhibition of gluconeogenesis in the liver brought about by a reduction of the $NAD:NADH_2$ ratio in the cytosol.[307] As had been found by Salaspuro and Maenpaefe[302] and Sholz et al.,[315] the administration of ethanol increases the concentration of lactate in the blood both in vitro[315] and in vivo.[316,317] The increased lactate permits the release of reducing equivalents into the bloodstream, while that of glycerophosphate provides the three-carbon skeleton for the synthesis of triglycerides, and the glyceraldehyde phosphate dehydrogenase shifts the equilibrium toward glyceraldehyde phosphate from the oxidation of glucose to gluconeogenesis.

Although this postulate seems quite plausible from the biochemical point of view, it does not fit in with the clinical side of the story. Brown and Harvey[318] found a clinical picture of hypoglycemia associated with alcoholic intoxication due to some form of denatured alcohol. Cummins[319] reported two cases of hypoglycemia in children following ingestion of alcohol. He suggests that alcohol utilizes glucose as a metabolic poison and thus interferes with gluconeogenesis or may inhibit glycogenolysis. Hypoglycemia is more pronounced when the liver is depleted of glycogen. Field et al.[320] consistently produced hypoglycemia in humans by the ingestion of 35 to 50 m$\ell$ of ethanol after a 2-day fast. Besides inhibiting glycogen synthesis, he also found that ethanol interfered with urea formation in the isolated, perfused rat liver, but not with amino acid mobilization. Freinkel et al.[321] drew that same conclusion after an overnight fast.

Henley and Scholz[322] suggest, however, that in the well-fed man or animal, the blood sugar may rise due to the action of catecholamines, which will promote the formation of glucose from glycogen stores.

Madison and Leonard[323] postulated that ethanol inhibits gluconeogenesis because of the marked increase in the ratio of $NADH_2:NAD$. This, he suggests, permits a unitary explanation for several puzzling phenomena. First, it will allow for the increased fatty acid synthesis. Second, it explains the decrease in hepatic conversion of galactose to glucose. Third, it justifies the decreased transformation of serotonin to 5-hydroxyindolacetic acid (5 HIAC). Fourth, it gives a logical rationale for the decreased glycerol metabolism, the depression of the Krebs cycle, and the decreased conversion of amino acids to glucose.

With regards to the effect of alcohol ingestion on protein synthesis in animals and man, the subject is too complex and not completely understood. While acute perfusion studies suggest that acetaldehyde is the main culprit, particularly in depressing cardiac protein synthesis in the subcellular microsomal system,[324,325] ethanol has not been shown to alter protein synthesis in vitro. Recent studies by Schreiber[326] in the guinea pig suggests that prolonged exposure to low levels of ethanol interferes with cardiac protein synthesis.

The effect of ethanol on protein synthesis may differ a great deal between acute and chronic administration as well as the level of ingestion and the route of administration. Amino acid deficiency caused by ethanol ingestion could play a major role in the inhibition of protein synthesis.[65,327-329] In support of these findings, the inhibition has been reversed by amino acid supplementation.[330,331] It must be mentioned also that chronic ethanol ingestion has been shown to enhance liver protein synthesis.[332,333]

As for the effect of a low-protein diet on the manifestation of ethanol toxicity, it has been shown by Best[62] and others[334-336] that a low-protein diet (6 to 30% casein) aggravated cirrhosis in rats but did not cause liver damage. Experimental evidence as to whether ethanol in the presence of a normal diet can bring about an increased triglyceride formation in the liver was provided by Lieber and DeCarli[303] and Lieber and Spritz,[337] who demonstrated that partial protection from alcoholic fatty liver can be achieved with large amounts of DL-methionine or choline with an average reduction of 80% of triglyceride accumulation in the liver.

Takada et al.[338] confirmed these results by providing further proof against the hypothesis that alcohol is cirrhogenic in humans by direct hepatotoxic action. They contended that this can afford hope for achieving some recovery of liver function if a high-protein diet is consumed containing abundant vitamins and essential food factors simultaneously with alcohol. Porta et al.[339] presented data to indicate that when alcohol provides less than 30% of the total caloric intake of rats, harmful dietary imbalances do not result as long as the basal food mixture is reasonably adequate. However, Rubin and Lieber[340] do not agree completely with this concept. Their data do not exclude the possibility that chronic malnutrition may exasperate alcohol-induced liver injury, but also demonstrate that alcohol itself is toxic to the liver, independent of nutritional factors. In a paired feeding experiment, Lowry et al.[341] supported this view that alcohol increases the severity of liver cirrhosis when rats are put on deficient diets. Furthermore, Kasliwal et al.[342] presented data to support the same view and concluded that undernutrition may hasten cirrhogenesis by rendering the liver more vulnerable to the ill effects of hepatotoxic factors.

If alcohol is isocalorically replaced by sucrose, however, sucrose-fed animals show significantly more hepatic lipid than alcohol-fed rats.[339] When sucrose-fed rats were put on a fortified diet ("Super Diet") they again showed a higher concentration of hepatic lipids than controls.[338] In the perfused liver the rat, unlike man, can make a histologic recovery from cirrhosis provided it receives an adequate diet.

The question that is raised by many is what would the effect of alcohol be on the liver that is already the site of cirrhosis. The answer was provided by Erenoglu et al.,[343] who gave alcohol daily to cirrhotic patients. One group received a nutritious diet and the other was given a protein-restricted diet. The first group of patients made steady improvement while in the second group there was definite deterioration. As indicated earlier, ethanol suppresses Krebs cycle activity in the liver. In man, unlike the rat, there is no way of reversing the lesions of cirrhosis once it sets in, although it has been shown that continued alcohol consumption with adequate diet such as in hospitalized patients is compatible with clinical improvement. According to Gyoergy and Goldblatt,[344] tocopherol is an additional protective dietary factor which may compensate for the absence of sulfur-containing amino acids. In a similar experiment, Gyoergy and Goldblatt[345] were able to reduce liver injury by giving from 10 to 20 mg choline daily, whereas addition of 50 mg cystine daily accentuated liver cirrhosis.

## X. ETHANOL AND NUTRIENT INTERACTION AS A FUNCTION OF AGE

We have seen in previous sections how the excessive use of ethanol in any form can be deleterious to vital organs of the body, particularly the liver, heart, and central nervous system. We have also seen that ethanol abuse can impair the metabolic function of major nutrients, the carbohydrates, proteins, and lipids, as well as minor nutrients, the vitamins, and trace metals. The question raised now is whether cellular impairment or organ dysfunction is accelerated during the aging processes or whether nutrient interaction is affected adversely by ethanol to the impairment of physical and mental well being of the elderly, and finally whether ethanol consumption is increased or decreased in our aging population and whether ethanol in moderation has any effect on longevity as such.

The effect of ethanol intake on the aging processes in man and animals has unfortunately received little attention by the scientific community. What is available in the literature seems to suggest that little effort has been made to explore the problem in greater depth, probably because such study does not lend itself to easy interpretation, or more likely, accessibility to human subjects is hard to attain. Recent surveys of the drinking habits of our elderly population indicate that there is a declining popularity of ethanol with age as well as a decline in excessive use.[346] Eckardt[346] suggests that the possibility exists that such a trend is probably due to the incorrect belief among the elderly that long-term ethanol abuse is correlated with decreased longevity or alternatively the inability of the elderly to gain access to ethanol because of their mental or physical handicaps. It is estimated, however, that in spite of this decrease, the number of alcoholics among the elderly is over 1 million. This number is likely to increase in future years due to the progressive increase in the elderly population.

Is ethanol intake good or bad for the elderly? In posing this question one would have to define the degree of intake. We have seen how excess intake of ethanol can be harmful to man and animals, to the young, to the fetus, and to the old by virtue of its toxic effect on the cellular components of the body. Such an effect can be compounded to the point of harmful synergism with preexisting old age diseases, malnutrition, or undernutrition due to social or economic stress. To compound the problem further, the elderly are more likely to use over-the-counter drugs and/or prescribed drugs than the young. In view of the adverse effects of combining ethanol with a variety of drugs used by the elderly, particularly the tranquilizers and antidepressants, it can be appreciated that side effects accrued therefrom could spell[349] disaster if not appropriately controlled. [347-349]

As for the effect of alcohol on longevity, it is agreed that heavy drinkers have a shorter life span than those who abstain or drink moderately. Nonetheless, there is no general agreement as to whether there is a difference between moderate drinkers and abstainers in longevity, although insurance company records show no significant difference.[252] Frequently cited are the findings of Pearl,[350] who presented data to demonstrate the shortest life expectancy for heavy drinkers, a somewhat higher expectancy for abstainers, and the highest for moderate drinkers.[351]

On the other hand, similar questions can be asked with regards to moderate drinking of ethanol and its influence on aging. If we make the assumption that the elderly live in a comfortable socioeconomic environment, moderate intake of alcohol daily is believed to affect longevity in a favorable way. Based on personal experience of the author (A. A. Yunice), during a trip to the Soviet Union in 1974 to study the factors that contribute to longevity in the mountain villages of the republics of Georgia, Armenia, and Azerbaidzhan, the observation was made by natives that the regular intake of homemade wine was more prevalent among the centurions, particularly in the republic of Georgia. This leads one to suspect that ethanol in small amounts could be beneficial to the elderly. It must be mentioned that this was a casual observation of the natives supported by testimony of the scientists who accompanied us, but it was not based on any scientific evidence or carefully documented by epidemiological studies. On the other hand, in other areas of the mountain villages of Adzharia and Dagestan, inhabited mostly by Moslems, old people never drank alcoholic beverages.[352]

## A. Studies in Animals

Animal models designed to study alcoholism in the aging animal have not been popular nor has sufficient evidence been presented in the literature to deny or confirm any association. Recently, Wood and Armbrecht[353] studied the behavioral effects of ethanol ingestion at different ages of mice. Although he did not utilize blood ethanol as a measure of the degree of intoxication, he concluded that differences exist between different age groups, particularly rate of ethanol metabolism, body water, and lean body mass, as well as increased neuronal sensitivity to ethanol. In both acute and chronic studies, ethanol ingestion, whether given i.p., by inhalation, or by liquid diet, induced more toxic effects in the older group as measured by the righting reflex, general motor activity, hypothermia, and withdrawal symptoms.

With regards to relative degree of ethanol consumption in different age groups, there is an agreement that old animals consume less if presented with ethanol on a voluntary basis.[354-356] These findings, however, have not been in total concert with other reports, which suggest that old animals consume more than the young.[357-358] Although this may not have any bearing on ethanol consumption in humans, it is interesting to note that very few studies measured blood-ethanol levels, the criteria acceptable by most investigators[205] for assessing the degree of intoxication.

Ethanol ingestion to the point of intoxication as reflected by $LD_{50}$ has been studied by several investigators in a variety of animal species. Most of these studies suggest that the dose of ethanol resulting in 50% mortality was less for old animals.[359,360] When other criteria are used such as righting reflex, sleeping time, blood ethanol clearance, or hypothermia, older animals were affected similarly but with less amount of ethanol, again suggesting that old animals are more sensitive to ethanol.[360,361]

Our laboratory has been interested in ethanol metabolism in the guinea pig as related to its interaction with other nutrients. In an unpublished report, we utilized a large number of animals (25 in each group) to determine the $LD_{50}$ in two different age groups. Ethanol was administered intraperitoneally to two groups, one group 2 months old (mean of 250 g body weight) and the other group, 6 months old (mean of 750 g body weight). The dose of ethanol resulting in 50% mortality was 4.25 g/kg in the old animals as compared to 5.35 g/kg in the young. In a separate study, we also observed that ethanol clearance from the blood as measured by blood ethanol levels was more rapid in the old than in the young. Similar findings have been obtained in the rat by other workers.[360]

Although these data seemingly suggest that a real difference exists between the old and young as far as how animal systems handle ethanol, this difference, if corrected for body water and lean body mass, becomes insignificant since it is known that these two parameters decline with increasing age.[362,363] Similar studies conducted on a chronic basis again suggest that old animals experience more detrimental effects than young ones.[364]

Impairment of learning and memory, though difficult to interpret in the animal model, has been demonstrated in rodents, the severity being more prominent with longer duration of ethanol consumption in a dose-response relationship.[283]

## B. Studies in Humans

If we accept the recent reports[365] which suggest that the elderly abuse alcohol, then it becomes imperative that clinical studies designed to answer some of these fundamental questions would be most urgently needed to fill the vacuum that exists in this area. If we envision an accelerated trend of alcohol abuse in the coming years as alluded to previously and if indeed the metabolic, pathological, and behavioral changes are a common denominator in both the aged and the alcoholic, then the efforts invested in this research area would be rewarding in gaining a better understanding of the aging processes. Unfortunately, most of the studies in humans are phenomenological at best. The recent symposium on the "Neurobiological Interactions between Aging and Alcohol Abuse" and the introductory title "Alcohol and Aging: Challenges for the Future" both attest to this fact.[366] There certainly is a great deal to be done in the future.

It is interesting to note that the future challenges Freund[366] puts forward are related to three major areas. First, altered pathways as reflected by biological organization should be compared between the alcoholic and the aged. Second, pathogenesis of impaired behavior, particularly CNS alterations at the molecular level, should be assessed in the two groups. Third, the individual variations that exist among the groups and the contributions of both genetic and environmental factors should be explored in great depth.

Much of the association between young alcoholics and older individuals that has recently been studied is related primarily to CNS alterations and cognitive deficit.[367] Perhaps the pathological alterations in the CNS could be exemplified by Wernickes' encephalopathy and Korsakov syndrome, and the cognitive deficit by impaired abstract-reasoning, learning, memory, and visuoperceptual skills. Impairment of these skills are most prominent among older alcoholics.[368] The term "premature aging" has been used by most investigators[369] to refer to the similarities in these altered parameters between chronic alcoholics and the nonalcoholic elderly. It appears from these data that, as has been found in animal studies, the aging brain in humans is also more sensitive to the deleterious effect of ethanol.

In support of these data, Ryan and Butters[370] compared cognitive deficits of younger and older alcoholics with those of age-matched nonalcoholic control subjects and concluded that of the former two groups, younger and older alcoholics performed more poorly than their counterpart controls. This data appears to be consistent with the premature aging hypothesis invoked by other investigators to explain why chronic alcohol intoxication induces such mental changes as to cause an individual to behave as if he were 10 years older.[369]

It is believed by most investigators that altered behavioral manifestations reflect altered pathological cellular brain damage as illustrated by the severe cases of chronic alcoholics who develop Wernickes' encephalopathy and Korsakov syndrome in terminal cases of the disease. We have previously alluded to nutrient deficiency in these diseases, particularly thiamin deficiency,[83,371] but whether there is a cause and effect relationship is not known. It is known, however, that thiamin supplementation reverses manifestations of the disease and that the disease occurs in nonalcoholic subjects as well.[372] Freund's conclusion,[366] however, emphasized the fact, as others have indicated, that ethanol itself is toxic to brain cells and that adequate thiamin supplementation is unlikely to prevent all CNS toxic effects in chronic alcoholism. Aging processes at the molecular level could interact synergistically or additively with the damage of molecular events due to ethanol, thus potentiating impairment to the cell more so than each alone.

Other theories have also been advanced to explain the ethanol-induced accelerated aging hypothesis. One theory is the autoimmune disorder hypothesis,[373] which postulated that changes in amino acid configurations may be the triggering mechanism since ethanol has been shown to alter RNA and protein synthesis in animals.[374] Reports by Porjesz et al.[375] could not confirm the accelerated aging hypothesis as demonstrated by studies of evoked brain potential deficit.

The hypothesis that premature aging in the alcoholic is a by-product of chronic alcoholism is, according to others, of limited scientific value because it does not lend itself to simple clinical modalities for scientific inquiry.[376] Although behavioral and morphological alterations due to aging processes and chronic alcoholism have been found to be identical, the conclusion cannot be made at this time that these similarities or differences are explainable on the basis of molecular mechanisms, but rather help project our thoughts and efforts toward a better understanding of aging.[377]

It should be noted here that most of the clinical studies performed to correlate accelerated aging with alcoholism have been of the longitudinal design and few of the cross-sectional type. Since in the cross-section study comparison is made between two or more different groups at the same time it lacks the followup required to observe the changes that occur in time if age changes are to be assessed and evaluated. The difficulty of testing these hypotheses in a longitudinal study would be due to ethical reasons. If these theories are to be tested in a clinical setting, studies of behavioral changes do not provide a complete answer to the problem. It would be necessary to examine the pathological alterations at the molecular level in the brain tissue in order to determine whether alcoholics perform like elderly nonalcoholics. In the words of Freund and Butters,[366] "Behavioral studies alone will never allow us to answer this question . . . only *neurohistological and biochemical studies will* produce the sort of data which can unequivocably ascertain the validity of a particular biological hypothesis."

And finally, a biochemical basis for the premature aging in chronic alcoholics can tentatively be put forward in spite of the many gaps that exist. Based on recent findings of depressed glutathione in magnesium-deficient animals,[378] we would like to propose an enzymic hypothesis that could explain premature aging in chronic alcoholics. We have alluded previously to the severe magnesium deficiency that exists in chronic alcoholics.[135-138] Glutathione plays a major role in protecting the cellular membrane by preventing free radical attack on the lipid molecule.[379] If indeed, glutathione is depressed in chronic alcoholics due to depressed magnesium, and if indeed, magnesium is essential in the maintenance of glutathione, then it is not unlikely to speculate that premature aging in chronic alcoholics is brought about by this cascading effect of deficiencies.

## XI. CONCLUSION

Alcoholism is one of the four major health problems in the U.S., if not worldwide, with only cardiovascular, mental disease, and malignancies sharing in importance. It has been a serious public health problem with adverse health consequences such as cardiovascular problems (cardiomyopathies), GI disorders, increased risk of certain cancers (mouth and throat), kidney failure, immunologic dysfunction, cirrhosis of the liver (post-alcoholic), neuropsychiatric problems, and malnutrition. About 7% of the adult population are believed to be alcohol abusers and 10 million of these are problem drinkers. In terms of fatal auto accidents, 40% are caused by alcohol abusers. Mortality rates which occur at the most productive age are two to four times higher than in the normal population. To further complicate the problem, the amount of alcohol consumed is on the increase in North America and in most other countries of the world. If the present trend continues, post-alcoholic cirrhosis is estimated to become the third leading cause of death after heart disease and cancer. Recent strengthening of federal research efforts and the enhancement of the intramural research program of the National Institute of Alcohol Abuse and Alcoholism (NIAAA) attest to the great national concern about this problem.

Although much information has now been gathered defining the biochemical metabolism of alcohol, little is known about the etiological factors responsible for the physiological and biochemical derangements (enzymatic, hormonal, nutritional). Research efforts have been directed from many disciplines of science to establish a rational approach to the prophylaxis and treatment of alcoholism and premature aging of the alcoholic. Speculation about the causes include the possibilities that they are genetically determined, that they are due to enzymatic polymorphism, or that they are due to a deficiency of certain nutrients or minerals.

Although many therapeutic model systems have been developed in recent years to attenuate or modulate the toxic effect of ethanol in humans, little has been done to define the relationship that exists between the premature aging processes of alcoholics and the aging processes that develop normally in the elderly nonalcoholic. The use of vitamins and minerals as therapeutic agents in the treatment of chronic alcoholism and aging is based on the rationale that a state of deficiency exists under these conditions and that supplementation of these nutrients is necessary to alleviate further exasperation of this deficiency. Whether the nutritional deficiency is a cause or effect of alcohol intoxication or premature aging of alcoholics has not yet been determined. Alcohol may interfere with absorption and utilization of nutrients by virtue of its toxic effect on cellular metabolism causing conditioned deficiencies, or alternatively, may lead to inadequate food intake due to its high content of "empty calories". In the elderly alcoholic, these conditions may be worsened further due to the synergistic effects of both alcohol toxicity and preexisting old age disease.

Although adequate replacement of essential nutrients can be useful to repair the injured hepatocyte, minimal amounts may not be sufficient in active post-alcoholic liver disease. During this phase of the disease, the liver cells lose many of the essential vitamins and minerals which are needed for repair of the necrotic cells and hence the patients' metabolic needs should be taken into account and nutrient therapy should be instituted accordingly.

While most investigators agree that it is unlikely that a single major causal factor can be attributed to alcoholism or premature aging of alcoholics, it is also becoming increasingly obvious that ethanol does not act by a single mechanism to produce the multiple physiological and biochemical effects; thus, there is a need for a trial of multifactorial elements, particularly nutritional elements, that are likely to give additive or synergistic effects. An understanding of the multiple causes of deficiency is essential if the mechanisms of ethanol toxicity and premature aging are to be understood and the needs for repair of tissue damage are to be met.

## ACKNOWLEDGMENTS

This research was supported by Hoffmann-La Roche, Inc. and the Medical Research Service of the Veterans Administration. The authors gratefully acknowledge the technical assistance of Ms. Sherry Henry and her help in the preparation of the manuscript, and to Darci Sargent and Shirley Logue of the Biochemistry Department for their secretarial assistance.

## REFERENCES

1 **Roueche, B.,** *Alcohol, Its History, Folklore & Effect on the Human Body,* Grove Press, New York, 1966, 22.
2. **Sinclair, H. M.,** Nutritional aspects of alcohol consumption, *Proc. Nutr. Soc.,* 31, 117, 1973
3. **Lieber, C. S., Jones, D. P., and DeCarli, L. M.,** Effects of prolonged ethanol intake on production of fatty liver despite adequate diets, *J. Clin Invest.,* 44, 1009, 1965.
4. **Block, A. M.,** *Alcoholism. Its Facet and Phases,* John Day Co., New York, 1965.
5. Alcohol and Alcoholism, *National Institute of Mental Health, Department of Health, Education and Welfare,* Publ NHSM 1640, 1967
6. **Krebs, H. A. and Perkins, J. R.,** The physiological role of liver alcohol dehydrogenase, *Biochemistry,* 118, 635, 1970.
7. **Lester, D.,** The concentration of apparent endogenous ethanol, *Q. J Stud. Alcohol,* 237, 125, 1962.
8. **Williams, R. J.,** *Alcoholism. The Nutritional Approach,* University of Texas Press, Austin, 1959.
9. *The Bible,* Leviticus 10:9, Old Testament, Revised Standard Version.
10. *The Bible,* Ephesians 5:18, New Testament, Revised Standard Version.

11 **Kelly, N. L.,** Social and legal programs of control in the United States, in *Alcohol Education for Classroom and Community. A Source Book for Educators,* McCarthy, R . Ed., McGraw/Hill, New York, 1964, chap 1.

12. **Greenberg, A. L.,** Alcohol in the body, *Sci. Am ,* 189, 86, 1953.

13. **Jacobson, E.,** The metabolism of ethyl alcohol, *Pharmacol Rev ,* 4, 109, 1952

14. **Korsten, M. A. and Lieber, C. S.,** Nutrition in the alcoholics, *Med Clin. N Am ,* 63, 963, 1979

15. **Thomson, A. D.,** Alcohol and nutrition, *Clin. Endocrinol Metab.,* 7, 405, 1978.

16. **Thomson, A. D. and Majumdar, S. K.,** The influence of ethanol on intestinal absorption and utilization of nutrients, *Clin. Gastroenterol.,* 10, 263, 1981

17. **Lieber, C. S. and DeCarli, L. M.,** Metabolic effects of alcohol on the liver, in *Metabolic Effects of Alcoholism,* Lieber, C. S., Ed., MTP Press, Lancaster, England, 1977, 31

18 **Badawy, A. A.-B.,** The metabolism of alcohol, *Clin Endocrinol Metab ,* 7, 247, 1978

19 **Smuckler, E. A.,** Alcoholic drink, its production and effects, *Fed Proc. Fed. Am. Soc. Exp Biol ,* 34, 2038, 1975

20 **Pawan, G. L. S.,** Metabolism of alcohol (ethanol) in man, *Proc Nutr. Soc.,* 31, 83, 1972

21. **Dubowski, K. M.,** Measurement of ethyl alcohol in breath, in *Diseases Caused by Toxic Agents,* Sunderman, F W. and Sunderman, F W , Jr., Eds , Warren H Green, St Louis, Mo., 1970, 316.

22 **Wilkinson, P. K.,** Pharmacokinetics of ethanol: a review, *Alcoholism Clin Exp. Res.,* 4, 6, 1980

23 **Wallgren, H. and Barry, H., III,** Drug actions in relation to alcohol effects, in *Actions of Alcohol,* Vol. 2, *Chronic and Clinical Aspects,* Elsevier, Amsterdam, 1971, 621

24 **Kalant, H., Mons, W., and Mahon, M. A.,** Acute effects of ethanol on tissue electrolytes in the rat, *Can. J. Physiol Pharmacol.,* 44, 1, 1966.

25 **Theorell, H.,** Function and structure of liver alcohol dehydrogenase, *Harvey Lect ,* 61, 17, 1967.

26 **Lester, D., Keskosky, W. Z., and Felzenberg, F.,** Effect of pyrazoles and other compounds on alcohol metabolism, *Q J. Stud Alcohol,* 29, 449, 1968.

27 **Lieber, C. S. and DeCarli, L. M.,** Hepatic microsomal ethanol-oxidizing system in vitro characteristics and adaptive properties in vivo, *J. Biol. Chem ,* 245, 2505, 1970.

28 **Blair, A. H. and Vallee, B. L.,** Some catalytic properties of human liver alcohol dehydrogenase, *Biochemistry,* 5, 2026, 1966.

29. **Harris, R. A.,** Metabolism of calcium and magnesium during ethanol intoxication and withdrawal, in *Biochemistry and Pharmacology of Ethanol,* Vol. 2, Majchrowicz, E and Noble, E P., Eds , Plenum Press, New York, 1979, 27

30. **von Wortburg, J. P., Papenberg, J., and Aebi, T. L.,** An atypical human ADH, *Can J. Biochem.,* 43, 889, 1965.

31. **Krebs, H. A.,** The effects of ethanol on the metabolic activities of the liver, *Adv Enzyme Regul.,* 6, 467, 1968.

32 **Lundquist, F., Tygstrup, N., Winkler, K.,** et al , Ethanol metabolism and production of free acetate in the human liver, *J Clin Invest.,* 41, 955, 1962

33. **Rewat, A. K.,** Effects of ethanol infusion on the redox state and metabolic levels in rat liver in vivo, *Eur J Biochem ,* 6, 585, 1968

34 **Krebs, H. A., Freedland, R. A., Hems, R.,** et al , Inhibition of hepatic gluconeogenesis by ethanol, *Biochem J ,* 112, 117, 1969.

35 **Larsen, J. A.,** Extrahepatic metabolism of ethanol in man, *Nature (London),* 184, 1236, 1959

36. **Videla, L. and Israel, Y.,** Factors that modify the metabolism of ethanol in rat liver and adaptive changes produced by its chronic administration, *Biochem. J ,* 118, 275, 1970.

37. **Lieber, C. S. and DeCarli, L. M.,** The role of the hepatic microsomal ethanol oxidizing system (MEOS) for ethanol metabolism in vivo, *J. Pharmacol. Exp. Ther.,* 181, 279, 1972

38 **Orme-Johnson, W. H. and Ziegler, D. M.,** Alcohol mixed function oxidase activity of mammalian liver microsomes, *Biochem. Biophys. Res. Commun.,* 21, 78, 1965.

39 **Lieber, C. S. and DeCarli, L. M.,** Ethanol oxidation by hepatic microsomes. adaptive increase after ethanol feeding, *Science,* 162, 917, 1968.

40. **Teschke, R., Hasamura, Y., and Lieber, C. S.,** Hepatic ethanol metabolism: respective roles of alcohol dehydrogenase, the MEOS and catalase, *Arch. Biochem. Biophys.,* 175, 635, 1976.

41 **Teschke, R., Matsuzaki, S., Ohnishi, K.,** et al., Microsomal ethanol oxidizing system (MEOS): current status of its characterization of its role, *Alcohol,* 1, 7, 1977.

42 **Cederbaum, A. I. and Rubin, E.,** Molecular injury to mitochondria produced by ethanol and acetaldehyde, *Fed Proc Fed. Am Soc. Exp. Biol ,* 34, 2045, 1975

43 **Pikkarainen, P. H. and Lieber, C. S.,** Concentration dependency of ethanol elimination rates in baboons. effects of chronic alcohol consumption, *Alcoholism Clin Exp. Res.,* 4, 40, 1980.

44. **Thieden, H. E. D. and Lundquist, F.,** The influence of fructose and its metabolites on ethanol metabolism in vitro, *Biochem. J.,* 102, 177, 1967.

45  **Badawy, A. A.-B. and Evans, M.,** The effect of ethanol on tryptophan pyrrolase activity and their comparison with those of phenobarbitone and morphine, *Adv. Exp. Med. Biol ,* 59, 229, 1975

46  **Israel, Y., Kalant, H., LeBlanc, A. E.,** et al., Changes in cation transport and (Na⁺-K⁺) activated adenosine triphosphatase produced by chronic administration of ethanol, *J Pharmacol Exp Ther.,* 174, 330, 1970.

47  **Sun, A. Y., Seaman, R. N., and Middleton, C. C.,** Effect of acute and chronic alcohol administration on brain membrane transport system, in *Alcohol Intoxication and Withdrawal III. Part 3A, Biological Effects of Ethanol,* Gross, M M , Ed , Plenum Press, New York, 1977, 123.

48. **Peng, T., Cooper, C. W., and Munson, P. L.,** The hypocalcemic effect of ethyl alcohol in rats and dogs, *Endocrinology,* 91, 586, 1972.

49. **Kastenbaum, R. and Ross, B.,** Psychosocial care of the aged: some historical perspectives, *Psychosocial Care of the Aged,* Howell, J G., Ed , Basic Books, New York, 1975

50  **Lucia, S. P.,** *A History of Wine as Therapy,* J. B Lippincott, Philadelphia, 1963

51. **Leake, C. D. and Silverman, M.,** *Alcoholic Beverages in Clinical Medicine,* Year Book Publishers, Chicago, 1966, 160.

52  **Chafetz, M.,** *Why Drinking Can Be Good For You,* Stein & Day, New York, 1976.

53  **Lolli, G.,** Assets outweigh liabilities, *Int. J Psychiatr ,* 9, 358, 1970—1971

54  **Burrill, R. H., McCourt, J. F., and Cutter, H. S. G.,** Beer: a social facilitator for PMI patients?, *Gerontology,* 14, 430, 1974.

55. **Mishara, B. L.,** et al , Alcohol effects in old age: an experimental investigation, *Soc. Sci. Med.,* 9, 535, 1975.

56. **Lolli, G.,** The role of wine in the treatment of obesity, *N.Y. State J Med.,* 62, 3438, 1962.

57  **Kastenbaum, R. and Slater, P. E.,** Effects of wine on the interpersonal behavior of geriatric patients' an exploratory study, in *New Thoughts On Old Age,* Kastenbaum, R , Ed., Springer, New York, 1964, 191

58  **Damrau, F., Liddy, E., and Damrau, A. M.,** Value of stout as a sedative and relaxing soporific, *J. Am Geriatr. Soc.,* 11, 238, 1963

59. **Sarley, V. C. and Tyndall, F. W.,** Wine is beneficial to geriatric nonpatients, too, *Nurs. Homes,* July, 73, 1971.

60. **Funk, L. P. and Prescott, J. H.,** Study shows wine patient attitudes, *Mod. Hosp.,* 108, 182, 1967.

61. **Black, A. L.,** Altering behavior of geriatric patients with beer, *Northwest Med.,* 456, 1969

62. **Best, C. H., Hartroff, W. S., Lucas, C. C., and Ridout, J. H.,** Liver damage produced by feeding alcohol or sugar and its prevention by choline, *Br Med. J.,* 2, 1001, 1949

63. **Lieber, C. S. and DeCarli, L. M.,** Animal models of ethanol dependence and liver injury in rats and baboons, *Fed. Proc. Fed. Am. Soc. Exp. Biol ,* 35, 1232, 1976.

64. **Janus, E. D. and Lewis, B.,** Alcohol and abnormalities of lipid metabolism, *Clin Endocrinol. Metab.,* 7, 321, 1978.

65. **Hsu, J. M.,** Ethanol consumption and free amino acids of rat plasma and liver, in *Alcohol and Nutrition,* NIAAA Research Monograph, DHEW Publ. No. ADM 79-790, Department of Health, Education and Welfare, Washington, D.C , 1979, 298

66  **Rothschild, M. A., Cratz, M., and Schreiber, S. S.,** Alcohol, amino acids and albumin synthesis, *Gastroenterology,* 65, 1200, 1971.

67. **Tomasula, P. A., Kater, R. M. H., and Iber, F. L.,** Impairment of thiamine absorption in alcoholism, *Am. J Clin Nutr ,* 21, 1340, 1968.

68  **Kalbfleish, J. M., Lindeman, R. D., Ginn, H. E.,** et al., Effects of ethanol administration on urinary excretion of magnesium and other electrolytes in alcoholic and normal subjects, *J. Clin. Invest ,* 42, 1471, 1963.

69. **Flink, E. B.,** Magnesium deficiency and magnesium toxicity in man, in *Trace Elements in Human Health and Disease,* Vol. 2, Prasad, A. S. and Oberleas, D., Eds., Academic Press, New York, 1976, 1.

70  **Sargent, W. Q., Simpson, J. R., and Beard, J. D.,** Twenty-four-hour fluid intake and renal handling of electrolytes after various doses of ethanol, *Alcoholism Clin. Exp. Res ,* 4, 74, 1980.

71  **Baines, M.,** Detection and incidence of B and C vitamin deficiency in alcohol-related illness, *Anal Clin. Biochem.,* 15, 307, 1978.

72. **Prasad, A. S.,** *Zinc Metabolism,* Charles C Thomas, Springfield, Ill., 1966.

73. **Lieber, C. S.,** Chronic alcoholism hepatic injury in experimental animals and man: biochemical pathway and nutritional factors, *Fed Proc. Fed Am. Soc Exp. Biol.,* 26, 1443, 1967.

74  **Hillman, R. W.,** Alcoholism and malnutrition, in *The Biology of Alcoholism,* Vol 3, Kissin, B and Beglieter, H., Eds., Plenum Press, New York, 1974, 515

75. **Leevy, C. M., Thompson, A., and Baker, H.,** Vitamins and liver injury, *Am J Clin. Nutr ,* 23, 493, 1970.

76. **Shaw, G. K. and Thomson, A. D.,** A joint psychiatric and medical outpatient clinic for alcoholics, in *Alcoholism· New Knowledge and New Responses,* Edwards, G. and Grant, M., Eds., Croom Helm, London, 1977, 328.

77 **Thomson, A. D., Rae, S. A., and Majumdar, S. K.,** Malnutrition in the alcoholic, in *Medical Consequences of Alcohol Abuse,* Clark, P. M S. and Kricka, L. J , Eds., Ellis Horwood, Chichester, England, 1980, 103.

78. **Vallee, B. L., Wacker, W. E. C., Bartholomay, D. D.** et al., Zinc metabolism in hepatic dysfunction. II. Correlation of metabolic patterns with biochemical findings, *N. Engl J Med ,* 257, 1055, 1957.

79 **Walravens, P. A.,** Nutritional importance of copper and zinc in neonates and infants, *Clin. Chem.,* 26, 185, 1980

80. **Prasad, S. A.,** *Zinc in Human Nutrition,* CRC Press, Boca Raton, Fla , 1979.

81. **Heaton, F. W., Pyrah, L. N., Beresford, C. C., Bryson, R. W., and Martin, D. F.,** Hypomagnesemia in chronic alcoholism, *Lancet,* 2, 802, 1962

82. **Nordin, B. E. C.,** *Calcium, Phosphate and Magnesium Metabolism: Clinical Physiology and Diagnostic Procedures,* Churchill Livingstone, Edinburgh, 1976, 186

83. **Victor, M., Adams, R. D., and Collins, G. H.,** *The Wernicke Korsakoff Syndrome,* Blackwell, Oxford, 1971.

84. **Graham, J. R., Woodhouse, D., and Read, F. H.,** Massive thiamine doses in an alcoholic with cerebellular cortical degeneration, *Lancet,* 2, 107, 1971.

85. **Perkoff, G. T., Dioso, M. M., Bleisch, V., and Klinkerfuss, G.,** A spectrum of myopathy associated with alcoholism. I. Clinical and laboratory features, *Ann. Intern. Med.,* 67, 481, 1967.

86. **Fennelly, J., Frank, O., Baker, H., and Leevy, C. M.,** Peripheral neuropathy of the alcoholic. I. Etiological role of anurin and other B-complex vitamins, *Br Med. J.,* 2, 1290, 1964.

87 **Erbslon, F. and Abel, M.,** *Handbook of Clinical Neurology,* Vol. 7, Vinken, V and Bruyn, T., Eds., North-Holland, Amsterdam, 1970.

88. **Hartroft, S. W. and Porta, E. A.,** Alcohol, diet and experimental hepatic injury, *Can. J. Physiol. Pharmacol.,* 46, 462, 1968.

89. **Lane, B. P. and Lieber, C. S.,** Ultrastructural alterations in human hepatocytes following ingestion of ethanol with adequate diets, *Am J. Pathol.,* 49, 593, 1966.

90. **Lieber, C. S. and DeCarli, L. M.,** An experimental model of alcohol feeding and liver injury in the baboon, *J. Med. Primatol.,* 3, 153, 1974.

91. **Lieber, C. S., DeCarli, L. M., and Rubin, E.,** Sequential production of fatty liver, hepatitis and cirrhosis in sub-human primates fed ethanol with adequate diets, *Proc Natl. Acad. Sci. U.S.A.,* 72, 437, 1975

92. **Goldberg, M., Hehir, R., and Hurowitz, M.,** Intravenous tri-iodothyronine in acute alcoholic intoxication, *N. Engl. J. Med.,* 263, 1336, 1966.

93. **Lowenstein, L. M., Somone, R., Boulter, P., and Nathan, P.,** *Effects of fructose on alcohol concentrations in the blood in man, JAMA,* 213, 1899, 1970.

94 **Breglia, R. J., Ward, C. O., and Jawoski, C. I.,** Effect of selected amino acids on ethanol toxicity in rats, *J. Pharm. Sci.,* 62, 49, 1973.

95. **Pawan, G. L. S.,** Vitamins, sugars and ethanol metabolism in man, *Nature (London),* 220, 374, 1968.

96. **Bonjour, J. P.,** Vitamins and alcoholism. I. Ascorbic acid, *Int. J. Vitam. Nutr. Res.,* 49, 434, 1979.

97. **Bonjour, J. P.,** Vitamins and alcoholism. II Folate and vitamin $B_{12}$, *Int. J. Vitam. Nutr. Res.,* 50, 96, 1980.

98. **Bonjour, J. P.,** Vitamins and alcoholism. IV Thiamin, *Int. J. Vitam. Nutr Res.,* 50, 321, 1980.

99. **Bonjour, J. P.,** Vitamins and alcoholism. V Riboflavin. VI. Niacin VII Pantothenic acid VIII. Biotin, *Int. J. Vitam. Nutr. Res.,* 50, 425, 1980.

100. **Sorrel, M. F., Baker, H., Barak, A. J., and Frank, O.,** Release by ethanol of vitamins in rat liver perfusate, *Am. J. Clin. Nutr.,* 27, 743, 1974.

101 **Leathen, J. H.,** *Nutrition in The Testes,* Vol 3, Johnson, A. D., Gomes, W R., and Vandermark, N. L., Eds., Academic Press, New York, 1970, 188.

102. **Van Thiel, D. H., Gavalet, J. S., and Lester, R.,** Ethanol inhibition of vitamin A metabolism in the testes. Possible mechanism for sterility in alcoholics, *Science,* 186, 941, 1974.

103. **Smith, J. C. J., Brown, E. D., McDaniel, E. G., and Chan, W.,** Alterations in vitamin A metabolism during zinc deficiency and food and growth restriction, *J. Nutr.,* 106, 569, 1976.

104. **Benhamou, J. P.,** Should chronic liver disease disqualify one for night driving? *Gastroenterology,* 65, 345, 1973.

105. **Van Thiel, D. H., Gavaler, J. S., Lester, R., and Goodman, M. D.,** Alcohol-induced testicular atrophy: an experimental model for hypogonadism occurring in chronic alcoholic men, *Gastroenterology,* 69, 326, 1975.

106. **Dergun, M. S., Fiabane, A., Paterson, C. K., and Zarembski, P.,** Vitamin and mineral nutrition in chronic alcoholics including patients with Korsakoff psychosis, *Br. J. Nutr.,* 45, 469, 1981.

107. **Schiff, E. R., Small, N. C., and Dietschy, J. M.,** Characterization of the kinetics of the passive and active transport mechanisms for bile acid absorption in the small intestine and colon of the rat, *J. Clin. Invest.,* 51, 1351, 1972.

108. **French, S. W. and Castagna, J.,** Some effects of chronic ethanol feeding on vitamin B$_6$ deficiency in the rat, *Lab. Invest ,* 16, 526, 1967.
109. **Lishman, W. A., Ron, M. A., and Acker, W.,** Computed tomography of the brain and psychometric assessment of alcoholic patients — a British Study, in *Addiction and Brain Damage,* Richter, D , Ed , Croom Helm, London, 1980, 215
110. **Centerwall, B. G. and Crigui, M. H.,** Prevention of Wernicke-Korsakoff syndrome, *N. Engl. J Med.,* 299, 285, 1978.
111. **Leevy, C. M., Thomson, A. D., and Baker, H.,** Vitamins and liver injury, *Am. J. Clin. Nutr ,* 23, 493, 1970.
112. **Desmond, P. V., Lowrenzs, C., and Breen, K. J.,** Thiamine hydrochloride absorption in man. normal kinetics and absence of acute effect of ethanol, *Austr. N.Z. J. Med.,* 6, 264, 1976.
113. **Eichner, E. R., Buchanan, B., Smith, J. W., and Hillman, R. S.,** Variations in the hematologic and medical status of alcoholics, *Am. J. Med. Sci.,* 263, 35, 1972.
114. **Lindenbaum, J., Saha, J. R., Shea, N., and Lieber, C. S.,** Mechanism of alcohol-induced malabsorption of vitamin B$_{12}$, *Gastroenterology,* 64, 762, 1973.
115 **Whitehead, T. P., Clarke, A. C., and Whitfield, A. G. W.,** Biochemical and haematological markers of alcohol intake, *Lancet,* 1, 978, 1978.
116. **Holick, M. F., Schnoes, H. K., and DeLuca, H. F.,** Identification of 1,25-dihydroxycholecalciferol, a form of vitamin D3 metabolically active in the intestine, *Proc Natl. Acad. Sci U.S.A.,* 68, 803, 1971.
117. **Dechavanne, M., Barbier, Y., Prost, G., Pehlivanian, E., and Tolot, E.,** Study of absorption of calcium in alcoholic cirrhosis: action of 25-hydroxycholecalciferol, *Nouv. Presse Med.,* 3, 2549, 1974
118. **Schwarz, K.,** Role of vitamin E, selenium, and related factors in experimental nutritional liver disease, *Fed. Proc. Fed. Am Soc. Exp. Biol.,* 24, 58, 1965.
119. **DiLuzio, N. R. and Hartman, A. D.,** Modification of acute and chronic ethanol-induced hepatic injury and the role of lipid peroxidation in the pathogenesis of the ethanol-induced fatty liver, in *Biochemical and Clinical Aspects of Alcohol Metabolism,* Sardesai, V. M., Ed., Charles C Thomas, Springfield, Ill , 1969, 133.
120. **Bunyan, J., Cawthorne, M. A., and Diplock, T. A.,** Vitamin E and hepatotoxic agents. I Carbon tetrachloride and lipid peroxidation in the rat, *Br. J Nutr.,* 23, 297, 1969.
121. **Levander, O. A., Morris, V. C., Higgs, D. J., et al.,** Nutritional interrelationships among vitamin E, selenium, antioxidants and ethyl alcohol in the rat, *J Nutr ,* 103, 536, 1973.
122 **Tappel, A. L. and Dillard, C. J.,** In vivo lipid peroxidation: measurement via exhaled pentane and protection by vitamin E, *Fed Proc. Fed Am. Soc. Exp. Biol.,* 40, 174, 1981.
123 **Chen, L. H. and Chang, M. L.,** Effect of dietary vitamin E and vitamin C on respiration and swelling of guinea pig liver mitochondria, *J. Nutr ,* 108, 1616, 1979.
124 **Chen, L. H., Lee, M. S., Hsing, W. F., and Chen, S. H.,** Effect of vitamin C on tissue antioxidant status of vitamin E deficient rats, *Int J Vitam Nutr. Res.,* 50, 156, 1980.
125. **Tamburo, C., Kirkland, M., Cabansag, C., and Leevy, C. M.,** Biochemical alterations in delirium tremens, *Clin. Res.,* 16, 353, 1968.
126. **Leevy, C. M. and Baker, H.,** Vitamins and alcoholism, *A. J. Clin. Nutr.,* 21, 1325, 1968.
127. **Knochel, J. P.,** The pathophysiology and clinical characteristics of severe hypo-phosphataemia, *Arch. Intern. Med.,* 137, 203, 1977.
128. **MacDonald, R. A., Jones, R. S., and Pechet, G. S.,** Folic acid deficiency and hemochromatosis, *Arch. Pathol.,* 80, 153, 1965.
129. **Aron, E., Paoletti, C., Jobard, P., and Gosse, C.,** *Arch. Mal. Appar. Dis.,* 50, 745, 1961.
130 **Charlton, R. W., Jacobs, P., Seftel, H., and Bothwell, T. H.,** Effect of alcohol on iron absorption, *Br. Med. J.,* 2, 1427, 1964.
131. **Miralles-Garcia, J. M. and De-Castro-del-Pozo, S.,** Iron deposits in chronic alcoholics: special studies in relation to the iron contained in red wine, *Acta Hepato-Gastroenterol.,* 23, 10, 1976.
132. **Cosnett, J. E.,** Miner's syndrome, *S. Afr. Med. J.,* 48, 2011, 1974.
133. **Chapman, D., Urbina, J., and Keough, K. M.,** Biomembrane phase transitions: studies of lipid-water systems using differential scanning colorimetry, *J. Biol. Chem.,* 249, 2512, 1974.
134. **Rasmussen, H. and Goodman, D. P. B.,** Relationships between calcium and cyclic nucleotides in cell activation, *Physiol. Rev.,* 57, 421, 1977.
135. **Aikawa, J. K.,** Biochemistry and physiology of magnesium, in *Trace Elements in Human Health and Disease,* Vol 11, Prasad, A. S and Oberleas, D., Eds., Academic Press, New York, 1976, 47.
136. **Wacker, W. E. C. and Parish, A. F.,** Magnesium metabolism, *N Engl. J. Med.,* 278, 712, 1968.
137. **Rubin, R. P.,** *Calcium and the Secretory Process,* Plenum Press, New York, 1974, 1
138 **Flink, E. B., Stutzman, F. L., Anderson, A. R., Konig, T., and Fraser, R.,** Magnesium deficiency after prolonged parenteral fluid administration and after chronic alcoholism complicated by delirium tremens, *J. Lab. Clin Med.,* 43, 169, 1954

139. **Ylikahri, R. H., Poso, A. R., Huttunen, M. D., and Hillbom, M. E.**, Alcohol intoxication and hangover effects on plasma electrolyte concentration and acid-base balance, *Scand. J. Clin. Lab Invest.*, 34, 327, 1974.

140. **Medalle, R. and Waterhouse, C.**, A magnesium-deficient patient presenting with hypocalcemia and hyperphosphatemia, *Ann. Intern. Med.*, 79, 76, 1973.

141. **Estep, H., Shaw, W. A., Watlington, C., Hobe, R., Holland, W., and Tucker, S. G.**, *J Clin. Endocrinol Metab.*, 29, 842, 1969.

142. **Moore, M. D.**, Studies with ion exchange calcium electrodes, *Gastroenterology*, 60, 43, 1971.

143. **D'Souza, A. and Floch, M. H.**, Calcium metabolism in pancreatic disease, *Am. J Clin. Nutr.*, 26, 352, 1973

144. **Smith, W. O. and Hamnarstein, J. F.**, Intracellular magnesium in delirium tremens and uremia, *Am. J. Med. Sci*, 237, 413, 1959.

145. **Wheeler, E. G., Smith, T., Gallindano, C., Alam, A. N., Wilkinson, S., Edmunds, C. J., and Williams, R.**, Potassium and magnesium depletion in patients with cirrhosis on maintenance diuretic regimes, *Gut*, 18, 683, 1977

146. **Whang, R., Ryan, M. P., and Aikawa, J. K.**, Delirium tremens; a clinical example of cation pump failure (Editorial), *Am. J. Clin. Nutr.*, 27, 447, 1974.

147. **Brooks, B. R. and Adams, R. D.**, Cerebrospinal fluid acid-base and lactate changes after seizures in unanesthesized man to alcohol withdrawal seizures, *Neurobiology*, 25, 943, 1975.

148. **Sargent, W. Q., Simpson, J. R., and Beard, J. D.**, The effect of acute and chronic alcohol administration on renal hemodynamics and monovalent ion excretion, *J. Pharmacol. Exp. Ther.*, 190, 501, 1974

149. **Peng, T. C. and Gitelman, H. J.**, Ethanol-induced hypocalcemia, hypermagnesemia and inhibition of the serum calcium-raising effect of parathyroid hormones in rats, *Endocrinology*, 94, 608, 1974.

150. **Krawitt, E. L.**, Effect of ethanol ingestion on duodenal calcium transport, *J Lab. Clin Med.*, 85, 665, 1975.

151. **Yunice, A. A.**, Altered tissue trace metals after chronic ingestion of ascorbic acid and/or ethanol, *Fed. Proc. Fed. Am. Soc. Exp Biol.*, 38, 449, 1979.

152. **Yunice, A. A., King, R. W., Kraikitpanitch, S.**, et al., Urinary zinc excretion following infusion of zinc sulfate, cysteine, histidine, or glycine, *Am J. Physiol.*, 235(1), F40, 1978.

153. **Yunice, A. A. and Haygood, C. C.**, Comparative effects of ascorbic acid and ethanol infusion on urinary excretion of cations in the dog, *Physiologist*, 22, 137, 1979.

154. **Yunice, A. A.**, Effect of ascorbic acid (ASA) loading on renal tubular transport of cations in the dog, *Fed. Proc. Fed. Am. Soc. Exp. Biol.*, 39, 3346, 1980

155. **Alexander, G. J. and Alexander, R. B.**, Alcohol consumption in rats treated with lithium carbonate or rubidium chloride, *Pharmacol Biochem. Behav.*, 8, 533, 1978.

156. **Messiha, F. S.**, Cesium and rubidium salts: effects on voluntary intake of ethanol by the rat, *Pharmacol. Biochem. Behav.*, 9, 647, 1978.

157. **Wren, J. C., Kline, N. S., Cooper, T. B.**, et al., Evaluation of lithium therapy in chronic alcoholism, *Clin. Med.*, 81, 33, 1974.

158. **Guerri, C. and Grisolla, S.**, Potentiation of ethanol toxicity by lithium and vanadate, in *Animal Models in Alcohol Research*, Erikson, K., Sinclair, J D., and Kiianmaa, D., Eds., Academic Press, New York, 1980, 323.

159. **Miller, W. J., Morton, D. J., Pitts, W. J.**, et al., Effect of zinc deficiency and restricted feeding on wound healing in the bovine, *Proc. Soc. Exp Biol. Med.*, 718, 427, 1965.

160. **Sandstead, H. H., Lanier, V. C., Shepherd, G. H.** et al , Zinc and wound healing effects of zinc deficiency and zinc supplementation, *Am J. Clin. Nutr.*, 23, 514, 1970

161. **O'Rian, S. H., Copenhagen, H. J., and Calman, J. S.**, The effect of sulfate on the healing of incised wounds in the rat, *Br J. Plast. Surg.*, 21, 240, 1968.

162. **Pories, W. J., Henzel, J. H., Rob, C. G.** et al., Acceleration of wound healing in man with zinc sulfate given by mouth, *Lancet*, 1, 121, 1967.

163. **Weser, U., Seeber, S., and Warneeke, P.**, Reactivity of zinc on nuclear DNA and RNA biosynthetic regenerating rat liver, *Biochem. Biophys. Acta*, 179, 442, 1969.

164. **Grey, P. C. and Dreosti, I. E.**, Deoxyribonucleic acid and protein metabolism in zinc-deficient rats, *J. Comp. Pathol.*, 82, 223, 1972.

165. **Yunice, A. A. and Lindeman, R. D.**, Effect of ascorbic acid and zinc sulfate on ethanol toxicity and metabolism, *Proc. Soc. Exp. Biol. Med.*, 154, 146, 1977

166. **Jamall, I. S., Mignano, J. E., Lynch, V. D.** et al., Protective effects of zinc sulfate and L-lysine on acute ethanol toxicity in mice, *Environ. Res.*, 19, 112, 1979

167. **Vallee, B. L., Wacker, W. E. C., Bartholomay, A. F., and Robin, E. D.**, Zinc metabolism in hepatic dysfunction. I. Serum zinc concentrations in Laennec's cirrhosis and their validation by sequential analysis, *N. Engl. J. Med.*, 255, 403, 1956

168. **Huber, A. M. and Gershoff, S. N.**, Effect of zinc deficiency on the oxidation of retinol and ethanol in rats, *J. Nutr.*, 105, 1486, 1975.
169. **Wang, J. and Pierson, R. N., Jr.**, Distribution of zinc in skeletal muscle and liver tissue in normal and dietary controlled alcoholic rats, *J. Lab. Clin. Med.*, 85, 50, 1975.
170. **Prasad, A. S.**, Effect of ethanol in zinc metabolism, in Alcohol and Nutrition, Li, T. K., Schenker, S., and Luming, L., Eds., Publ. No. ADM 79-780, Department of Health Education and Welfare, Washington, D.C., 1979, 157.
171 **Sullivan, J. F.**, Effect of alcohol on urinary zinc excretion, *Q. J. Stud Alcohol*, 23, 216, 1962.
172. **Reinhold, J. G., Pascol, E., Arscanian, M.** et al., Relation of zinc metallic enzyme activities to zinc concentration in tissues, *Biochem. Biophys Acta*, 215, 430, 1970.
173 **Dreosti, I. E. and Record, I. R.**, Superoxide dismutase, zinc status and ethanol consumption in maternal and foetal rat livers, *Br. J. Nutr.*, 41, 399, 1979.
174 **Smith, J. J.**, The effects of alcohol on the adrenal ascorbic acid and cholesterol of the rat, *J. Clin Endocrinol.*, 11, 792, 1951.
175. **Chvapil, M., Ryan, J. H., Elias, S. I.** et al., Protective effect of zinc on carbon tetrachloride-induced liver injury in rats, *Exp Mol. Pathol.*, 19, 186, 1973.
176. **Chin, J. H. and Goldstein, D. B.**, Drug tolerance in biomembranes; a spin label study of the effects of ethanol, *Science*, 196, 684, 1977.
177. **Fredericks, R. E., Tanaka, K. R., and Valentine, W. N.**, Zinc in human blood cells. normal values and abnormalities associated with liver disease, *J. Clin. Invest.*, 39, 1651, 1960.
178. **Sullivan, J. F. and Lankford, H. G.**, Urinary excretion of zinc in alcoholism and post-alcoholic cirrhosis, *Am. J. Clin. Nutr.*, 10, 153, 1962
179. **Prasad, A. S., Oberleas, D., and Halsted, J. A.**, Determination of zinc in biological fluids by atomic absorption spectrophotometry in normal and cirrhotic patients, *J. Lab Clin. Med.*, 66, 508, 1965.
180. **Kahn, A. M., Helwig, H. L., Redeker, A. G.** et al., Urine and serum zinc abnormalities in disease of the liver, *Am. J. Clin. Pathol.*, 44, 426, 1965.
181 **Halsted, J. A. and Smith, J. C., Jr.**, Plasma zinc in health and disease, *Lancet*, 1, 322, 1970.
182. **Sinha, S. N. and Gabrielli, E. R.**, Serum copper and zinc levels in various pathological conditions, *Am J. Clin. Pathol.*, 54, 570, 1970.
183. **Lindeman, R. D., Baxter, D. J., Yunice, A. A.** et al., Serum concentrations and urinary excretions of zinc in cirrhosis, nephrotic syndrome and renal insufficiency, *Am. J. Med. Sci.*, 275, 17, 1978.
184. **Neville, J. N., Eagles, J. A. Samson, G.** et al., Nutritional status of alcoholics, *Am. J. Clin Nutr.*, 21, 1329, 1968.
185. **Halsted, C. H., Robles, E. A., and Mezey, E.**, Intestinal malabsorption in folate-deficient alcoholics, *Gastroenterology*, 64, 526, 1973.
186. **Krasner, N., Dow, J., and Goldberg, A.**, Re-evaluation of Vitamin C, in *Ethanol Metabolism and Vitamin C*, Hanck, A and Ritzel, C., Eds., Hans Huber Verlag, Berlin, 1977, 139.
187. **Busnel, R. G. and Lehmann, A. G.**, Antagonistic effect of sodium ascorbate on ethanol-induced changes in swimming of mice, *Behav. Brain Res*, 1, 351, 1980.
188. **Sprince, H., Parker, C. M., and Smith, G. C.**, L-Ascorbic acid in alcoholism and smoking. protection against acetaldehyde toxicity as an experimental model, *Int. J. Vitam. Nutr Res.*, 16(Suppl.), 185, 1977.
189. **Sprince, H., Parker, C. M., Smith, G. L.** et al., Protective action of ascorbic acid and sulfur compounds against acetaldehyde toxicity: implication in alcohol and smoking, *Agents Actions*, 5, 164, 1975
190. **Undefriend, S.**, Formation of hydroxyproline in collagen, *Science*, 152, 1335, 1965.
191. **Stone, N. and Meister, A.**, Function of ascorbic acid in the conversion of proline to collagen hydroxyproline, *Nature (London)*, 194, 555, 1962.
192. **Harwood, R., Grant, M. E., and Jackson, D. S.**, Influence of ascorbic acid on ribosomal patterns and collagen biosynthesis in healing wounds of scorbutic guinea pigs, *Biochem. J*, 142, 641, 1974.
193. **Forbes, J. C. and Duncan, G. M.**, Effect of vitamin C intake on the adrenal response of rats and guinea pigs to alcohol administration, *Q. J. Stud. Alcohol*, 22, 22, 1953.
194 **Forbes, J. C. and Duncan, G. M.**, Effect of repeated alcohol administration on adrenal ascorbic acid and on development of scurvy in the guinea pig, *Q. J. Stud. Alcohol*, 14, 540, 1953.
195. **Vanha-Pertulla, T. P. J.**, The influence of vitamin C on eosinophil response to acute alcohol intoxication in rats, *Acta Endocrinol*, 35, 585, 1960.
196. **Krasner, N., Dow, J., Moore, M. R.** et al., Ascorbic acid saturation and ethanol metabolism, *Lancet*, 2, 693, 1974.
197. **Dow, J., Krasner, N., and Goldberg, A.**, Relation between hepatic alcohol dehydrogenase activity and the ascorbic acid in leucocytes of patients with liver disease, *Clin. Sci. Mol Med.*, 49, 603, 1975
198. **Dow, J. and Goldberg, A.**, Ethanol metabolism in the vitamin C deficient guinea pigs, *Biochem. Pharmacol.*, 24, 863, 1975.
199. **Beattie, A. D. and Sherlock, S.**, Ascorbic acid deficiency in liver disease, *Gut*, 17, 571, 1976

200. **Lester, D., Buccino, R., and Bizzoco, D.,** Vitamin C status of alcoholism, *J. Nutr*, 70, 278, 1960.

201. **Zannoni, V. G. and Sato, P. H.,** Interaction with drugs and environmental chemicals. Effect of ascorbic acid on microsomal drug metabolism, *Ann N.Y. Acad. Sci.*, 258, 119, 1975.

202. **Hsu, J. M. and Hsieh, H. S.,** Ethanol increases urinary and tissue ascorbic acid concentrations in rats, *Proc. Soc Exp. Biol Med.*, 170, 448, 1982.

203. **Jellinek, E. M.,** *Disease Concept of Alcoholism,* Hillhouse Press, New Haven, Conn., 1960.

204. **Clinebell, H. J., Jr.,** *Understanding and Counseling the Alcoholic,* Abington Press, Nashville, 1956.

205. **Cicero, T. J.,** A critique of animal analogues of alcoholism, in *Biochemistry and Pharmacology of Ethanol,* Vol 2, Majchrowicz, E. and Nobel, E. P., Eds., Plenum Press, New York, 1979, 533.

206. **Turner, T. B., Mezey, E,. and Kimball, A. W.,** Measurement of alcohol-related effects in man. Chronic effect in relation to levels of alcohol consumption, *Johns Hopkins Med. J*, 141(Parts A and B), 235, 141, 273, 1977.

207. **Terris, M.,** Epidemiology of cirrhosis of the liver: national mortality data, *Am. J. Public Health*, 57, 2076, 1967.

208. **Brenner, M. H.,** Trends in alcohol consumption and associated illnesses, *Am. J Public Health*, 65, 1279, 1975.

209. **Mezey, E.** et al., Pancreatic function and intestinal absorption in chronic alcoholism, *Gastroenterology,* 59, 657, 1970.

210. **Haggard, H. W., Greenberg, L. A., and Lolli, G.,** The absorption of alcohol with special reference to its influence on the concentration of alcohol appearing in the blood, *Q. J. Stud. Alcohol*, 1, 684, 1941.

211 **Hawkins, R. D. and Kalant, H.,** The metabolism of ethanol and its metabolic effects, *Pharmacol. Rev.*, 24, 67, 1972.

212. **Popper, H.,** Pathological aspects of cirrhosis: a review, *Am. J. Pathol.*, 81, 206, 1917.

213. **Lieber, C. S. and DeCarli, L. M.,** Metabolic effects of alcohol on the liver, in *Metabolic Aspects of Alcoholism,* Lieber, C. S., Ed., MTP Press, Lancaster, England, 1977, 135.

214. **Rubin, E. and Lieber, C. S.,** Early fine structural changes in the human liver induced by alcohol, *Gastroenterology,* 52, 1, 1967.

215. **Moon, V. H.,** Experimental cirrhosis in relation to human cirrhosis, *Arch. Pathol. (Chicago),* 18, 381, 1934.

216. **Boyd, W.,** *Pathology for the Physician,* 7th ed., Lea & Febiger, Philadelphia, 1965, 461.

217. **Popper, H., Rubin, E., Krus, S., and Schaffner, F.,** Postnecrotic cirrhosis in alcoholics, *Gastroenterology,* 39, 669, 1960

218. **Gomez-Dumm, C. L. A., Porta, M. D., and Hartroft, W. S.,** A new experimental approach in the study of chronic alcoholism. II. Effects of high alcohol intake in rats fed diets of various adequacies, *Lab. Invest.,* 18, 365, 1968.

219. **Leevy, C. M., Tamburro, C., and venHove, W.,** Ethanol oxidizing enzymes and response to liver injury, *Liver Res.,* Suppl 10b, 71, 1967.

220. **Popper, H. and Schaffner, F.,** *Structural Studies in Alcohol and Drug Induced Liver Injury in Cirrhosis and Other Toxic Hepatopathies,* Nordiska Bokhandelno Forlag, Sweden, 1968, 15.

221. **Lieber, C. S. and DeCarli, L. M.,** Metabolic effects of alcohol on the liver, in *Metabolic Aspects of Alcoholism,* Lieber, C. S., Ed., MTP Press, Lancaster, England, 1977, 31.

222. **DeCarli, L. and Lieber, C. S.,** Fatty liver in the rat after prolonged intake of ethanol with a nutritionally adequate new liquid diet, *J. Nutr.,* 91, 331, 1967.

223. **Popper, H.,** Morphologic and biochemical aspects of fatty liver, *Acta Hepato-Splenol.,* 8, 279, 1961.

224. **Wands, J. R., Center, E. A., Bucher, N. L. R., and Isselbacher, K. S.,** Inhibition of hepatic regeneration in rats by acute and chronic ethanol intoxication, *Gastroenterology,* 77, 528, 1979.

225. **Suematsu, T. and Matsumura, T.** et al., Lipid peroxidation in alcoholic liver disease in humans, *Alcoholism Clin. Exp. Res.,* 5, 427, 1981

226. **Waesm, K. C. and Lieber, C. S.,** Glutamate dehydrogenase, a reliable marker of liver cell necrosis in the alcoholics, *Br. Med. J.,* 2, 1508, 1977.

227 **Gillespie, J. A.,** Vasodilator properties of alcohol, *Br. Med. J.,* 2, 274, 1967.

228. **Eliaser, M., Jr. and Giensircusa, F. J.,** Heart and alcohol, *Calif. Med.,* 84, 234, 1956.

229. **Regan, T. J., Khan, M. I., Ettinger, P. O., Haider, B., Lyons, M. M., and Oldewurtel, H. A.,** Myocardial function and lipid metabolism in the chronic alcoholic animals, *J. Clin. Invest.,* 54, 740, 1974.

230. **Perkoff, G. T., Hardy, P., and Velez-Garcia, E.,** Reversible acute muscular syndrome in chronic alcoholism, *N. Engl. J. Med.,* 274, 1277, 1966.

231. **Wendt, V. E., Wu, C., Balcon, R., Dotty, G., and Bing, R. J.,** Hemodynamic and metabolic effects of chronic alcoholism in man, *Am. J. Cardiol.,* 15, 175, 1965.

232. **Evans, W.,** Electrocardiogram of alcoholic cardiomyopathy, *Br. Heart J.,* 21, 445, 1959.

233. **Bridgen, W.,** Uncommon myocardial disease, non-coronary cardiomyopathies, *Lancet,* 2, 1179, 1957.

234. **Burch, G. E. and Walsh, J. J.,** Cardiac insufficiency in chronic alcoholism, *Am. J. Cardiol.,* 6, 864, 1960.

235. **Ferrans, V. J., Ribbs, R. G., and Weilbaecher, D. G.,** Alcoholic cardiomyopathy: a histochemical and electron microscopic study, *Am. J. Cardiol.,* 13(Abstr.), 106, 1964.

236. **Rubin, E.,** Alcohol: toxic or tonic? *Cardiovasc. Rev. Rep.,* 2, 23, 1981.

237. **Demakis, J. G.** et al., The natural course of alcoholic cardiomyopathy, *Ann. Intern. Med.,* 80, 293, 1974.

238. **Levi, G. F.** et al., Preclinical abnormality of left ventricular function in chronic alcoholics, *Br. Heart J.,* 39, 35, 1977.

239. **Svob, A.** et al., Cardiovascular diseases in chronic alcoholics, *Alcoholism Zagreb.,* 10, 81, 1974.

240. **Stason, W. B.** et al , Alcohol consumption and nonfatal myocardial infarction, *Am. J. Epidemiol.,* 104, 603, 1976.

241. **Yano, K., Rhoads, G. G., and Kangan, A.,** Coffee, alcohol and risk of coronary heart disease among Japanese men living in Hawaii, *N. Engl. J. Med.,* 297, 405, 1977.

242. **Barboriak, J. J.** et al., Coronary artery occlusion and alcohol intake, *Br. Heart J.,* 39, 289, 1977.

243. **Klatsky, A. L.** et al., Alcohol consumption and blood pressure: Kaiser-Permanente multiphasic health examination data, *N Engl. J. Med.,* 296, 1194, 1977.

244. **Barboriak, J. G., Anderson, A. J., and Hoffmann, R. G.,** Interrelationship between coronary artery occlusion, high-density lipoprotein, cholesterol and alcohol intake, *J. Lab. Clin. Med.,* 94, 348, 1979.

245 **Hennekens, C. H., Willett, W., Rooner, B., Cole, D. S., and Mayrent, S. L.,** Effects of beer, wine and liquor in coronary deaths, *JAMA,* 242, 1973.

246. **Hennekens, C. H., Rosner, B., and Cole, D. S.,** Daily alcohol consumption and fatal coronary heart disease, *Am. J Epidemiol.,* 107, 116, 1978.

247. **Epstein, M., Berk, D. P., and Hollenberg, H. K.,** et al., Renal failure in the patient with cirrhosis: the role of active vasoconstriction, *Am. J. Med.,* 49, 175, 1970.

248. **Papper, S., Belsky, J. L., and Bleifer, K. H.,** Renal failure in Laennec's cirrhosis of the liver, I. Description of clinical and laboratory features, *Ann. Intern. Med.,* 51, 759, 1959

249. **Laube, H., Norris, H. T., and Robbins, S. L.,** The nephromegaly of chronic alcoholics with liver disease, *Arch. Pathol.,* 84, 290, 1967

250. **Van Thiel, D. H., Gavaler, J. S., Little, J. M., and Lester, F. C.,** Alcohol its effect on the kidney, *Metabolism,* 26, 857, 1977.

251. **Scheriker, V. J.,** Adrenal hormones and amine metabolism in alcoholism, *Psychosom. Med.,* 28, 564, 1966.

252 Alcohol and the Impaired Driver, A Manual on the Medicolegal Aspects of Chemical Tests for Intoxication, American Medical Association, Chicago, 1968

253. **Davis, V. E., Brown, H., Huff, A. J., and Cashaw, L. J.,** Ethanol-induced alterations of norepinephrine metabolism in man, *J. Lab. Clin. Med.,* 69, 787, 1967.

254. **Davis, V. E., Brown, H., Huff, A. J., and Cashaw, L. J.,** The alteration of serotonin metabolism to 5-hydroxytryptamine by ethanol ingestion in man, *J. Lab. Clin. Med.,* 69, 132, 1967.

255. **Duritz, G. and Truitt, E. B., Jr.,** Importance of acetaldehyde in the action of ethanol on brain norepinephrine and 5-hydroxytryptamine, *Biochem Pharmacol ,* 15, 711, 1966

256 **Towne, J. C.,** Effect of ethanol and acetaldehyde on liver and brain monoamine oxidase, *Nature (London),* 201, 709, 1964.

257. **Lieber, C. S. and Davidson, C. S.,** Some metabolic effects of ethyl alcohol, *Am. J. Med ,* 33, 319, 1962.

258. **Ridge, J. W.,** The metabolism of acetaldehyde by the brain in vivo, *Biochem J ,* 88, 95, 1963

259 **Truitt, E. B., Jr. and Duritz, G.,** The role of acetaldehyde in the actions of ethanol, in *Biochemical Factors in Alcoholism,* Maickel, R. P., Ed., Pergamon Press, New York, 1967

260 **Sutherland, V. C.** et al., Cerebral metabolism in problem drinkers under the influence of alcohol and chlorpromazine hydrochloride, *J Appl. Physiol.,* 15, 189, 1960

261. **Edwards, J. A. and Price, J.,** Metronidazole and human alcohol dehydrogenase, *Nature (London),* 214, 190, 1967.

262 **Sarles, H.,** Alcohol and the pancreas, *Nutr Metab.,* 21, 175, 1977.

263. **Strum, W. B. and Spiro, H. M.,** Chronic pancreatitis, *Ann Intern. Med.,* 74, 264, 1971

264. **Lemire, S. and Iber, F. L.,** Pancreatic secretion in rats with protein malnutrition, *Johns Hopkins Med. J.,* 120, 21, 1967.

265. **Mezey, E. and Tobonyl, E.,** Rates of ethanol clearance and activities of the ethanol-oxidizing enzymes in chronic alcoholic patients, *Gastroenterology,* 61, 707, 1971

266 **Cameron, J. L., Zuideman, G. D., and Margolis, S.,** A pathogenesis for alcoholic pancreatitis, *Surgery,* 77, 754, 1975.

267. **Zieve, F. G.,** Jaundice, hyperlipemia and hemolytic anemia a heretofore unrecognized syndrome associated with alcoholic fatty liver and cirrhosis, *Ann. Intern. Med ,* 48, 471, 1958

268. **Blass, J. P. and Dean, H. M.,** The relation of hyperlipemia to hemolytic anemia in an alcoholic patient, *Am. J. Med.,* 40, 283, 1966.

269. **Brewster, A. C., Lankford, H. G., Schwartz, M. G., and Sullivan, F. J.,** Ethanol and alimentary lipemia, *Am. J Clin. Nutr.,* 19, 255, 1966.

270. **Sobel, H. J. and Waye, J. D.**, Pancreatic changes in various types of cirrhosis in alcoholics, *Gastroenterology*, 45, 341, 1963
271. **Beard, J. D. and Knott, D. H.**, Hematopoietic response to experimental chronic alcoholism, *Am J Med. Sci.*, 252, 518, 1966
272 **Pierce, H. I., McGuffin, R. G., and Hillman, R. S.**, Clinical studies in alcoholic sideroblastosis, *Arch Intern. Med.*, 136, 283, 1976
273 **Sullivan, L. W. and Herbert, V.**, Suppression of hematopoiesis by ethanol, *J. Clin. Invest.*, 43, 2048, 1964.
274 **Eichner, E. R. and Hillman, R. S.**, The evolution of anemia in alcoholic patients, *Am J. Med*, 50, 218, 1971.
275 **Lindenbaum, J. and Lieber, C. S.**, Effect of ethanol on the blood and bone marrow, *Am. J. Clin. Nutr.*, 21, 533, 1968
276. **Herbert, V., Zalisky, R., and Davidson, C. S.**, Correlation of folate deficiency with alcoholism and associated macrocytosis, anemia and liver disease, *Ann Intern. Med.*, 58, 977, 1963
277 **Lemoine, P., et al**, Les enfants de parents alcooliques: anomalies observees a propos de 127 cas, *Quest. Med*, 25, 477, 1968
278. **Jones, K. L. and Smith, D. W.**, The fetal alcohol syndrome, *Teratology*, 12, 1, 1975
279. **Mulvihill, J. H. et al**, Fetal alcohol syndrome, *Am. J. Obstet. Gynecol.*, 125, 937, 1976
280 **Cahalan, D., Cisin, I. H., and Crossley, H. M.**, *American Drinking Practices. A National Study of Drinking Behavior and Attitudes*, Rutgers Center of Alcohol Studies, New Brunswick, N J, 1969.
281. **Van Thiel, D. H. and Lester, R.**, Alcoholism: its effect on hypothalamic pituitary gonadal function, *Gastroenterology*, 71, 318, 1976
282 **Van Thiel, D. H., Lester, R., and Sherin, R. J.**, Hypogonadism in alcoholic liver disease: evidence for a double defect, *Gastroenterology*, 67, 1188, 1974
283 **Kent, J. R., Scaramuzzi, R. J., Lammer, W. et al**, Plasma testosterone estradiol and gonadotropins in hepatic insufficiency, *Gastroenterology*, 64, 111, 1973
284. **Mendelson, J. H. and Mello, N. K.**, Alcohol aggression and androgens, *Res Public Assoc. Res Nutr. Mental Dis*, 52, 225, 1974
285. **Kreuz, L. E., Rose, R. M., and Jennings, J. R.**, Suppression of plasma testosterone levels and psychological stress, *Arch. Gen Psychiatr*, 26, 479, 1972
286 **West, C. D. et al**, Simultaneous measurement of multiple plasma steroids by radioimmunoassay demonstrating episodic secretion, *J. Clin. Endocrinol Metab*, 36, 1236, 1973
287 **Saville, P. D. and Lieber, C. S.**, Increases in skeletal calcium and femur cortex thickness produced by undernutrition, *J. Nutr.*, 99, 141, 1969
288. **Saville, P. D.**, Alcohol and skeletal disease, in *Metabolic Aspects of Alcoholism*. Leiber, C S., Ed, MTP Press, Lancaster, England, 1977, 135.
289 **Saville, P. D.**, Changes in bone mass with age and alcoholism, *J. Bone J. Surg*, 47A, 492, 1965
290. **Nilsson, B. E. and Westlin, N. E.**, Femure density in alcoholism and after gastrectomy, *Calcium, Tissue Res.*, 10, 167, 1972
291 **Schottenfeld, D.**, Alcohol as a co-factor in the etiology of cancer, *Cancer*, 43, 1962, 1979
292. **Tuyns, A. J. and Audiger, J. C.**, Double wave cohort increase for oesophageal and laryngeal cancer in France in relation to reduced alcohol consumption during the Second World War, *Digestion*, 14, 197, 1976
293 **Marks, V.**, Biochemical and metabolic basis of alcohol toxicity, *Adv Biol Psychiatr*, 3, 88, 1979
294 **Bouchier, I. A. D. and Dawson, A. M.**, The effect of infusion of ethanol on the plasma free fatty acids in man, *Clin Sci*, 26, 47, 1964
295 **Brown, D. F.**, The effect of ethyl α-p-chloro-phenoxyisobutyrate on ethanol-induced hepatic steatosis in the rat, *Metabolism*, 15, 868, 1966.
296 **DiLuzio, N. R. and Poggi, M.**, Abnormal lipid tolerance and hyperlipidemia in acute ethanol-treated rats, *Life Sci.*, 2, 751, 1963
297 **Elko, E. E., Wooles, W. R., and DiLuzio, N. R.**, Alteration and mobilization of lipids in acute ethanol-treated rats, *Am J. Physiol*, 201, 923, 1961
298. **Lieber, C. S., Jones, P. D., Losowsky, S. M., and Davidson, S. C.**, Interrelation of uric acid and ethanol metabolism in man, *J. Clin. Invest*, 41, 1863, 1962
299. **Thorpe, M. E. C. and Shorey, C. D.**, Long-term alcohol administration: its effects on the ultrastructure and lipid content of the rat liver cell, *Am J. Pathol*, 48, 557, 1966.
300. **Feinman, L. and Lieber, C. S.**, Effect of ethanol on plasma glycerol in man, *Am. J Clin Nutr.*, 20, 400, 1967.
301. **French, S. W.**, Effect of chronic ethanol feeding on rat liver phospholipid, *J. Nutr.*, 91, 292, 1967
302. **Salaspuro, M. P. and Maenpaefe, P. H.**, Influence of ethanol on the metabolism of perfused normal, fatty and cirrhotic rat livers, *Biochem J*, 100, 768, 1966
303 **Lieber, C. S. and DeCarli, L. M.**, Study of agents for the prevention of the fatty liver produced by prolonged alcohol intake, *Gastroenterology*, 50, 316, 1966.

304 **Barona, E. and Lieber, C. S.,** Effect of ethanol and lipoprotein metabolism, *Clin Res ,* 16 (Abstr ), 279, 1968

305. **Crouse, J. R., Gerson, C. D., DeCarli, L. M., and Lieber, C. S.,** Role of acetate on the reduction of plasma free fatty acids produced by ethanol in man, *J Lipid Res.,* 9, 509, 1968.

306. **Lieber, C. S.,** Chronic alcoholic hepatic injury in experimental animals and man. Biochemical pathway and nutritional factors, *Fed. Proc Fed Am Soc Exp Biol.,* 26, 1443, 1967

307. **Krebs, H. A.,** The effects of ethanol on the metabolic activities of liver, *Adv. Enzyme Regul ,* 6, 467, 1968.

308 **Dajani, M. R. and Kouyoumjian, C.,** Alcohol dehydrogenase: in vitro studies of its relationship to hepatic lipid synthesis in the rat, *Q. J. Stud. Alcohol,* 30, 295, 1969

309. **French, S. W.,** Effect of acute and chronic ethanol ingestion on rat liver ATP, *Proc Soc. Exp. Biol. Med ,* 121, 681, 1966.

310 **Kiessling, K. H. and Tilander, K.,** Biochemical changes in rat tissue after prolonged alcohol consumption, *Q J. Stud. Alcohol,* 22, 535, 1961

311. **Kiessling, K. H. and Pilstrom, L.,** Effect of ethanol on rat liver. II. Number, size and appearance of mitochondria, *Acta Pharmacol Toxicol.,* 24, 103, 1966.

312. **Kiessling, K. H. and Pilstrom, L.,** The effect of bile obstruction on the oxidation rate of ethanol in the rat, *Acta Physiol Scand.,* 691, 187, 1967

313 **Marks, V.,** Alcohol and carbohydrate metabolism. *Clin Endocrinol. Metab.,* 7, 333, 1978

314. **Arky, R. A., Veverbrants, E., and Abramson, E. A.,** Irreversible hypoglycemia· a complication of alcohol and insulin, *JAMA,* 206, 575, 1968.

315. **Scholz, R., Grunst, J., Buecher, T., Henley, K. S., and Hendeiman, L. U.,** Metabolic studies with perfused cirrhotic liver, *Gastroenterology,* 54, 348, 1968.

316. **Lieber, C. S. et al ,** Interrelation of uric acid and ethanol metabolism in man, *J Clin. Invest.,* 41, 1863, 1962.

317. **Lieber, C. S., Leevy, C. M., Stein, S. W., George, W. S., Cherrick, G. R., Abelmann, W. H., and Davidson, C. S.,** Effect of ethanol on plasma free fatty acids in man, *J. Lab Clin. Med.,* 59, 826, 1962

318 **Brown, T. M. and Harvey, A. M.,** Spontaneous hypoglycemia in "Smoke" drinkers, *JAMA,* 117, 12, 1941.

319. **Cummins, L. H.,** Hypoglycemia and convulsion in children following alcohol ingestion, *J. Pediatr ,* 58, 23, 1961.

320 **Field, J. B., Williams, H. E., and Mortimove, G. E.,** Studies on the mechanisms of ethanol induced hypoglycemia, *J Clin Invest ,* 42, 497, 1963.

321. **Freinkel, N., Singer, D. L., Arky, R. A., Bleicher, S. J., Anderson, J. B., and Silbert, C. K.,** Alcohol hypoglycemia. I. Carbohydrate metabolism of patients with clinical alcohol hypoglycemia and the experimental production of the syndrome with pure ethanol, *J Clin. Invest ,* 42, 1112, 1963.

322 **Henley, S. K. and Scholz, R.,** Ethanol: drug, "Hepatotoxin", and modifier of intermediary metabolism, *Med. Clin N. Am.,* 53, 1413, 1969.

323. **Madison, L.,** Ethanol-induced hypoglycemia, *Adv. Metab. Dis ,* 3, 85, 1968

324. **Schreiber, S. S., Briden, K., Oratz, M., and Rothchild, M. A.,** Ethanol, acetaldehyde and myocardial protein synthesis, *J. Clin. Invest ,* 51, 2820, 1972.

325 **Schreiber, S. S., Oratz, M., Rothchild, M. A., Reff, F., and Evans, C. D.,** Alcoholic cardiomyopathy. II. The inhibition of cardiac microsomal protein synthesis by acetaldehyde, *J Mol. Cell Cardiol.,* 6, 207, 1974

326 **Schreiber, S. S., Evans. C. D., Reff, F., Rothchild, M. A., and Oratz, M.,** Prolonged feeding of ethanol to the young growing guinea pig. I. The effect on protein synthesis in the afterloaded right ventricle measured in vitro, *Alcoholism. Clin. Exp. Res ,* 6, 384, 1982

327. **Kirsch, R. E., Frith, L. O'C., Stead, R. H., and Saunders, S. J.,** Effect of alcohol on albumin synthesis by the isolated perfused rat liver, *Am J Clin Nutr ,* 26, 1191, 1973.

328. **Oratz, M. and Rothschild, M. A.,** The influence of alcohol and altered nutrition on albumin synthesis, in *Alcohol and Abnormal Protein Biosynthesis,* Rothschild, M. A , Oratz, M., and Schreiber, S S., Eds., Pergamon Press, New York, 1975, 343.

329 **Perin, A., Scalabrino, G., Sessa, A., and Arnaboldi, A.,** In vitro inhibition of protein synthesis in rat liver as a consequence of ethanol metabolism, *Biochim Biophys. Acta,* 366, 101, 1974.

330. **Rothschild, M., Oratz, M., Mongelli, J., and Schreiber, S.,** Alcohol-induced depression of albumin synthesis. reversal by tryptophan, *J. Clin Invest ,* 50, 1812, 1971.

331. **Jeejeebhoy, K. N., Phillips, M. J., Bruce-Robertson, A., Ho, J., and Sodtke, U.,** The acute effect of ethanol on albumin, fibrinogen and transferrin synthesis in the rat, *Biochem. J.,* 126, 1111, 1972

332. **Kuriyama, K., Sze, P. T., and Rauscher, G. E.,** Effects of acute and chronic ethanol administration on ribosomal protein synthesis in mouse brain and liver, *Life Sci ,* 10, 181, 1971.

333. **Renis, M., Giovinc, A., and Bertolino, A.,** Protein synthesis in mitochondrial and microsomal fractions from rat brain and liver after acute or chronic ethanol administration, *Life Sci.,* 16, 1447, 1975.

334. **Blumbert, H. and Grady, H. S.,** Production of cirrhosis of the liver in rats by feeding low protein, high fat diets, *Arch. Pathol.*, 34, 1035, 1942.

335. **Bunyan, J., Cawthorne, M. A., Diplock, A. T., and Green, J.,** Vitamin E and hepatotoxic agents, II Lipid peroxidation and poisoning with orotic acid, ethanol and thioacetamide in rats, *Br. J. Nutr.*, 23, 309, 1969

336. **Patek, A. J., Plough, I. C., and Bevans, M.,** Relative effects of protein and lipotropic substances in the treatment of nutritional cirrhosis in rats, *Ann. N.Y. Acad. Sci.*, 57, 772, 1954.

337. **Lieber, C. S. and Spritz, N.,** Effects of prolonged ethanol intake in man: role of dietary, adipose and endogenously synthesized fatty acids in the pathogenesis of the alcoholic fatty liver, *J. Clin. Invest.*, 45, 1400, 1966.

338. **Takada, A., Porta, E. A., and Hartroft, W. S.,** Regression of dietary cirrhosis in rats fed alcohol and a "Super Diet", *Am. J. Clin. Nutr.*, 20, 213, 1967.

339. **Porta, E. A., Hartroft, W. H., Cesar, L. A., Gomez-Dumm, C. L. A., and Koch, O. R.,** Dietary factors in the progression and regression of hepatic alterations associated with experimental chronic alcoholism, *Fed. Proc. Fed. Am. Soc. Exp. Biol.*, 26, 1449, 1967.

340. **Rubin, E. and Lieber, C. S.,** Early fine structural changes in the human liver induced by alcohol, *Gastroenterology*, 52, 1, 1967.

341. **Lowry, J. V., Ashburn, L. L., Daft, F. S., and Sebrell, W. H.,** Effect of alcohol in experimental liver cirrhosis, *Q. J. Stud. Alcohol*, 3, 168, 1942.

342. **Kasliwal, R. M., Sharma, B. M., and Agarwal, M. P. et al.,** A study of the etiological factors of cirrhosis of liver with special reference to the role of undernutrition as compared to hepatotoxic factors, as seen in India, *India J. Med. Res.*, 53, 1148, 1965.

343. **Erenoglu, E., Edreira, J. C., and Patek, A. J., Jr.,** Observation on patients with Laennec's cirrhosis receiving alcohol while on controlled diets, *Ann. Intern. Med.*, 60, 814, 1964.

344. **Gyoergy, P. and Goldblatt, H.,** Experimental production of dietary liver injury (necrosis, cirrhosis) in rats, *Proc. Soc. Exp. Biol. Med.*, 46, 492, 1941.

345. **Gyoergy, P. and Goldblatt, H.,** Further observations on the production and prevention of dietary hepatic injury (cirrhosis) in rats, *J. Exp. Med.*, 89, 245, 1949.

346. **Eckardt, M. J.,** Consequences of alcohol and other drug use in the aged, in *Biology of Aging*, Behnke, J. A., Finch, C. E., and Moment, G. B., Eds., Plenum Press, New York, 1978, 191.

347. **Klot, V., Avant, G. R., Hoyumpa, A., Schenker, S., and Wilkinson, G. R.,** The effects of age and liver disease on the dispositions and elimination of diazepam in adult man, *J. Clin. Invest.*, 55, 347, 1975.

348. **Roberts, R. K., Wilkinson, G. R., Branch, R. A., and Schenker, S.,** Effect of age and parenchymal liver disease on the disposition and elimination of chlordiazepoxide (Librium), *Gastroenterology*, 75, 479, 1978.

349. **Sellers, E. M., Naranjo, C. A., Giles, H. G., Frecher, R. C., and Beeching, M.,** Intravenous diazepam and oral ethanol interaction, *Clin. Pharmacol. Ther.*, 29, 638, 1980.

350. **Pearl, R.,** *Alcohol and Longevity*, Knopf, New York, 1926.

351. **Pearl, R.,** The search for longevity, *Sci. Monthly*, 46, 461, 1938

352. **Pitskhelaure, G. T.,** *Personal communication*, 1974.

353. **Wood, W. G. and Armbrecht, H. J.,** Behavioral effects of ethanol in animals: age difference and age changes, *Alcoholism Clin. Exp. Res.*, 6, 3, 1982.

354. **Goodrick, C. L.,** Behavioral differences in young and aged mice strain differences for activity measures, operant learning, sensory discrimination and alcohol preference, *Exp. Aging Res.*, 1, 191, 1975.

355. **Wood, W. G.,** Ethanol preference in the C57BL/6 and BALB/c mice at three ages and eight ethanol concentrations, *Exp. Aging Res.*, 2, 425, 1976.

356. **Abel, E. L.,** Effects of ethanol and pentobarbital in mice of different ages, *Physiol. Psychol.*, 6, 366, 1978.

357. **Parisella, R. M. and Pritham, G. H.,** Effect of age on alcohol preference by rats, *Q. J. Stud. Alcohol*, 25, 248, 1964.

358. **Wallgren, H. and Forsander, O.,** Effect of adaptation to alcohol and of age on voluntary consumption of alcohol by rats, *Br. J. Nutr.*, 17, 453, 1963.

359. **Wiberg, G. S., Trenholm, H. L., and Coldwell, B. B.,** Increased ethanol toxicity in old rats: changes in LD 50, in vivo and in vitro metabolism and liver alcohol dehydrogenase activity, *Toxicol. Appl. Pharmacol.*, 16, 718, 1970.

360. **Abel, E. L. and York, J. L.,** Age-related differences in response to ethanol in the rat, *Physiol. Psychol.*, 7, 391, 1979.

361. **Ritzmann, R. F. and Springer, A.,** Age-differences in brain sensitivity and tolerance to ethanol in mice, *Age*, 3, 15, 1980.

362. **Edelman, I. S., Haley, H. B., Schloerb, P. R., Sheldon, D. B., Friis-Hansen, B. J., Stoll, G., and Moore, F. D.,** Further observations on total body water. I. Normal values throughout the life span, *Surg. Gynecol. Obstet.*, 95, 1, 1952.

363 **Timiras, P. S.,** *Developmental Physiology and Aging,* Macmillan, New York, 1972.

364. **Rider, A. A.,** Growth and survival in a rat colony maintained on moderate ethanol intake, *Nutr. Rep. Int ,* 22, 57, 1980

365. **Eckardt, M. J.,** Consequences of alcohol and other drug use, in the aged, in *Biology of Aging,* Behnke, J A., Finch, C. E , and Moment, G B , Eds., Plenum Press, New York, 1978, 191.

366. **Freund, G. and Butters, N.,** Alcohol and aging: challenges for the future, *Alcohol,* 6, 1, 1982

367. **Parker, E. S. and Nobel, E. P.,** Alcohol consumption and cognitive functioning in social drinkers, *Q. J. Stud. Alcohol,* 38, 1224, 1977.

368. **Jones, B. and Parsons, O. A.,** Specific vs. generalized deficits of abstracting ability in chronic alcoholics, *Arch Gen. Psychiatr.,* 26, 380, 1972.

369. **Ryan, C.,** Alcoholism and premature aging: a neuropsychological perspective, *Alcoholism Clin. Exp. Res.,* 6, 22, 1982.

370. **Ryan, C. and Butters, N.,** Further evidence for a continuum-or-impairment encompassing male alcoholic, Korsakoff patients and chronic alcoholic men, *Alcoholism: Clin. Exp. Res.,* 4, 111, 1980.

371. **Phillips, G. B., Victor, M., Adams, R. D., and Davidson, C. S.,** A study of the nutritional defect in Wernicke's syndrome. The effect of a purified diet, thiamine and other vitamins on the clinical manifestations, *J. Clin. Invest.,* 31, 859, 1952.

372. **Lopez, R. I. and Collins, G. H.,** Wernicke's encephalopathy A complication of chronic hemodialysis, *Arch. Neurol.,* 18, 248, 1968.

373. **Blumenthal, H. T.,** Immunological aspects of the aging brain, in *Neurobiology of Aging,* Terry, R. D. and Gershon, S., Eds., Raven Press, New York, 1976, 313.

374. **Tabakoff, B., Noble, E. B., and Warren, K. R.,** Alcohol, nutrition and the brain, in *Nutrition and the Brain,* Vol. 4, Wurtman, R. J. and Wurtman, J. J., Eds., Raven Press, New York, 1979, 159.

375. **Porjesz, B., Begleiter, H., and Samuelly, I.,** Cognitive deficits in chronic alcoholics and elderly subjects assessed by evoked brain potential, *Acta Psychiatr. Scand ,* 62, 15, 1980.

376. **Parsons, O. A. and Leiber, W. R.,** Premature aging, alcoholism and recovery, in *Alcoholism and Aging: Advances in Research,* Wood, W. G and Elias, M F., Eds., CRC Press, Boca Raton, Fla , 1982, 91

377 **Freund, G.,** Interactions of aging and chronic alcohol consumption on the central nervous system, in *Alcoholism and Aging. Advances in Research,* Wood, W. G. and Elias, M. F , Eds., CRC Press, Boca Raton Fla., 1982, 145

378. **Hsu, J. M., Rubenstein, B., and Paleker, A. G.,** Role of magnesium in glutathione metabolism of rat erythrocyte, *J Nutr.,* 112, 488, 1982.

379. **Rotruck, J. T., Pope, A. L., Ganther, H. E., Swanson, A. B., Hafeman, D. G., and Hoekstra, W. G.,** Selenium: biochemical role as component of glutathione peroxidase, *Science,* 179, 588, 1973.

380. **Yunice, A. A., Hsu, J. M., Fahmy, A., and Henry, S.,** Ethanol-ascorbate interrelationship in acute and chronic alcoholism in the guinea pig, *Proc. Soc. Exp. Biol Med.,* 77, 262, 1984.

*Section II*
*Nutritional Basis for Diseases in Aging*

Chapter 3

# CARDIOVASCULAR DISEASES

**Kang-Jey Ho**

## TABLE OF CONTENTS

## I. INTRODUCTION

Cardiovascular disease in the elderly might belong to one of the following three categories: (1) physiologic changes related to the aging process per se, (2) a specific pathologic process which is increased in frequency and severity with age, and (3) a combination of the above two. In fact, the distinction among these three categories for a particular entity of cardiovascular disease is often extremely difficult, if not impossible. Lipofucsin accumulation in the myocardium, an essential feature of so-called senile cardiomyopathy, is definitely related to the aging process.[1] The amount of lipofucsin pigment increases progressively with age at an accumulation rate of 0.3% of the total heart volume per decade.[2] By the time a person reaches 90 years of age, lipofucsin pigment may account for as much as 30% of the total solids of myocardial fibers. Amyloid deposit in the myocardium and basophilic degeneration of myocardial fibers are common but not universal findings in elderly persons. Endocardial fibrous thickening, valvular ring calcification, arrhythmia, cardiac enlargement, atrophy, and fibrosis cannot be considered as universal or even common results of aging. With advancing age, most vessels increase in both diameter and wall thickness, with fragmentation of elastic fibers and proliferation of collagen resulting in less extensibility of the vessel wall. The incidence and degree of hyalinization of arterioles also increase with age. However, it remains uncertain whether such changes are related to the aging process alone or its interaction with environmental factors.[1,3,4]

Regardless of how much aging per se contributes to the disease process, the most common and important cardiovascular diseases in the elderly patients are coronary heart disease, cerebrovascular accident, aortic aneurysm, occlusive peripheral vascular disease, and phlebothrombosis.[4] All of these diseases, with the exception of venous thrombosis, are based primarily upon a single common pathologic process, atherosclerosis. Hypertension and its role in cerebrovascular accidents and congestive heart failure are discussed elsewhere in this book and will not be repeated here. The discussion in this chapter is focused on the basic disease process of atherosclerosis.

Atherosclerosis is a multifactorial disease; aging is only one of the many factors attributing to such a process. First reviewed is our current understanding of the pathogenesis of atherosclerosis, followed by identification of its risk factors, particularly those pertinent to the nutritional aspects of aging.

## II. PATHOGENESIS OF ATHEROSCLEROSIS

The current concept of the pathogenesis of atherosclerosis is briefly summarized in Figure 1. The primary event initiating atherosclerosis is generally believed to be endothelial injury.[5] Since naturally occurring atherosclerosis develops slowly over a period of decades, such endothelial injury must be subtle, repetitive, and chronic. Endothelium can be damaged by excessive shear stress of hemodynamic forces, particularly those associated with hypertension.[8-15] Low-density lipoproteins (LDL) may cause functional endothelial injury as evidenced by both morphologic and physiologic alterations in the endothelium.[17-21] Other chemicals and toxins, such as homocystine and bacterial endotoxins, can produce vascular injury and arterial thrombosis.[22-24] Endothelium can also be damaged by immunologic means,[25-28] viruses,[29] hypoxemia,[30] and many other factors yet to be identified.

The damaged endothelium becomes dysfunctional, disrupted, and even desquamated.[6] Endothelial dysfunction leads to the loss of its thromboresistant mechanisms, such as the synthesis and secretion of glycocalyx,[31] $\alpha_2$-macroglobulin,[32] nonthrombotic type V surface collagen,[33] prostacyclin,[34,35] and plasminogen activator,[36] and also the metabolism of thrombin and vasoactive amines,[37,38] Endothelial denudation results in the direct exposure of subendothelial matrix, such as collagen (Types I and III), to the circulating blood elements.

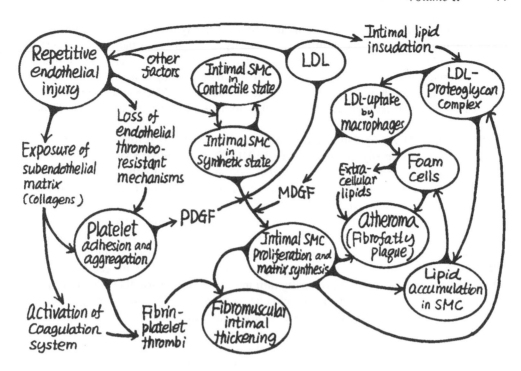

FIGURE 1.   Current concept of the pathogenesis of atherosclerosis. LDL: Low-density lipoprotein. SMC: Smooth muscle cell. PDGF: Platelet-derived growth factor. MDGF: Macrophage-derived growth factor

Adherence of platelets to the collagen takes place; the process is enhanced by the presence of von Willebrand factor.[39-43] Platelets in contact with collagen swell and undergo degranulation, releasing adenosine diphosphate (ADP) and other platelet constituents. ADP induces further platelet aggregation and degranulation. Degranulation in turn releases more ADP. A platelet thrombus is thus kept growing. The activated platelets also synthesize and release thromboxane $A_2$, which causes vasoconstriction and further platelet aggregation.[35,40] Among other constituents released by platelets during degranulation, platelet factor 4 (PF4), a glycoprotein, counteracts the anticoagulant effect of heparin,[44] and platelet factor 3 (PF3), a phospholipoprotein, together with the exposed subendothelial collagen, initiates the intrinsic coagulation mechanism, leading to the formation of fibrin-platelet thrombi.[45]

The aortic arterial intima of children contains very few smooth muscle cells (SMC). The number of the intimal SMC increases with age. It has been shown that SMC from the media respond to vascular injury by migration from the media to the intima.[46] Such migration is influenced by certain factors released from the platelets.[47-51] The intimal SMC exist in two distinct phenotypes: contractile and synthetic.[50-53] SMC in contractile state manifest myosin and contractility and are not stimulated to divide by various mitogenic factors such as LDL,[54,55] platelet-derived growth factor (PDGF),[50,51,56,57] and macrophage-derived growth factor (MDGF).[58,59] The contractile state is maintained by high cellular density in tissue culture and also by endothelium-derived inhibitor.[52,53] Endothelial injury with subsequent loss of the endothelium-derived inhibitor and low cellular density in the intima should favor the modulation of SMC from the contractile state to the synthetic state. The intimal SMC in the synthetic state have lost myosin and contractility but are capable of multiplication and synthesis of connective tissue elements, namely collagen and proteoglycans, in response to the stimulation by LDL, PDGF, and MDGF. SMC in the synthetic state may return to the contractile state. However, some of them do not regain the contractile phenotype and appear permanently in the synthetic state.[52]

LDL, PDGF, and MDGF stimulate intimal SMC in the synthetic state to proliferate and synthesize arterial matrix through different mechanisms. PDGF is of particular interest and importance. PDGF is a highly stable cationic protein contained in the α-granules of platelets and is released when platelets aggregate.[57,60] It is a potent mitogen and is bound to SMC through a specific, high-affinity receptor on the cell surface.[61] PDGF also stimulates SMC to increase their LDL-receptor activity as well as to increase uptake of LDL via bulk-phase endocytosis.[62,63]

In the absence of hyperlipidemia, the repetitive endothelial injury is followed by a repetitive repairing process, primarily involving intimal SMC. Such repetitive SMC proliferation and its matrix synthesis, and also organization of the fibrin platelet thrombi, over a long period of time result in the formation of a diffuse fibromuscular intimal thickening, invariably present in the adult and elderly population.[64-66]

In the presence of hyperlipidemia, on the other hand, endothelium is further injured and intimal SMC proliferation and matrix synthesis are enhanced by LDL.[54,55] Furthermore, LDL forms a complex with proteoglycans synthesized by the stimulated intimal SMC.[64,65,67-70] The latter also increase the uptake of LDL through both receptor-mediated and bulk-phase endocytosis and eventually turn into foam cells.[71] Another source of intimal foam cells is macrophages derived from circulating monocytes.[72-75] Macrophages bind and uptake both normal and altered LDL such as acetylated, malondialdehyde-conjugated, endothelial cell-modified, and smooth muscle cell-modified LDL, by a high-affinity receptor different from that for LDL.[76-81] The extracellular lipids in the intima are derived from LDL-proteoglycan complex and from breakdown of the foam cells. The intimal lesions are gradually turning into fibrofatty plaques or atheromas.[66]

The age-associated progressive diffuse fibromuscular intimal thickening is rather harmless because it rarely causes luminal occlusion. Atheromas, on the contrary, can narrow and occlude the arterial lumen by their growing size, intramural hemorrhage, or thrombosis developed on the rough surface of ulcerative plaques. The consequence of luminal occlusion is infarction, commonly occurring in heart, brain, intestine, and other organs. The ulceration of atheromatous plaques also gives rise to atheromatous embolism. Weakening of the vascular wall by atheromas is the mechanism for the formation of atherosclerotic aneurysms. Since hyperlipidemia is essential for the development of atheromas, it plays the central role in atherosclerosis and its grave complications were just described. At this point, it seems appropriate to briefly review lipoprotein and cholesterol metabolism before discussing the individual risk factors for atherosclerosis and ischemic heart disease.

## III. LIPOPROTEIN AND CHOLESTEROL METABOLISM

The liver and intestine are the primary organs for the synthesis of circulating lipoproteins (Figure 2).[82] Liver synthesizes apolipoproteins (apo) A-I, A-II, B-100, C-I, C-II, C-III, D, and E, while intestine synthesizes apo A-I, A-II, and B-48.[82,83] Apo B-100 is essential for the transport of endogenous lipids as very low-density lipoprotein (VLDL) from the liver and apo B-48 for transport of exogenous lipids as chylomicrons from the intestine. Other than apo B-100, VLDL also contains apo C-I, C-II, C-III, and E. By the action of lipoprotein lipase existing on the endothelium of the extrahepatic or peripheral tissue and its cofactor, apo C-II, the triglyceride in the VLDL undergoes lipolysis,[84] VLDL is thus progressively decreasing in size due to selective unloading of its triglyceride to the tissue and becomes intermediate-density lipoprotein (IDL). At the same time, the redundant surface membrane, which contains unesterified cholesterol and apo C and E, is removed by high-density lipoproteins (HDL) in exchange for esterified cholesterol. IDL undergoes similar progressive lipolysis by lipoprotein lipase. By the time that IDL loses all its apo C and E, no further lipolysis is possible because of the lack of the lipoprotein lipase cofactor, apo C-II. The remnants containing all the apo B-100 in the original VLDL and no other apolipoproteins

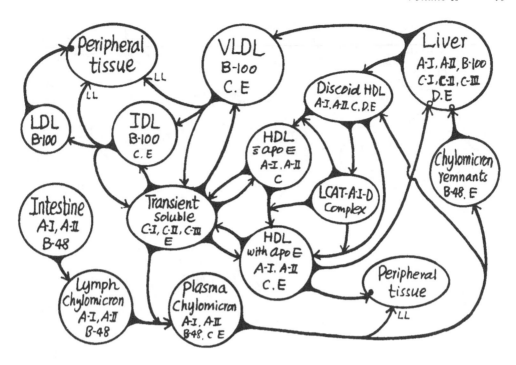

FIGURE 2    Lipoprotein metabolism and their interrelationship. VLDL, IDL, LDL, and HDL. Very low, intermediate, low, and high density lipoproteins, respectively. LL: Lipoprotein, lipase  Open circle. Apo-E receptor.
Closed circle· Apo B-100:E receptor. LCAT: Lecithin.cholesterol acyl transferase.

become LDL. LDL is removed by cells in the liver and peripheral tissue through an apo B-
100:E receptor and also by the mechanism of receptor-independent bulk-phase
endocytosis.[71,85,86]

The chylomicrons newly secreted from the intestine to the mucosal lacteals lack apo C
and E. They gain apolipoproteins quickly from HDL as soon as they enter the systemic
circulation. The chylomicrons are then subjected to lipolysis by lipoprotein lipase. With
progressive loss of triglyceride the chylomicrons decrease in size and their redundant surface
membrane, which contains apo A-I, A-II, C, and E, is detached from the parent particle to
form discoid HDL.[87] The chylomicron remnants, which retain all the original apo B-48 and
some of the acquired apo E, are rapidly removed by the hepatocyte through a high-affinity
apo E receptor.[86,88]

The nascent HDL synthesized in and secreted by the liver is also discoid in shape. By
the action of a cholesteryl ester transfer complex consisting of lecithin, cholesterol acyl
transferase (LCAT), apo A-I, and apo D (the transfer protein), the free cholesterol in the
discoid HDL is converted to esterified cholesterol.[89,90] The nonpolar esterified cholesterol
enters the space between the lipid bilayers; in such a way the discoid HDL is converted into
spherical HDL. There are two functional species of HDL: HDL without apo E and HDL
with apo E.[86] The smaller and denser HDL without apo E is converted to HDL with apo E
by obtaining apo E and free cholesterol from VLDL, IDL, and tissue. During the process,
free cholesterol is esterified by the cholesteryl ester transfer complex.[89,90] HDL with apo E,
but not HDL without apo E, is removed from the circulation by the liver through the apo
E receptors, and also by the peripheral tissue by apo B-100:E receptors.[86,88,91–94] HDL with
apo E, therefore, plays dual roles: feeding the peripheral tissue with cholesterol when needed
by the tissue and transporting excessive cholesterol in the tissue and lipoproteins to the liver
for disposal.

In terms of cholesterol transport and metabolism, the primary role of chylomicrons is to
carry the absorbed dietary cholesterol to the liver in the form of esterified cholesterol in

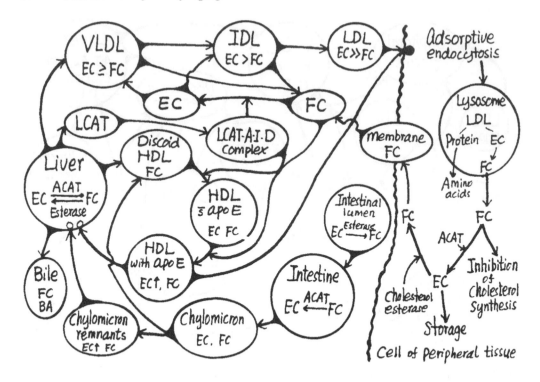

FIGURE 3.    Cholesterol metabolism and its relationship to lipoproteins. VLDL, IDL, LDL, and HDL. Very low, intermediate, low, and high-density lipoproteins, respectively  EC: Esterified cholesterol. FC: Free cholesterol  LCAT: Lecithin-cholesterol acyl transferase. ACAT. Acyl-CoA:cholesterol acyl transferase  Open circle. Apo E receptor  Closed circle: Apo B-100 E receptor.

chylomicron remnants, while that of VLDL is to transport cholesterol stored or newly synthesized in the liver to the peripheral tissue (Figure 3). The cholesterol ester transfer complex in the plasma constantly converts free cholesterol to esterified cholesterol among HDL, VLDL, and IDL. The ratio of esterified cholesterol to free cholesterol thus increases progressively when VLDL proceeds through IDL to form LDL. This complex also converts discoid HDL to spherical HDL, and HDL without apo E to HDL with apo E. Both LDL and HDL with apo E are picked up by the cells in the peripheral tissue through apo B-100:E receptors by adsorptive endocytosis. Newly arrived esterified cholesterol inside the cell is hydrolyzed in lysosomes by cholesterol esterase to form free cholesterol.[71] The free cholesterol, in turn, exerts the following three regulatory activities: (1) inhibition of β-hydroxy-β-methylglutaryl CoA reductase, the rate-limiting enzyme for endogenous cholesterol synthesis; (2) diminution of the number of apo B-100:E receptors on the cell surface, and (3) stimulation of acyl-CoA: cholesterol acyl transferase for conversion of free cholesterol to esterified cholesterol, mainly cholesteryl oleate, for storage.[71] On the other hand, excessive free cholesterol on lipoprotein and on cell membrane is transported to HDL and esterified by cholesteryl ester transfer complex.[95] Esterified cholesterol carried on HDL with apo E is then taken up by the liver cells through apo E receptors. Inside the liver cells, esterified cholesterol is again hydrolyzed to free cholesterol by cholesterol esterase. The free cholesterol is excreted in the bile as such or after conversion to bile acids, or re-esterified by acyl-CoA: cholesterol acyl transferase for storage or synthesis of VLDL.

## IV. RISK FACTORS FOR ATHEROSCLEROSIS AND ISCHEMIC HEART DISEASE

The "established" risk factors for atherosclerosis and ischemic heart disease (IHD) are

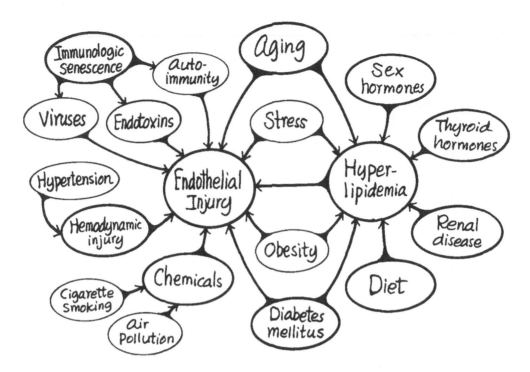

FIGURE 4    Various risk factors for atherosclerosis relating to endothelial injury and/or hyperlipidemia

summarized in Figure 4. Each of these risk factors is related to either one or both of the following two basic pathogenetic mechanisms of atherosclerosis: endothelial injury and hyperlipidemia. Those risk factors which are applicable to elderly persons will be discussed in more detail in the following sections.

## A. Age as a Risk Factor

The death rates of the U.S. population as a whole during the past 10 years have been fixed at around 880/100,000.[96] However, a steady across-the-board decline of the IHD mortality at all age groups has been noted.[96,97] Yet the major cardiovascular diseases today still account for 49% of the total deaths, including 38% caused by heart disease, 8% by cerebrovascular accidents, and 3% by peripheral vascular disease. The age-adjusted death rates of IHD increase rapidly after 55 years of age in males and after 65 in females (Figure 5).[96] For those 65 years of age and older, 2.6 and 1.8% of males and females, respectively, die of IHD each year. For those 85 years of age and older, the figures were 7.7 and 6.3% per year, respectively. These vital statistics clearly show that the relative risk of dying of IHD increases with age in both males and females. Aging alone is thus an independent risk factor for atherosclerosis.[98]

The mechanisms of the effect of aging on the atherosclerotic process are multiple and complicated by interactions among many factors.[98,99] Some of these factors might in turn be age-related. The plasma lipid level, particularly cholesterol, is elevated with age. The elevated serum lipid levels are also contributed to by age-related obesity, diabetes mellitus, sex-hormone status, thyroid function, and renal disease all to be discussed later. According to the clonal-senescence hypothesis of aging, when endothelial cells exhaust their finite replicative life span, they become more vulnerable to injury and less capable of repair when injured.[100,101] Endothelial injury is related to immunologic senescence, cumulative effects of long-term hypertension, tobacco smoking, and other environmental factors to be discussed

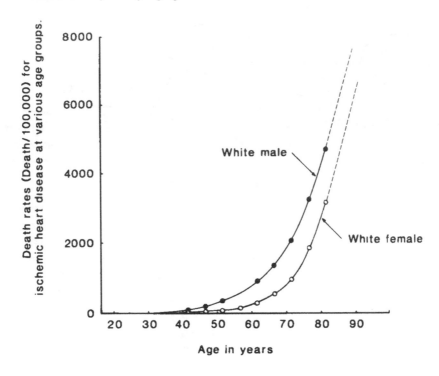

FIGURE 5.    Age-adjusted death rates of ischemic heart disease among male and female white population in the U S  in 1978 [96]

later. With increased age the arterial intima contains more collagen, especially thrombogenic types I and III collagen, and more high molecular weight elastin, which binds LDL.[70,102-104] Moreover, aging also modifies the composition of intimal glycosaminoglycans in such a way that their lipoprotein-complexing ability is enhanced.[105,106]

## B. Male Predominance

The death rate of IHD at any adult age group is always higher in males than in females (Figure 5).[96] Before 75 years of age the female lags behind the male in the IHD death rate for about 10 years. Such a gap becomes somewhat narrower after 75 years of age. The percentage of total death caused by IHD increases linearly in females from age 30 through over 85, whereas in males it increases sharply and linearly after age 30, reaches a plateau at age 55, and rises again after age 80 (Figure 6). The two lines cross each other at age 75. The steady increase in both the absolute and relative death rates caused by IHD with age in females strongly indicates that atherosclerosis, the underlying pathologic process of IHD, is indeed age related. In males the rapid increase in the relative death rate of IHD before age 55 is reflected by the clinical evidence of premature or accelerated atherosclerosis.[107] The fact that the relative rate becomes steady in spite of the continuous increase in the absolute rate in males between age 55 and 80 indicates that there is also a simultaneous increase in the absolute death rate caused by diseases other than IHD. Cerebrovascular accidents, aortic aneurysms, and occlusive peripheral vascular disease are some of the examples in this category. After age 80 the relative death rate of IHD in males resumes its up-rising trend. Here again aging is proven to be a time-tested risk factor for atherosclerosis.

The premature development of IHD in males is reflected by the high male to female ratio of the IHD death rate in the white population, which is particularly striking in the younger age group (Figure 7).[96] The ratio is as high as 5.7 in the 35 to 39 age group and declines steadily and linearly from 5.2 in the 45 to 49 age group to 1.2 in patients over 85 years of

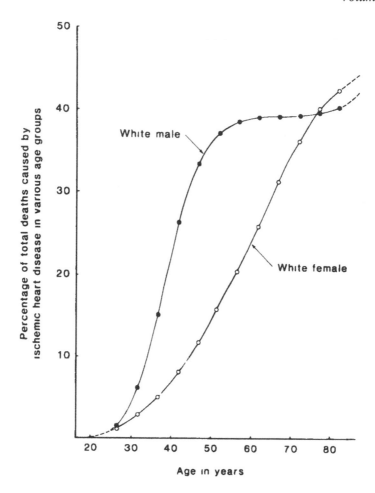

FIGURE 6. Age-adjusted percentage of total death caused by ischemic heart disease among male and female white population in the U.S. in 1978.[96]

age. The male to female ratio of all deaths is much lower when IHD is excluded (Figure 7). A similar male to female relationship is also observed in the U.S. black population.[96]

The male predominance of IHD must stem from differences in sex hormones. This might be directly related to male sex hormone or deficit of female sex hormone, or an imbalance of both.[108] In elderly persons the circulating sex hormone levels decrease in both sexes; the male predominance of IHD also becomes less obvious.[107] The higher plasma total cholesterol and LDL-cholesterol levels, and lower HDL-cholesterol level in the adult male than in the premenopausal female, might be the major factors accounting for the "premature" atherosclerosis in the male (Figure 8).[109-111] At the tissue level the male human aorta displays significantly higher activities of several enzymes than the female aorta.[112,113] Such enzymes include NADP-dependent enzymes, reduced NADP-producing enzymes, and glycero-3-phosphate dehydrogenase, suggesting an increased tendency toward triglyceride synthesis. Such sexual differences are less obvious among elderly persons.

## C. Obesity and Diabetes Mellitus

Obesity is often associated with aging.[114] As age advances, fat accumulates in the body while the overall mass of bone and muscle is reduced.[115] Therefore, despite the absence of gross obesity in some elderly persons, they do contain more fat than younger ones. It has been estimated that a 20% rise in body weight of a subject weighing 70 kg represents a 2-

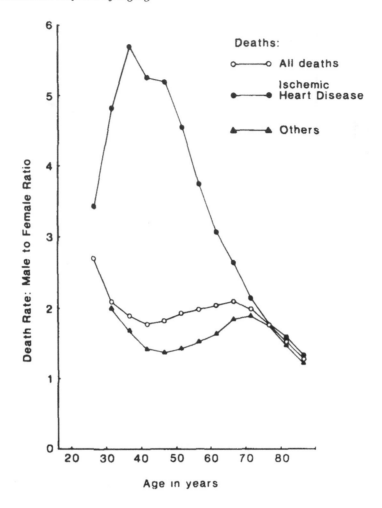

FIGURE 7    Male to female ratio of total and ischemic heart disease death rates and death rate due to causes other than ischemic heart disease in U S white population in 1978.[96]

fold increase in adipose tissue.[116] According to Dilman,[114 117] such age-related gross or relative obesity is secondary to the decline in the hypothalamic sensitivity to regulation by glucose. In other words, the threshold of satiety center is elevated. The process is accelerated in elderly persons due to the reduction in physical activity.

Both insulin-dependent and noninsulin-dependent diabetes mellitus (IDDM and NIDDM, respectively) are associated with accelerated atherosclerosis.[118] IDDM is more common in children, whereas NIDDM is more common in adults.[119,120] The age of onset is peaked at 12 years for IDDM and in the 5th and 6th decades for NIDDM. Although only 2 to 4% of the population has documented diabetes, the incidence increases with age. As high as 40 to 60% of persons in the 9th decade have obvious abnormal glucose tolerance.[120] In fact, every normal individual over 50 years of age develops a certain degree of chemical diabetes.[121] This could be simply related to the age-associated obesity (diabetes of obesity) or related to prediabetic state or overt diabetes. Patients with NIDDM have a very high frequency of obesity.[120] The interrelationship among aging, NIDDM, obesity, and atherosclerogenesis is summarized in Figure 9.

Insulin resistance is present in both obesity and NIDDM.[122-124] Such insulin resistance could be the result of the combined effects of (1) an increase in glucocorticoid output

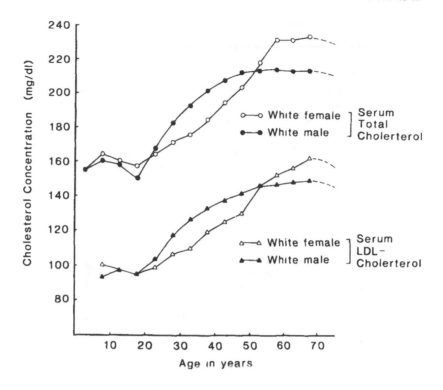

FIGURE 8. Plasma total cholesterol and low density lipoprotein-cholesterol levels of U S. male and female population at various ages [109]

frequently observed in normal aging, diabetes, and obesity,[114,115] (2) a decrease in insulin receptors in obesity,[126-128] (3) an increase in growth hormone level in NIDDM,[129] and (4) down regulation of the insulin receptors by hyperlipidemia.[130] Insulin resistance leads to compensatory hyperinsulinemia. NIDDM patients, even if not obese, may also be resistant to insulin and hence develop hyperinsulinemia.[131,132] Hyperinsulinemia in turn exerts the following effects: (1) enlargement of fat depots through activation of the insulin-sensitive lipoprotein lipase, leading to further obesity,[133] (2) stimulation of intimal SMC proliferation,[134,135] and (3) increase in the hepatic synthesis and secretion of VLDL, resulting in hypertriglyceridemia.[136-139] Although there is ample evidence to support such an insulin-mediated rise in hepatic synthesis and secretion of VLDL,[136-139] some NIDDM patients may have normal circulating insulin levels, yet still develop hypertriglyceridemia.[138,140] Such patients usually have an elevated plasma concentration of free fatty acids which acts on a ''normally insulinized'' liver to increase VLDL synthesis and secretion as the primary cause for their hypertriglyceridemia.[138] The elevated plasma VLDL level or hypertriglyceridemia impairs or down regulates insulin receptor function,[130] which in turn leads to greater cellular insulin resistance. Although the epidemiologic evidence indicating hypertriglyceridemia as an independent risk factor in IHD is still in some doubt,[141,142] hypertriglyceridemia is associated with low HDL levels[143-145] and may also lead to an elevated lipoprotein remnant level. Both are considered to be atherogenic.[146]

The plasma HDL cholesterol levels are inversely related to both relative body weight and plasma triglyceride or VLDL concentration.[143-145] In other words, obesity and NIDDM are associated with low plasma HDL level. The endogenous hyperinsulinemia in obesity and NIDDM is regarded as ''central'' hyperinsulinemia in contrast to the ''peripheral'' hyperinsulinemia due to administration of exogenous insulin.[145] In central hyperinsulinemia the insulin level is higher in the liver than in the extrahepatic tissue. Therefore, it is directed

FIGURE 9   The proposed mechanism for aging, obesity, and noninsulin dependent diabetes mellitus as risk factors for atherosclerogenesis.

to hepatic metabolism while peripheral tissues may be metabolically insulin deficient. Since both lipoprotein lipase and hepatic triglyceride lipase are insulin dependent,[147,148] the central hyperinsulinemia results not only in enhanced VLDL synthesis and secretion, but also in high hepatic triglyceride lipase activity, which may accelerate the removal of HDL from the plasma and thereby contribute to the low HDL levels.[145,149] On the other hand, the relative insulin deficiency in the peripheral tissue reduces the activity of lipoprotein lipase and hence reduces the rate of removal of VLDL from the plasma.[145,149] Hypertriglyceridemia is thus the result of slower catabolism coupled with enhanced synthesis of VLDL. Hypertriglyceridemia accompanied by hypo-high-density lipoproteinemia is believed to be one of the major contributory factors for increased atherosclerosis and the excessive morbidity and mortality from cardiovascular disease in obesity and NIDDM.[118]

## D. Diet and Other Factors Affecting Plasma Lipid Levels

### 1. Diet

Atherosclerosis may be considered as a proliferative disease of the arterial intima.[17] As mentioned earlier, in the absence of hyperlipidemia, the intimal proliferative process results in a nonocclusive diffuse fibromuscular thickening.[64-66] Some degree of hyperlipidemia is thus essential for the development of full-blown features of atherosclerosis and its grave consequences. Plasma total cholesterol levels indeed correlate positively with the severity of atherosclerosis and clinical IHD.[150-153] Among various serum lipoproteins, LDL is considered to be atherogenic whereas HDL is considered protective from atherogenesis.[143,144,150-153] Therefore, LDL- or HDL-cholesterol provides better prediction for IHD than plasma total cholesterol alone. The ratio of LDL- or HDL-cholesterol has even better predictive value than LDL- and HDL-cholesterol alone. In the future, the ratio of apo-B to apo-A (A-I,A-II), could be proven to be very useful as an indicator for monitoring plasma lipids in the prevention of IHD.

Plasma cholesterol levels are affected by lipid composition of the diet.[154-159] The change in plasma cholesterol level (c) in milligrams per deciliter can be predicted from the dietary intake of polyunsaturated (P) and saturated fatty acids (S) expressed as a percentage of total energy of the diet, dietary cholesterol content (C) in milligrams per day, and total energy of the diet (E) in kilocalories per day, by the following equation:[159]

$$c = 1.26(2S - P) + 1.5(100 \ C/E)^{1/2}$$

A recent diet-cholesterol study reveals that 64% of 83 males responded within 30% of prediction calculated from the above equation, 82% within 50%, and only 3% were labeled as nonresponders, even though they show some responses.[158]

A statistically significant positive correlation between the percentage of calories from saturated fats and the level of plasma total cholesterol and LDL-cholesterol has been repeatedly observed.[151-159] Polyunsaturated fatty acids of the ω-6 and ω-3 families present in vegetable oil and in fish oil, on the other hand, are hypocholesteremic and reduce drastically the plasma LDL-cholesterol level.[160] The mechanism of the hypocholesteremic effect of the dietary polyunsaturated fatty acids has been proposed to be through a change in the fatty acid composition of cell membrane and lipoproteins, resulting in an increase in the biliary and fecal excretion of cholesterol and bile acids and a simultaneous decrease in hepatic VLDL synthesis.[160]

Humans in general have a rather limited capacity for the intestinal absorption of dietary cholesterol.[161] The individual responses of plasma cholesterol to the dietary intake of cholesterol vary greatly. In large-scale population studies, a positive correlation between the dietary cholesterol intake and plasma cholesterol level often exists.[154,157] The plasma levels of HDL-cholesterol have been found not to be associated with dietary cholesterol and polyunsaturated and saturated fats, but rather with the female sex hormone, higher alcohol and lower carbohydrate intake, regular enduring exercise, and familial hyper α-lipoproteinemia.[110,162,163] The last entity is associated with longevity,[163]

Figure 10 briefly summarizes the proposed atherosclerogenetic mechanism of a high-fat, high-cholesterol diet. Such a diet induces the following changes in plasma lipoproteins both in humans and in animals:[86,164,165] (1) presence of a cholesterol ester-rich VLDL that floats at a density less than 1.006 g/mℓ and has β-electrophoretic mobility (β-VLDL), (2) an increase in cholesterol ester-rich LDL that is larger than ordinary LDL, (3) a reduction in the level of typical HDL or HDL without apo E, and (4) an increase in the level of HDL with apo E. The sequence of events is the following. A high-fat, high-cholesterol diet leads to the synthesis and secretion of cholesterol ester-rich chylomicrons in the intestine. After lipolysis at the peripheral tissue the cholesterol ester-rich chylomicron remnants are cleared by the liver through apo E receptors. The liver then synthesizes and secretes cholesterol ester-rich β-VLDL. β-VLDL is either picked up by macrophages though a specific receptor[166-168] or metabolized by lipoprotein lipase to form cholesterol ester-rich LDL.[84] It has been proposed that such LDL can be modified chemically by certain local factors such as malondialdehyde release from platelets or by lipid peroxidation.[76-81] The modified LDL

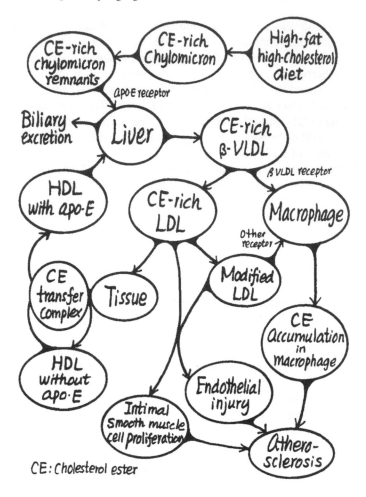

FIGURE 10. The proposed mechanism for a high-fat high cholesterol diet as a risk factor for atherosclerogenesis

is removed by macrophages through a specific independent receptor and increases accumulation of cholesterol ester in macrophages.[76-81] Furthermore, such modified or unmodified LDL can induce endothelial injury and stimulate intimal SMC proliferation.[54,55] Therefore, dietary cholesterol is more atherogenic than endogenous cholesterol. At the same time, excessive tissue cholesterol is cleared by HDL without apo E, which is converted to HDL with apo E by cholesteryl ester transfer complex as described before.[86,88-94] Cholesterol carried on HDL with apo E is then transferred to hepatocytes through apo E receptor for either storage or excretion in the bile. The fast formation of HDL without apo E and fast clearance of HDL with apo E, coupled with slow conversion of the former to the latter, are reflected by their plasma levels observed after the high-fat high-cholesterol diet just described.[86,164,165]

### 2. Chronic Renal Failure

Other than diabetes mellitus, the two most common causes of secondary hyperlipidemia in elderly persons are chronic renal failure and hypothyroidism. The normal aging process of the kidney contributes to a progressive reduction in the glomerular filtration rate and in renal blood flow.[169,170] With cumulative damage by hypertension, glucose intolerance, atherosclerosis, arteriosclerosis, and pyelonephritis, elderly persons are more prone to develop chronic renal failure than young ones. The usual type of hyperlipidemia in chronic renal failure is a mild to moderate degree of hypertriglyceridemia.[171,172] The key alterations of lipid metabolism in chronic renal failure are enhanced intestinal synthesis of triglyceride-rich lipoproteins, namely chylomicrons and VLDL B-48, and diminished activity of lipoprotein lipase and hepatic triglyceride lipase.[171-175] Such a combination of overproduction and slow catabolism results in the accumulation of VLDL B-48 and triglyceride-enriched IDL and LDL in the circulation. The IDL and LDL are atypical not only because of their increased triglyceride content, but also because of their abnormal apoprotein patterns, especially the presence of apo C and E in LDL. Even HDL in chronic renal failure is also enriched with triglyceride. For unknown reasons, HDL levels are reduced. One possible explanation is the rapid clearance of HDL by peripheral tissue since HDL with apo E can serve as a cholesterol supplier to the tissue when catabolism of VLDL through IDL to LDL is impaired. Such a moderate hypertriglyceridemia would ordinarily be considered at most to be a marginal risk for an increase in IHD. However, the atypical lipoproteins in chronic renal failure seem to have greater atherogenic potential than the VLDL found in otherwise healthy persons with moderate hypertriglyceridemia. This might account for the higher incidence of IHD in patients with chronic renal failure and in patients who undergo chronic dialysis.[176]

### 3. Hypothyroidism

Hypothyroidism is characteristically associated with elevated plasma cholesterol levels. The changes in plasma lipoproteins can mimic type II-a, type II-b, or type III familial hyperlipidemia.[177] It is customarily believed that aging is associated with the development of a hypothyroid state. The overall incidence of hypothyroidism in geriatric patients is, however, only about 5%.[178] The oxygen utilization and thus the basal metabolic rate decreases with age, but when related to the active cell mass, the oxygen utilization per unit mass of cells does not change with age. Although both plasma concentration of triiodothyronine $(T_3)$[179] and plasma thyroxine $(T_4)$[180] turnover rates decrease with age, the response of thyroid to thyroid-stimulating hormone and the thyroxine turnover during acute febrile illness appear to be within normal limits in elderly persons.[181] The ability of the aged thyroid to respond normally to stress implies that the age-related decrease in the thyroxine turnover is due to the reduced peripheral need rather than to the failure of thyroid response. Therefore, clinical hypothyroidism among elderly persons is not as common as generally believed.

Furthermore, the atherogenic effect of hypercholesteremia in hypothyroidism is questionable. We have observed a paradox in animal experimental atherosclerosis.[182] Thiouracil-treated rabbits have higher plasma cholesterol levels but less cholesterol accumulation in the aorta than the euthyroid controls, while thyroxine-treated rabbits tend to have plasma cholesterol levels lower than those of the control during the period of cholesterol feeding but exhibit a greater degree of cholesterol accumulation in the aorta. In humans, the extent and severity of atherosclerosis observed at autopsy does not seem to have been influenced by the presence of either hypo- or hyperthyroidism.[183,184] The relationship among aging, thyroid state, hyperlipidemia, and atherosclerosis remains to be further clarified.

### E. Immunologic Senescence

A progressive decline in immune function with age has been well documented.[185,186] The decline involves both humoral and cell-mediated immunity. There is a progressive shift in balance between idiotypic and autoanti-idiotypic reactions with age as manifested by the increasing incidence of autoantibodies among elderly persons.[187] Although there have been conflicting reports and the entire immune system could be much more complicated, I have briefly summarized the mechanism of immunologic senescence and its relationship to atherosclerogenesis in Figure 11.

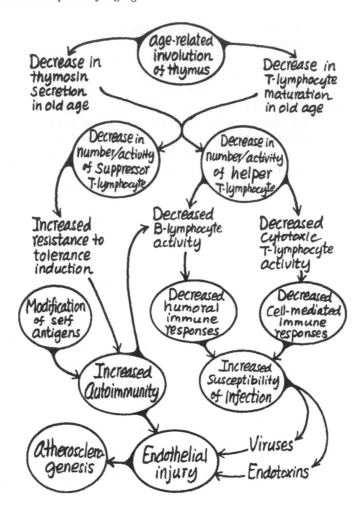

FIGURE 11.   The proposed mechanism for immunologic senescence as a risk factor for atherosclerogenesis.

The involution of the thymus gland in humans begins at sexual maturity and is complete by 45 to 50 years of age.[188] With such involution, the plasma concentration of thymic hormone or thymosin, which consists of a family of polypeptides,[189] declines progressively and becomes undetectable in normal humans over 60 years of age.[190] The aging thymus gland also loses its capacity to differentiate immature lymphocytes, resulting in an increase in immature T lymphocytes in the thymus and in blood.[191] The decrease in number and/or activity of helper and suppressor T lymphocytes becomes evident. As a consequence of the decrease in helper T lymphocyte function, both cell-mediated immunity and humoral immunity are impaired. The number of mitogen-responsive T lymphocytes and the proliferative capacity of these mitogen-responsive T lymphocytes decline.[192] The delayed hypersensitivity reactions to common skin-testing antigens in elderly persons is less vigorous than in young persons.[186] The evidence for impaired humoral immunity in elderly persons includes the decreased serum levels of natural antibodies, such as isoagglutinins, and the decreased responses to foreign antigens, such as tetanus toxoid.[186] The decreased suppressor T lymphocyte function leads to increased resistance to tolerance induction and thus to the increase in the incidence of autoantibodies and monoclonal immunoglobulins.[187,193,194] One class of autoantibodies, the autoanti-idiotypic antibody, can block lymphocytic secretion of specific antibodies and, therefore, further impairs the humoral immune responses.[187] Another reason for increasing autoimmunity is the modification of self-antigens, which occurs more commonly in aging tissue than in young tissue. Although the presence of autoantibodies may not be necessarily associated with clinical autoimmune diseases, humans with autoantibodies have a shorter survival than do age-matched persons without autoantibodies.[186] Clinico-pathologic evidence suggests that immunologically induced arterial injury leads to accelerated atherosclerosis in humans.[24-28] It is conceivable that the increasing severity of atherosclerosis with age might be at least in part related to direct endothelial injury mediated through autoantibodies or immune complexes. Moreover, the suppression of both humoral and cell-mediated immunities in elderly persons renders them susceptible to infection. Subsequently, endothelial injury by viral infection or bacterial endotoxin can also contribute to the process of atherosclerogenesis.

## F. Other Factors Inducing Endothelial Injury

### 1. Hemodynamics

Atheromatous plaques have a certain predilection for locations in the arterial system: arteries of relatively substantial calibers and orifices of the major branches of the aorta, particularly the coronary, renal, and iliac arteries.[11,12 15 16] This is based on the fact that flowing blood affects the vascular wall principally through shear stress, the force due to its viscosity applied parallel to the surface of the vessel, and the fact that excessive shear stress occurs near the orifice of branches arising from the aorta, the predilection areas of atheromatous plaque formation.[11,12,15,16] The excessive wall shear stress disrupts the endothelial lining of the artery, allowing entry of plasma proteins and lipids and also activating the platelets and coagulation cascade. The consequence of the cumulative effect of such minor but constant endothelial injury is the progression of atherosclerosis with age. Such mechanical vascular injury is exaggerated in patients with hypertension.[10,13-15] The frequency of blood pressure above 160/95 mmHg rises progressively with age so that by age 70 more than one third of the population is affected.[195] Therefore, higher incidence of hypertension and prolonged hemodynamic injury contribute in part to the severity of atherosclerosis in elderly persons.

### 2. Chemicals

The chemicals that are capable of causing endothelial injury are either endogenous or exogenous. Examples of endogenous chemicals are lipoproteins (LDL)[17-21] and homocystine.[22] Among various exogenous chemicals, those related to smoking and pollution are most important. Such environmental factors are not directly related to aging but their effects on the arterial wall are cumulative with time. The following short account is limited to tobacco smoking only (Figure 12).

Tobacco, after combustion during smoking, yields countless chemical compounds; among them, nicotine and carbon monoxide have a profound effect on the cardiovascular system.[30] Nicotine is readily absorbed through the lungs and oral mucosa.[196] Nicotine is a stimulator of both sympathetic and parasympathetic ganglia and thus increases heart rate, cardiac output, blood pressure, myocardial oxygen consumption, and coronary blood flow in normal subjects. In patients with damaged myocardium, however, nicotine may cause a falling of cardiac output and predispose a person to the initiation of arrhythmia.[30] Release of catecholamines by nicotine stimulation enhances lipolysis, leading to an increase in plasma free fatty acid.[197] It has been well established that long-chain saturated fatty acids, such as stearic acid, are thrombogenic and may indeed predispose to coronary thrombosis and enhancement of the atherosclerotic process.[160,198] A recent study suggests that cigarette smoking may attenuate the anti-atherogenic properties of HDL by altering the surface phospholipid components.[199]

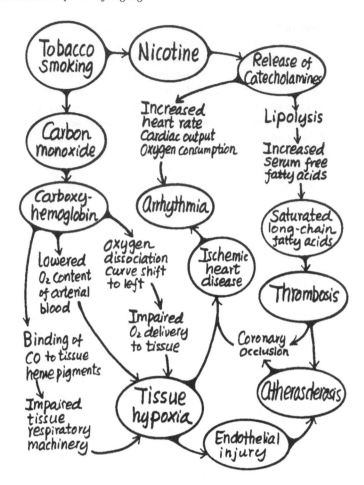

FIGURE 12    The proposed mechanism for tobacco smoking as a risk factor for atherosclerogenesis

Carbon monoxide produced by tobacco combustion is absorbed through the lung. Carbon monoxide has an affinity for hemoglobin about 250 times higher than the affinity of oxygen. The carboxyhemoglobin may reach a high plasma level up to 15 to 20% of the total hemoglobin in heavy smokers. Carboxyhemoglobin not only lowers the oxygen content of arterial blood but also shifts the oxygen dissociation curve to the left, and thus impairs oxygen delivery to the tissues. Moreover, carbon monoxide also binds to the heme pigments in the tissue and thus impairs the respiratory machinery of the tissues. The net effect of carbon monoxide inhalation is tissue hypoxia as reflected by myocardial degeneration and endothelial injury.[30] The relationship between carboxyhemoglobin and atherosclerosis has been established experimentally and epidemiologically.[30]

## V. PREVENTION AND MANAGEMENT

The atherosclerotic process starts early in life; by age 65 and over, everyone has developed a considerable degree of atherosclerosis. The diffuse fibromuscular intimal thickening seems to be an inevitable process related to aging but rarely produces serious clinical consequences.[64-66] Exuberant atherosclerosis associated with hyperlipidemia, on the other hand, often leads to early death due to its grave complications. Therefore, the prevention of atherosclerosis should begin early in life and continue for as long as one lives. Regression of the atherosclerotic plaques is difficult and limited to a certain degree, but is not impossible.[200,201] Even though the regression is slow, the preventive and therapeutic measures, when implemented, should at least slow down or hold steady the atherosclerotic process.

The prevention and treatment aim at the elimination of the risk factors described in this chapter or alleviation of their harmful effects. The following are the general principles applicable to the elderly population:

1. Maintenance of ideal body weight or avoidance of excessive weight gain by dietary control and adequate exercise.[202]
2. Detection and evaluation of the diabetic state and control by dietary means with medication if indicated.
3. Detection and evaluation of hyperlipidemic state and control by dietary measures with medication if indicated.[203,204] A high-fat high-cholesterol diet, by all means should be avoided.
4. Detection and evaluation of abnormal renal function, particularly in regard to the abnormal lipoproteins associated with chronic renal failure, and consultation with a nephrologist for preservation of renal functions.
5. Detection and evaluation of immunologic status and prevention of viral and bacterial infection by vaccination if indicated.
6. Detection and evaluation of hypertensive state and control by medication if indicated. Special caution must be exercised with administration of diuretics, adrenergic inhibitors, and vasodilators to elderly patients. Whenever possible, polypharmacy should be avoided to prevent adverse interactions and side effects.[205]
7. Tobacco smoking should be discontinued or reduced to a minimum.

In view of the fact that the great majority of elderly persons have his or her well-established personal lifestyle and dignity, all the evaluative, preventive, and therapeutic measures should be conservative instead of aggressive and should be persuasive instead of mandatory. Very often, to do more is more harmful than to do less, particularly among elderly persons.

# REFERENCES

1. **Burch, G.,** Senile cardiomyopathy, *J. Am. Geriatr. Soc.,* 30, 642, 1982.
2. **Strehler, B. L., Mark, D. D., Mildvan, A. S., and Gee, M. V.,** Rate and magnitude of age pigment accumulation in the human myocardium, *J. Gerontol.,* 14, 430, 1959.
3. **Weisfeldt, M. L.,** *The Aging Heart. Its Function and Response to Stress,* Raven Press, New York, 1980
4. **Exton-Smith, A. N. and Overstall, P. W.,** Cardiovascular system, in *Geriatrics,* Exton-Smith, A. N. and Overstall, P. W., Eds , University Park Press, Baltimore, 1980, 117.
5. **Ross, R. and Glomset, J. A.,** The pathogenesis of atherosclerosis, *N. Engl. J. Med.,* 295, 369, 420, 1976.
6. **Ross, R.,** Atherosclerosis: a problem of the biology of arterial wall cells and their interaction with blood components, *Arteriosclerosis,* 1, 293, 1981.
7. **Schwartz, S. M., Gajdusek, C. M., and Selden, S. C., III,** Vascular wall growth control: the role of the endothelium, *Arteriosclerosis,* 1, 107, 1981.
8. **Taylor, C. B., Hass, G. H., and Ho, K. J.,** The role of repair of arterial injury and degeneration in arteriosclerosis, in *Le Role de la Paroi Arterielle Dans Latherogenese,* Centre National de la Recherche Scientifique, Ed., Paris, 1967, 853.
9. **Fry, D. L.,** Responses of the arterial wall to certain physical factors, *Ciba Found. Symp.,* 12, 93, 1973.
10. **Schwartz, S. M. and Benditt, E. P.,** Aortic endothelial cell replication. I. Effects of age and hypertension in the rat, *Circ. Res.,* 41, 248, 1977.

11. **Texon, M.**, Atherosclerosis, its hemodynamic basis and implications, *Med. Clin. N. Am.*, 58, 257, 1974.

12. **McMillan, D. E.**, Physical factors important in the development of atherosclerosis in diabetes, *Diabetes*, 30(Suppl, 2), 97, 1981.

13 **Christlieb, A. R., Warram, J. H., Krolewski, A. S., Busick, E. J., Ganda, O. P., Asmal, A. C., Soeldner, J. S., and Bradley, R. F.**, Hypertension: the major risk factor in juvenile-onset insulin-dependent diabetes, *Diabetes*, 30(Suppl.2), 90, 1981

14. **Chobanian, A. V.**, Hypertension: major risk factor for cardiovascular complications, *Geriatrics*, 31, 87, 1976.

15 **Towne, J. B., Quinn, K., Salles-Cunha, S., Bernhard, V. M., and Clowry, L. J.**, Effect of increased arterial blood flow on localization and progression of atherosclerosis, *Arch. Surg.*, 117, 1469, 1982.

16. **Friedman, M. H., Deters, O. J., Mark, F. F., Bargeron, C. B., and Hutchins, G. M.**, Arterial geometry affects hemodynamics. A potential risk factor for atherosclerosis, *Atherosclerosis*, 46, 225, 1983.

17. **Thomas, W. A. and Kim, D. N.**, Biology of disease: atherosclerosis as a hyperplastic and/or neoplastic process, *Lab. Invest.*, 48, 245, 1983.

18 **Nelson, E., Gertz, S. D., Forbes, M. S., Rennels, M. L., Heaod, F. P., Kahn, M. A., Farber, T. M., Miller, E., Husain, M. M., and Earl, F. L.**, Endothelial lesions in the aorta of egg yolk-fed miniature swine: a study by scanning and transmission electron microscopy, *Exp. Mol. Pathol.*, 25, 208, 1976.

19 **Weber, G., Fabbrini, I., and Resi, L.**, Scanning and transmission electron microscopy observation on the surface lining of aortic intimal plaques in rabbits on a hypercholesterolemic diet, *Virchow's Arch. A:*, 364, 325, 1974.

20. **Henriksen, T., Evensen, S. A., and Carlander, B.**, Injury to human endothelial cells in culture induced by low density lipoproteins, *Scand. J. Clin. Lab. Invest.*, 39, 361, 1979.

21. **Gimbrone, M. A., Jr.**, Endothelial dysfunction and the pathogenesis of atherosclerosis, in *Atherosclerosis: Proc 5th Int. Symp. Atherosclerosis*, Gotto, A. M., Smith, L. C., and Allen, B., Eds , Springer-Verlag, New York, 1980, 415.

22. **Harker, L., Ross, R., Slichter, S., and Scott, C.**, Homocystine-induced arteriosclerosis: the role of endothelial cell injury and platelet response in its genesis, *J. Clin. Invest.*, 58, 731, 1976.

23. **Reidy, M. A., and Bowyer, D. E.**, Distortion of endothelial repair. The effect of hypercholesteremia on regeneration of aortic endothelium following injury by endotoxin. A scanning electron microscope study, *Atherosclerosis*, 29, 459, 1978.

24. **Gaynor, E.**, Increased mitotic activity in rabbit endothelium after endotoxin: an autoradiographic study, *Lab. Invest.*, 24, 318, 1971.

25. **Minick, C. R.**, Immunologic arterial injury in atherogenesis, *Ann. N.Y. Acad. Sci.*, 275, 210, 1976.

26. **Minick, C. R., Alonso, D. R., and Rankin, L.**, Role of immunologic arterial injury in atherosclerosis, *Thromb. Haemostas. (Stuttgart)*, 39, 304, 1978.

27. **Alonso, D. R., Starek, P. K., and Minick, C. R.**, Studies on the pathogenesis of atherogenesis of athero-arteriosclerosis induced rabbit cardiac allografts by the synergy of graft rejection and hypercholesterolemia, *Am. J. Pathol.*, 84, 415, 1977.

28. **Vlaicu, R., Niculescu, F., Rus, H. G., and Cristea, A.**, Immune deposits in human aortic atherosclerotic wall, *Rev. Roum. Med.*, 21, 3, 1983.

29. **Fabricant, C. G.**, Herpes virus-induced atherosclerosis, *Diabetes*, 30(Suppl.2), 29, 1981.

30. **Astrup, P. and Kjeldsen, K.**, Carbon monoxide, smoking, and atherosclerosis, *Med. Clin. N Am.*, 58, 323, 1973.

31. **Danon, D. and Skutelsky, E.**, Endothelial surface charge and its possible relationship to thrombogenesis, *Ann. N.Y. Acad. Sci.*, 275, 47, 1976.

32. **Becker, C. G. and Harpel, P. C.**, Alpha$_2$-macroglobulin on human vascular endothelium, *J. Exp. Med.*, 144, 1, 1976.

33. **Madri, J. A., Dreyer, B., Pitlick F. A., and Furthmayr, H.**, The collagenous components of the subendothelium. *Lab. Invest.*, 43, 303, 1980.

34. **Moncada, S., Herman, A. G., Higgs, E. A., and Vane, J. R.**, Differential formation of prostacyclin (PGX or PGI$_2$) by layers of the arterial wall. An explanation for the anti-thrombotic properties of vascular endothelium, *Throm. Res.*, 11, 323, 1977.

35. **Moncada, S.**, Prostacyclin and arterial wall biology, *Arteriosclerosis*, 2, 193, 1982.

36. **Loskutoff, D. J. and Edgington, T. E.**, Synthesis of a fibrinolytic activator and inhibitor by endothelial cells, *Proc. Natl. Acad. Sci. U.S.A.*, 74, 3903, 1977.

37. **Owen, W. G. and Esmon, C. T.**, Functional properties of an endothelial cell cofactor for thrombin-catalyzed activation of protein C, *J Biol. Chem.*, 256, 5532, 1981.

38. **Gimbrone, M. A., Jr.**, Culture of vascular endothelium, in *Progress in Hemostasis and Thrombosis*, Spaet, T. H., Ed., Grune & Stratton, New York, 1976, 3, 1.

39. **Lewis, J. C. and Kottke, B. A.**, Endothelial damage and thrombocyte adhesion in pigeon atherosclerosis, *Science*, 196, 1007, 1977.

40. **Weksler, B. B. and Nachman, R. L.,** Platelets and atherosclerosis, *Am. J Med.,* 71, 331, 1981.

41 **Fuster, V., Bowie, E. J. W., Lewis, J. C., Fass, D. N., Owen, C. A., Jr., and Brown, A. J., Jr.,** Resistance to arteriosclerosis in pigs with von Willebrand's disease, *J. Clin Invest.,* 61, 722, 1978

42. **Sakariassen, K. S., Bolhuis, P. A., and Sixma, J. J.,** Human blood platelet adhesion to artery subendothelium is mediated by factor VIII-von Willebrand factor bound to the subendothelium, *Nature (London),* 279, 636, 1979.

43. **Bowie, E. J. W., Fuster, V., Fass, D. N., and Owen, C. A., Jr.,** The role of Willebrand factor in platelet-blood vessel interaction, including a discussion of resistance to atherosclerosis in pigs with von Willebrand's disease, *Phil. Trans. R. Soc. London B,* 294. 267. 1981.

44. **Nath, N., Niewiarowski, S., and Joist, J. H.,** Platelet factor 4: antiheparin protein releasable from platelets. Purification and properties, *J. Lab. Clin. Med.,* 82, 754, 1973.

45. **Schiffman, S., Rapaport, S. I., and Chong, M. M. Y.,** Platelets and initiation of intrinsic clotting, *Br. J. Haematol.,* 24, 633, 1973.

46 **Geer, J. C. and Haust, M. D.,** Smooth muscle cells in atherosclerosis, in *Monographs on Atherosclerosis,* Vol. 2, Pollack, O. J., Sims, H. S., and Kirk, S. E., Eds., S. Karger, Basel, 1972.

47. **Thorgeirsson, G., Roberston, A. L., Jr., and Cowan, D. H.,** Migration of human vascular endothelial and smooth muscle cells, *Lab. Invest.,* 41, 51, 1979.

48. **Grotendorst, G. R., Seppa, H. E. J., Kleinman, H. K., and Martin G. R.,** Attachment of smooth muscle cells to collagen and their migration toward platelet-derived growth factor, *Proc. Natl. Acad. Sci. U.S.A.,* 78, 3369, 1981.

49. **Bernstein, L. R., Antoniades, H. N., and Zetter, B. R.,** Migration of cultured vascular cells in response to plasma and platelet-derived factors, *J. Cell Sci.,* 56, 71, 1982.

50. **Chamley-Campbell, J., Campbell, G. R., and Ross, R.,** The smooth muscle cell in culture, *Physiol. Rev.,* 59, 1, 1979

51. **Chamley-Campbell, J., Campbell, G. R., and Ross, R.,** Phenotype-dependent response of culture aortic smooth muscle to serum mitogens, *J. Cell Biol.,* 89, 378, 1981.

52. **Campbell, G. R. and Chamley-Campbell, J.,** Smooth muscle phenotypic modulation: role in atherogenesis, *Med. Hypoth.,* 7, 729, 1981.

53. **Chamley-Campbell, J. H. and Campbell, G. R.,** What controls smooth muscle phenotype?, *Atherosclerosis,* 40, 347, 1981.

54. **Fischer-Dzoga, K., Fraser, R., and Wissler, R. W.,** Stimulation of proliferation in stationary primary cultures of monkey and rabbit aortic smooth muscle cells. I. Effects of lipoprotein fractions on hyperlipidemic serum and lymph, *Exp. Mol. Pathol.,* 24, 346, 1976.

55. **Fless, G. M., Kirchhausen, T., Fischer-Dzoga, K., Wissler, R. W. and Scanu, A. M.,** Relationship between the properties of the apo B containing low-denisty lipoproteins (LDL) of normolipidemic rhesus monkeys and their mitogenic action on arterial smooth muscle cells grown in vitro, in *Atherosclerosis V,* Grotto, A. M., Jr., Smith, L. D., and Allen, B., Eds., Springer-Verlag, New York, 1980, 607.

56. **Ross, R., Glomset, J., Kariya, B., and Harker, L.,** A platelet-dependent serum factor that stimulates the proliferation of arterial smooth muscle cells in vitro, *Proc. Natl. Acad. Sci. U.S.A.,* 71, 1207, 1974.

57. **Antoniades, N. H. and Williams, L. T.,** Human platelet-derived growth factor: structure and function, *Fed. Proc. Fed. Am. Soc. Exp. Biol.,* 42, 2630, 1983.

58. **Liebovich, S. J. and Ross, R.,** A macrophage-dependent factor that stimulates the proliferation of fibroblasts in vitro, *Am. J. Pathol.,* 84, 501, 1976.

59. **Martin, B. M., Gimbrone, M. A., Jr., Unanue, E. R., and Cotran, R. S.,** Stimulation of non-lymphoid mesenchymal cell proliferation by a macrophage-derived growth factor, *J. Immunol.,* 126, 1510, 1981.

60. **Witte, L. D., Kaplan, K. L., Nossel, H. L., Lages, G. A., Weiss, H. J., and Goodman, D. S.,** Studies of the release from human platelets of the growth factor for cultured human arterial smooth muscle cells, *Circ. Res.,* 42, 402, 1978.

61. **Williams, L. T., Tremble, P., and Antoniades, H. N.,** Vascular smooth muscle cell receptors for platelet-derived growth factor: formation of nondissociated state of binding, *Proc. Natl. Acad. Sci. U.S.A.,* 79, 5867, 1982.

62. **Chait, A., Ross, R., Albers, J. J., and Bierman, E. L.,** Platelet-derived growth factor stimulates activity of low density lipoprotein receptors, *Proc. Natl. Acad. Sci. U.S.A.,* 77, 4084, 1980.

63. **Witle, L. D. and Cornicelli, J. A.,** Platelet-derived growth factor stimulates low density lipoprotein receptor activity in cultured human fibroblasts, *Proc. Natl. Acad. Sci. U.S.A.,* 77, 5962, 1980.

64. **Movat, H. Z., More, R. H., and Haust, M. D.,** The diffuse intimal thickening of the human aorta with aging, *Am. H. Pathol.,* 34, 1023, 1958.

65. **Wilens, S. L.,** The nature of diffuse intimal thickening of arteries, *Am. J. Pathol.,* 28, 825, 1951.

66. **Tracy, R. E., Toca, V. T., Lopez, C. R., Kissling, G. E., and Devaney, K.,** Fibrous intimal thickening and atherosclerosis of the thoracic aorta in coronary heart disease, *Lab. Invest.,* 48, 303, 1983.

67. **Hollander, W.,** Unified concept on the role of acid mucopolysaccharides and connective tissue proteins in the accumulation of lipids, lipoproteins, and calcium in the atherosclerotic plaque. *Exp. Mol. Pathol.,* 25, 106, 1976.

68. **Avila, E. M., Lopez, F., and Camejo, G.,** Properties of low density lipoprotein related to its interaction with arterial wall components — in vitro and in vivo studies, *Artery,* 4, 36, 1978.

69 **Camejo, G., Lalaguna, F., Lopez, F., and Starosta, R.,** Characterization and properties of a lipoprotein - complexing proteoglycan from human aorta, *Atherosclerosis,* 35, 307, 1980

70. **Camejo, G.,** The interaction of lipids and lipoproteins with the intercellular matrix of arterial tissue: its possible role in atherogenesis, *Adv. Lipid Res.,* 19, 1, 1982.

71. **Brown, M. S., Kovanen, P. T., and Goldstein, J. L.,** Regulation of plasma cholesterol by lipoprotein receptors, *Science,* 212, 627, 1981.

72. **Gerrity, R. G.,** Transition of blood-borne monocytes into foam cells in fatty lesions, *Am. J. Pathol.,* 103, 181, 1981.

73. **Gerrity, R. G.,** Migration of foam cells from atherosclerotic lesions, *Am. J. Pathol.,* 103, 191, 1981.

74. **Gaton, E. and Wolman, M.,** The role of smooth muscle cells and hematogenous macrophages in atheroma, *J. Pathol.,* 123, 123, 1977.

75. **Taylor, K., Schaffner, T., Wissler, R. W., and Glagov, S.,** Immunomorphologic identification and characterization of cells derived from experimental atherosclerotic lesions, *SEM,* 111, 815, 1978.

76. **Goldstein, J. L., Ho, Y. K., Basu, S. K., and Brown, M. S.,** Binding site on macrophages that mediates uptake and degradation of acetylated low density lipoprotein, producing massive cholesterol deposition, *Proc Natl. Acad. Sci. U.S.A.,* 76, 333, 1979

77. **Schechter, I., Fogelman, A. M., Haberland, M. E., Scager, J., Hokom, M., and Edwards, P. A.,** The metabolism of native and malondialdehyde altered low density lipoproteins by human monocyte-macrophages, *J. Lipid Res.,* 22, 63, 1981.

78. **Henriksen, T., Mahoney, E. M., and Steinberg, D.,** Enhanced macrophage degradation of low density lipoprotein previously incubated with cultured endothelial cells: recognition by the receptor for acetylated low density lipoproteins, *Proc. Natl. Acad. Sci. U.S A.,* 78, 6499, 1981.

79. **Henriksen, T., Mahoney, E. M., and Steinberg, D.,** Interaction of plasma lipoproteins with endothelial cells, *Ann. N.Y. Acad. Sci.,* 401, 102, 1983.

80. **Henriksen, T., Mahoney, E. M., and Steinberg, D.,** Enhanced macrophage degradation of biologically modified low density lipoprotein, *Arteriosclerosis,* 3, 149, 1983.

81. **Steinberg, D.,** Lipoproteins and atherosclerosis: a look back and a look ahead, *Arteriosclerosis,* 3, 383, 1983.

82. **Schaefer, E. J., Eisenberg, S., and Levy, R. I.,** Lipoprotein apoprotein metabolism, *J. Lipid Res.,* 19, 667, 1978.

83. **Kane, J. P., Hardman, D. A., and Paulus, H. E.,** Heterogeneity of apolipoprotein B: isolation of a new species from human chylomicrons, *Proc. Natl Acad. Sci. U.S A.,* 77, 2465, 1980.

84. **Reardon, M. F., Sakai, H., and Steiner, G.,** Roles of lipoprotein lipase and hepatic triglyceride lipase in the catabolism in vivo of triglyceride-rich lipoproteins, *Arteriosclerosis,* 2, 396, 1982.

85 **Basu, S. K., Goldstein, J. L., and Brown, M. S.,** Characterization of the low density lipoprotein receptor in membrane prepared from human fibroblasts, *J. Biol Chem.,* 253, 3852, 1978.

86. **Mahley, R. W.,** Atherogenic hyperlipoproteinemia: the cellular and molecular biology of plasma lipoproteins altered by dietary fat and cholesterol., *Med. Clin. N. Am.,* 66, 375, 1982.

87. **Tall, A. R. and Small, M. D.,** Plasma high-density lipoprotein, *N. Engl. J. Med.,* 299, 1232, 1978.

88. **Mahley, R. W., Hui, D. Y., Innerarity, T. L., and Weisgraber, K. H.,** Two independent lipoprotein receptors on hepatic membranes of dog, swine, and man: apo-B,E and apo-E receptors, *J. Clin. Invest.,* 68, 1197, 1981.

89. **Fielding, P. E. and Fielding, C. J.,** A cholesteryl ester transfer complex in human plasma, *Proc. Natl. Acad. Sci. U.S.A.,* 77, 3327, 1980.

90. **Fielding, C. J. and Fielding, P. E.,** Cholesterol transport between cells and body fluids: role of plasma lipoproteins and the plasma cholesterol esterification system, *Med. Clin. N. Am.,* 66, 363, 1982.

91. **Innerarity, T. L. and Mahley, R. W.,** Enhanced binding by cultured human fibroblasts of apo-E-containing lipoproteins as compared with low density lipoproteins, *Biochemistry,* 17, 1440, 1978.

92. **Sherrill, B. C., Innerarity, T. L., and Mahley, R. W.,** Rapid hepatic clearance of the canine lipoproteins containing only the E apoprotein by a high affinity receptor: identify with the chylomicron remnant transport process, *J. Biol. Chem.,* 255, 1804, 1980.

93. **Blum, C. B., Deckelbaum, R. J., Witte, L. D., Tall, A. R., and Cornicelli, J.,** Role of apolipoprotein E-containing lipoproteins in a betalipoproteinemia, *J. Clin. Invest.,* 70, 1157, 1982.

94. **Angelin, B., Raviola, C. A., Innerarity, T. L., and Mahley, R. W.,** Regulation of hepatic lipoprotein receptors in the dog: rapid regulation of apolipoprotein B, E receptors but not of apolipoprotein E receptors by intestinal lipoproteins and bile acids, *J. Clin. Invest.,* 71, 816, 1983.

95. **Heideman, C. I. and Hoff, H. F.,** Lipoproteins containing apolipoprotein A-I extracted from human aortas, *Biochim. Biophys. Acta,* 711, 431, 1982.

96 Health United States 1980, DHHS Publication No. (PHS)81-1232, U. S Department of Health and Human Services, Hyattsville, Md., *1980.*

97. **Levy, R. I.,** Declining mortality in coronary heart disease, *Arteriosclerosis,* 1, 312, 1981.

98. **Blumenthal, H. T.,** Athero-arteriosclerosis as an aging phenomenon, in *The Physiology and Pathology of Human Aging,* Goldman, R. and Rockstein, M., Eds., Academic Press, New York, 1975, 123.

99. **Hazzard, W. R.,** Aging and atherosclerosis interaction with diet, heredity, and associated risk factors, in *Nutrition, Longevity, and Aging,* Rockstein, M and Sussman, M. L., Eds., Academic Press, New York, 1976, 143.

100 **Martin, G. M. and Sprague, C. A.,** Symposium on in vitro studies related to atherogenesis: life histories of hyperplastoid cell lines from aorta and skin, *Exp Mol. Pathol.,* 18, 125, 1973.

101. **Duthu, G. S. and Smith, J. R.,** In vitro proliferation and lifespan of bovine aorta endothelial cells: effect of culture conditions and fibroblast growth factor, *J. Cell Physiol.,* 103, 385, 1980.

102. **Schwarz, K. O. and Levy, G. L.,** Species and age difference in the collagen types of the arterial wall — a possible explanation for differences in susceptibility to atherosclerosis, *Med. Hypoth.,* 8, 619, 1982.

103. **Toledo, O. M. S. and Nourao, P. A. S.,** Sulfated glycosaminoglycans in normal aortic wall of different mammals, *Artery,* 6, 341, 1980.

104. **Noma, A., Takahashi, T., and Wada, T.,** Elastin-lipid interaction in the arterial wall. II In vitro binding of lipoprotein-lipids to arterial elastin and the inhibitory effect of high density lipoproteins on the process, *Atherosclerosis,* 38, 373, 1981.

105. **Kumar, V., Berenson, G. S., Ruiz, H., Dalferes, E. R., and Strong, J. P.,** Acid mucopolysaccharides of human aorta. I. Variation with maturation, *J Atheroscler. Res ,* 7, 573, 1967.

106. **Wagner, W. D. and Nohlgren, S. R.,** Aortic glycosaminoglycans in genetically selected WC-2 pigeons with increased atherosclerosis susceptibility, *Arteriosclerosis,* 1, 192, 1981.

107. **Kannel, B. W., Hjortland, M. C., McNamara, P. M., and Gordon, T.,** Menopause and risk of cardiovascular disease. The Framingham Study, *Ann. Intern. Med ,* 85, 447, 1976.

108. **Velonakis, L. E., Oekonomakos, P., Laskaris, J., and Katsimades, D.,** Serum sex hormones in patients with coronary disease and their relationship to known factors causing atherosclerosis, *Cardiology,* 69, 98, 1982.

109. U S. Department of Health and Human Services, Public Health Service, National Institute of Health, Lipid Metabolism Branch, NHLBI, *The Lipid Research Clinics,* Population Studies Data Book, Vol. 1., NIH Publ. No. 80-1527, NIH, Bethesda, Md., 1980

110. **Mjos, O. D., Thelle, D. S., Forde, O. H., and Vik-Mo, H.,** Family study of high density lipoprotein cholesterol and the relation to age and sex, *Acta Med Scand.,* 201, 323, 1977.

111. **Castelli, W. P., Cooper, G. R., Doyle, J. T., Garcia-Palmieri, M., Gordon, T., Hames, C., Hulley, S. B., Kagan, A., Kuchmak, M., McGee, D., and Vicic, W. J.,** Distribution of triglyceride and total LDL and HDL cholesterol in several populations: a cooperative lipoprotein phenotyping study, *J Chron Dis.,* 30, 147, 1977.

112. **Kirk, J. E.,** Comparison of enzyme activities of arterial samples from sexually mature men and women, *Clin Chem ,* 10, 184, 1964.

113. **Kirk, J. E.,** *Enzymes of the Arterial Wall,* Academic Press, New York, 1969.

114. **Dilman, V. M.,** *The Law of Deviation of Homeostasis and Diseases of Aging,* Blumenthal, H T , Ed., John Wright, PSG, Boston, 1981 (translation).

115. **Dudl, R. T. and Ensinck, J. W.,** Insulin and glucagon relationships during aging in man, *Metabolism,* 26, 33, 1977

116. **Keys, A. and Brozek, J.,** Body fat in adult man, *Physiol Rev.,* 33, 245, 1953

117. **Dilman, V. M.,** On the age-associated elevation of activity of certain hypothalamic centers, *Trans Inst. Physiol. Acad Sci U.S.S.R.,* 326, 1958.

118 **West, K. W.,** *Epidemiology of Diabetes and its Vascular Lesions,* Elsevier, New York, 1978, 354.

119 WHO Expert Committee on Diabetes Mellitus, Second report, *WHO Tech. Rep. Ser.,* 646, 1980.

120. **Wyngaarden, J. B.,** Metabolic diseases, in *Cecil Textbook of Medicine,* Wyngaarden, J. B and Smith, L. S., Jr., Eds., W B. Saunders, Philadelphia, 1982.

121 **Vinik, A. I. and Jackson, W. P. U.,** Diabetes in the elderly. Is it a disease? *Excerpta Med.,* 140, 195, 1967.

122. **Bagdade, J. D., Bierman, E.L., and Porte, D., Jr.,** Influence of obesity on the relationship between insulin and triglyceride levels in endogenous hypertriglyceridemia, *Diabetes,* 20, 664, 1971.

123. **Genuth, S.,** Classification and diagnosis of diabetes mellitus, *Med Clin N. Am.,* 66, 1191, 1982

124. **Reaven, G. M. and Olefsky, J. M.,** Role of insulin resistance in the pathogenesis of diabetes mellitus, *Adv. Metab. Res ,* 9, 312, 1978.

125. **Goth, M. and Goenczi, J.,** Effect of insulin-induced hypoglycemia on the plasma corticol and growth hormone levels in obese and diabetic persons, *Endocrinology,* 60, 8, 1972

126. **Bar, R. S., Gordon, P., Roth, J., Kahn, C. R., and de Meyts, P.,** Fluctuations in the affinity and concentration of insulin receptors on circulating monocytes of obese patients: effects of starvation, refeeding, and dieting, *J. Clin Invest.,* 58, 1123, 1976.

127. **Bar, R. S., Harrison, L. C., Muggeo, M., Gorden, P., Kah, C. R., and Roth, J.,** Regulation of insulin receptors in normal and abnormal physiology in humans, *Adv. Intern. Med.,* 24, 23, 1979.

128. **Taylor, A. and Blackard, W. G.,** Role of insulin receptors in obesity-related diabetes, *Horm. Metab. Res.,* 14, 623, 1982.

129 **Luft, R. and Guillemin, R.,** Growth hormone and diabetes, *Diabetes,* 23, 783, 1974.

130. **Steiner, G. and Vranic, M.,** Hyperinsulinemia and hypertriglyceridemia, a vicious cycle with atherogenic potential, *Int. J. Obesity,* 6 (Suppl.1), 117, 1982.

131. **Kolterman, O. G., Gray, R. S., Griffin, J., Burstein, P., Insel, J., Scarlett, J. A., and Olefsky, J. M.,** Receptor and post-receptor defects contribute to the insulin resistance in non-insulin dependent diabetes mellitus, *J. Clin. Invest ,* 68, 957, 1981.

132. **Ciaraldi, T. P., Kolterman, O. G., Scarlett, J. A., Kao, M., and Olefsky, J. M.,** Role of glucose transport in the postreceptor defect of non-insulin-dependent diabetes mellitus, *Diabetes,* 31, 1016, 1982.

133. **Bjoerntorp. P.,** The role of adipose tissue in human obesity, in *Obesity,* Greenwood, M R C , Ed., Churchill Livingstone, New York, 1983.

134 **Stout, R. W., Bierman, E. L., and Ross, R.,** Effect of insulin on the proliferation of cultured primate arterial smooth muscle cells, *Circle Res.,* 36, 319, 1975.

135. **Taggart, H. and Stout, R. W.,** Control of DNA synthesis in cultured vascular endothelial and smooth muscle cell — response to serum platelet-deficient serum, lipid-free serum, insulin, and estrogens, *Atherosclerosis,* 37, 549, 1980

136. **Bagdade, J. D., Bierman, E. L., and Porte, D., Jr.,** Influence of obesity on the relationship between insulin and triglyceride levels in endogenous hypertriglyceridemia, *Diabetes,* 20, 664, 1971.

137. **Eaton, P. and Nye, W. H. R.,** The relationship between insulin secretion and triglyceride concentrations in endogenous lipemia, *J. Lab. Clin Med ,* 81, 682, 1975

138. **Reaven, G. M. and Greenfield, M. S.,** Diabetic hypertriglyceridemia: evidence for three clinical syndromes, *Diabetes,* 30(Suppl. 2), 66, 1981.

139. **Greenfield, M., Kolterman, O., Olefsky, J., and Reaven, G.,** Mechanism of hypertriglyceridemia in diabetic patients with fasting hyperglycaemia, *Diabetologia,* 18, 441, 1980

140. **Olefksy, J. M., Farquhar, J. W., and Reaven, G. M.,** Reappraisal of the role of insulin in hypertriglyceridemia, *Am. J. Med.,* 57, 551, 1974.

141. **Hulley, S. B., Rosenman, R. H., Bawol, R. D., and Brand, R. J.,** Epidemiology as a guide to clinical decisions: the association between triglyceride and coronary heart disease, *N. Engl. J. Med ,* 302, 1383, 1980

142. **Lippel, K., Tyroler, H., Eder, H., Gotto, A. J., and Vahouny, G.,** Relationship of hypertriglyceridemia to atherosclerosis, *Arteriosclerosis,* 1, 406, 1981.

143. **Miller, G. J.,** High density lipoproteins and atherosclerosis, *Ann. Rev. Med.,* 31, 97, 1980.

144 **Glomset, J. A.,** High-density lipoproteins in human health and disease, *Adv. Intern. Med.,* 25, 91, 1980

145. **Nikkilae, E. A.,** High density lipoproteins in diabetes, *Diabetes,* 30(Suppl. 2), 82, 1981

146 **Zilversmith, D. B.,** Atherogenesis: a postprandial phenomenon, *Circulation,* 60, 473, 1979.

147. **Robinson, D. S.,** The function of the plasma triglycerides in fatty acid transport, in *Comprehensive Biochemistry,* Vol. 18, Florkin, M. and Stotz, E. H., Eds., Elsevier, Amsterdam, 1970, 51.

148. **Elkeles, R. S. and Hambley, J.,** The effects of fasting and streptozotocin diabetes on the hepatic triglyceride lipase activity in the rat, *Diabetes,* 26, 58, 1977

149. **Taskinen, M. R., Nikkilae, E. A., Kuusi, T., and Harno, K.,** Lipoprotein lipase activity and serum lipoproteins in untreated type-2 (insulin-dependent) diabetes associated with obesity, *Diabetologia,* 22, 46, 1982.

150. **McGill, H. J.,** Introduction to the geographic pathology of atherosclerosis, *Lab. Invest.,* 18, 465, 1968.

151. **Sternby, N. H.,** Atherosclerosis, smoking and other risk factors, in *Atherosclerosis,* Gotto, J. M., Jr., Smith, L. C., and Allen, B., Eds , Springer-Verlag, New York, 1980, 67.

152. **Holme, I., Enger, S. C., Helgeland, A., Hjerman, I., Leren, P., Lund-Larsen, P. G., Solberg, L. A., and Strong, J. P.,** Risk factors and raised atherosclerotic lesions in coronary and cerebral arteries: statistical analysis from the Oslo study, *Arteriosclerosis,* 1, 250, 1981

153. **Kannel, W. B.,** The role of cholesterol in coronary atherogenesis, *Med. Clin. N. Am.,* 58, 363, 1974.

154. **Kato, H., Tillotson, J., Nichaman, M. Z., Rhoads, G. G., and Hamilton, H. B.,** Epidemiologic studies of coronary heart disease and stroke in Japanese men living in Japan, Hawaii, and California: serum lipids and diet, *Am. J. Epidemiol.,* 97, 372, 1973.

155. **Connor, W. E., Cerqueira, M. T., Connor, R. W., Wallace, R. B., Malinow, M. R., and Casdorph, H. R.,** The plasma lipids, lipoproteins, and diet of the Tarahumara Indians of Mexico, *Am. J. Clin. Nutr.,* 31, 1131, 1978.

156. **Ho, K. J., Mikkelson, B., Lewis, L. A., Feldman, S. A., and Taylor, C. B.**, Alaskan Arctic Eskimo: responses to the customary high fat diet, *Am. J. Clin. Nutr.*, 25, 737, 1972

157. **Shekelle, R. B., Shryock, A. M., Paul, O., Lepper, M., Stamler, J., Liu, S., and Raynor, W. J., Jr.**, Diet, serum cholesterol, and death from coronary heart disease· the Western Electric Study, *N. Engl. J. Med.*, 304, 65, 1981.

158. **Jacobs, D. R., Anderson, J. T., Hannan, P., Keys, A., and Blackburn, H.**, Variability in individual serum cholesterol response to change in diet, *Arteriosclerosis*, 3, 349, 1983.

159. **Keys, A., Anderson, J. T., and Grande, F.**, Prediction of serum cholesterol responses of man to changes in fats in the diet, *Lancet*, 2, 959, 1957.

160. **Goodnight, S. H. J., Harris, W. S., Connor, W. E., and Illingworth, D. R.**, Polyunsaturated fatty acids, hyperlipidemia, and thrombosis, *Arteriosclerosis*, 2, 87, 1982.

161. **Ho, K. J., Biss, K., and Taylor, C. B.**, Serum cholesterol levels in U. S. males: their correlation with various kinetic parameters of cholesterol metabolism, *Arch. Pathol.*, 97, 306, 1974.

162. **Ernst, N., Fisher, M., Smith, W., Gordon, T., Rifkind, B., Little, A., and Williams, O. D.**, The association of plasma high-density lipoprotein cholesterol with dietary intake and alcohol consumption The Lipid Research Clinics Program Prevalence Study, *Circulation*, 62(Suppl. IV), 41, 52, 1980

163. **Glueck, C. T., Fallat, R. W., Millett, F., Gartside, P., Elston, R. C., and Go, R. C. P.**, Familial hyperalphalipoproteinemia — studies of 18 kindreds, *Metabolism*, 24, 1243, 1975

164. **Mahley, R. W.**, Alterations in plasma lipoproteins induced by cholesterol feeding in animals including man, in *Disturbances in Lipid and Lipoprotein Metabolism*, Dietschy, J. M., Gotto, A. M., Jr., and Ontko, J. A., Eds., American Physiological Society, Bethesda, Md., 1978, 181

165. **Noel, S., Wong, L., Dophin, P. J., Dory, L., and Rubinstein, D.**, Secretion of cholesterol-rich lipoproteins by perfused livers of hypercholesteremic rats, *J. Clin. Invest.*, 64, 674, 1979.

166. **Goldstein, J. L., Ho, Y. K., Brown, M. S., Innerarity, T. L., and Mahley, R. W.**, Cholesterol ester accumulation in macrophages resulting from receptor-mediated uptake and degradation of hypercholesterolemic containing β-very low density lipoproteins, *J. Biol. Chem.*, 255, 1839, 1980.

167. **Mahley, R. W., Innerarity, T. L., Brown, M. S., Ho, Y. K., and Goldstein, J. L.**, Cholesterol ester synthesis in macrophages: stimulation by β-very low density lipoproteins from cholesterol-fed animals of several species, *J. Lipid Res.*, 21, 970, 1980.

168. **Pitas, R. E., Innerarity, T. L., and Mahley, R. W.**, Foam cells in explants of atherosclerotic rabbit aortas have recpetors for β-very low density lipoproteins and modified low density lipoproteins, *Arteriosclerosis*, 3, 2, 1983.

169. **Bell, E. T.**, *Renal Diseases*, Henry Kimpton, London, 1946

170. **Hollenberg, N. K., Adams, D. F., Solomon, H. S., Rachid, A., Abrams, H. L., and Merrill, J. P.**, Senescence and the renal vasculature in normal man, *Circ. Res.*, 34, 307, 1974

171. **Nestel, P. J., Fidge, N. H., and Tan, M. H.**, Increased lipoprotein-remnant formation in chronic renal failure, *N Engl. J. Med.*, 307, 329, 1982.

172 **Bagdade. J., Casaretto, A., and Albers, J.**, Effects of chronic uremia, hemodialysis and renal transplantation on plasma lipids and lipoproteins in man, *J Lab Clin. Med.*, 87, 37, 1976.

173. **Huttunen, J. K., Pasternack, A., Vaenttinen, T., Ehnholm, C., and Nikkilae, E. A.**, Lipoprotein metabolism in patients with chronic uremia, *Acta Med. Scand.*, 204, 211, 1978.

174. **Mordasini, R., Frey, F., Flury, W., Klosse, G., and Greten, H.**, Selective deficiency of hepatic triglyceride lipase in uremic patients, *N. Engl. J. Med.*, 297, 1362, 1977.

175. **Bolzano, K., Krempler, F., and Sandhofer, F.**, Hepatic and extrahepatic triglyceride lipase activity in uremic patients on chronic hemodialysis, *Eur. J. Clin. Invest.*, 8, 289, 1978.

176. **Lindner, A., Charra, B., Sherrard, D. J., and Scribner, B. H.**, Accelerated atherosclerosis in prolonged maintenance hemodialysis, *N. Engl. J. Med.*, 290, 697, 1974

177. **Havel, R. J.**, Approach to the patient with hyperlipidemia, *Med. Clin. N. Am.*, 66, 319, 1982.

178. **Jefferyo, P. M., Farran, H. E. A., Hoffenberg, R., Fraser, P. M., and Hodkinson, H. M.**, Thyroid-function tests in the elderly, *Lancet*, 1, 924, 1972.

179. **Rubenstein, H. A., Butler, V. P., and Werener, S. C.**, Progressive decrease in serum triiodothyronine concentrations with human aging: radioimmunoassay following extraction of serum, *J. Clin. Endocrinol. Metab.*, 37, 247, 1973

180. **Gregerman, R. I., Graffney, G. W., Shock, N. W., and Crowder, S. E.**, Thyroxine turnover in euthyroid man with special reference to changes with age, *J. Clin. Invest.*, 41, 2065, 1962.

181. **Gregerman, R. I. and Solomon, N.**, Acceleration of thyroxine and triiodothyronine turnover during bacterial pulmonary infections and fever: implications for the functional state of the thyroid during stress and in senescence, *J. Clin. Endocrinol.*, 27, 93, 1967

182. **Fu, S. T., Ho, K. J., Taylor, C. B., and Manalo-Estrella, P.**, Thyroid hormones: effect on aortic and mucopolysaccharides and atherosclerosis in rabbits, *Proc. Soc. Exp. Biol. Med.*, 142, 607, 1973

183. **Jones, R. J., Cohen, L., and Corbus, H.**, The serum lipid pattern in hyperthyroidism and coronary atherosclerosis, *Ann. N.Y. Acad. Sci.*, 72, 980, 1959.

184 **Blumgart, H. L., Freedberg, A. S., and Kurland, G. S.,** Hypercholesterolemia, myxederma, and atherosclerosis, *Am J Med ,* 14, 665, 1985

185. **Doggert, D. L., Chang, M. P., Makinodan, T., and Strehler, B. L.,** Cellular and molecular aspects of immune system aging, *Mol Cell Biochem ,* 37, 137, 1981.

186. **Weksler, M. E.,** Age-associated changes in the immune response, *Am. Geriatr. Soc.,* 30, 718, 1982

187. **Goidl, E. A., Thorbecke, G. J., Weksler, M. E., and Siskind, G. W.,** Production of auto-anti-idiotypic antibody during the normal immune response changes in the auto-anti-idiotypic antibody response and the idiotype repertoire associated with aging, *Proc Natl Acad. Sci U S.A ,* 77, 6788, 1980.

188. **Singh, F. and Singh, A. K.,** Age-related changes in human thymus, *Clin Exp Immunol ,* 37, 507, 1979

189. **Low, T. L. K. and Goldstein, A. L.,** Thymosin and other thymic hormones and their synthetic analogues, *Springer Semin Immunopathol.,* 2, 169, 1979

190 **Hirokawa, K. and Makinodan, T.,** Thymic involution effect of T cell differentiation, *J Immunol ,* 114, 1661, 1975

191. **Weksler, M. E.,** The senescence of the immune system, *Hosp Pract.,* 16, 53, 1981.

192. **Inkeles, B., Innes, J. B., Kuntz, M. M., Kadish, A. S., and Weksler, M. E.,** Immunological studies of aging III Cytokinetic basis for the impaired response of lymphocytes from aged humans to plant lectins, *J Exp Med.,* 145, 1176, 1977

193. **Weksler, M. E., Dekruyff, R., Dobken, J., and Siskind, G. W.,** Ease of tolerance induction in mice of different ages, in *Immunological Aspects of Aging,* Segre, D. and Smith, L., Eds , Marcel Dekker, New York, 1981, 119

194 **Ho, K. J., DeWolfe, V. G., Siler, W., and Lewis, L. A.,** Cholesterol dynamics in autoimmune hyperlipidemia, *J. Lab Clin Med ,* 88, 769, 1976

195 **Kaplan, N. M.,** *Clinical Hypertension,* Medcom Press, New York, 1973, 1

196. **Armitage, A. K. and Turner, D. M.,** Absorption of nicotine in cigarette and cigar smoke through the oral mucosa, *Nature (London),* 226, 1231, 1970

197 **Kershbaum, A., Bellet, S., Hirabayashi, M., Feinberg, L. J., and Eilberg, R.,** Effect of cigarette, cigar, and pipe smoking on nicotine excretion: the influence of inhalation, *Arch Intern Med.,* 120, 311, 1967

198. **Connon, W. E.,** The acceleration of thrombus formation by certain fatty acids, *J. Clin. Invest ,* 41, 1199, 1962

199. **Hegarty, K. M., Turgiss, L. E., Mulligan, J. J., Cluette, J. E., Kew, R. R., Stack, D. J., and Hojnacki, J. L.,** Effect of cigarette smoking on high density lipoprotein phospholipids, *Biochem. Biophys. Res. Commun.,* 104, 212, 1972.

200. **Ho, K. J. and Taylor, C. B.,** Comparative studies on tissue cholesterol, *Arch. Pathol.,* 122, 585, 1968.

201. **Malinow, M. R.,** Regression of atherosclerosis in human. a new frontier? *Postgrad Med.,* 73, 231, 239, 1983.

202. **Feldman, E. B.,** Does nutrition play a role in cardiovascular disease? *Geriatrics,* 35, 65, 1980

203. **Grundy, S. M.,** Questions and answers on hyperlipidemia, *Geriatrics,* 36, 101, 1981

204 **Rees, E. D.,** Serum lipid disorders: what levels should you aim for? *Geriatrics,* 36, 77, 1981

205. **Moser, M.,** The management of cardiovascular disease in the elderly, *J Am. Geriatr. Soc.,* 30, S29, 1982.

Chapter 4

CANCER

Part I

NUTRITIONAL BIOCHEMISTRY OF CANCER

**Betty J. Mills**

TABLE OF CONTENTS

## I. INTRODUCTION

### A. The Nature of Cancer

The origins of human cancer remain obscure, largely because of what appears to be a long latent period, possibly as long as 30 years, between the initiating events and overt symptoms of malignant disease. It is well documented that the incidence and mortality rates from cancer increase precipitously in the later years of life. This fact has been depicted graphically from statistics compiled by the U.S. Department of Health and Human Services (Figure 1).[1]

It is important to understand that cancer is not a single disease entity and may arise from numerous causes, not yet all understood. In spite of the multiplicity of originating factors, most cancers develop through a series of defined steps. It is necessary to know these progressive stages in cancer development in order to discriminate at which points it may be possible to intervene to prevent, arrest, or reverse the neoplastic growth.

This sequence of stages developing from the molecular to the tissue level of biological organization has been identified in vivo and in vitro using culture systems of both animal and human cell lines.[2] The first stage, *initiation*, occurs at the molecular level and involves an irreversible alteration or damage to the DNA of the cell by some substance called an "initiator". Such an alteration in DNA does not inevitably lead to cancer. However, the second stage of carcinogenesis, *promotion*, leads to the transformation of normal cells to tumor cells by affecting cell differentiation and phenotypic expression, and increases cell proliferation. Compounds that are not in themselves carcinogenic may be metabolized to ultimate active carcinogens by other compounds called "promoters". This is a critical stage that may be reversible by deactivating mechanisms. Various aspects of promotion and the molecular events leading to cell transformation have been discussed in detail by others.[3,4]

The final stage in the process of carcinogenesis is *progression*, which is the uncontrolled proliferation of transformed cells to the clinical entity called cancer. It is usually at this point that some intervention in the process is initiated, whether it is surgery, radiation, or chemotherapy. The role of nutrition at this stage has generally been limited to the use of hyperalimentation or total parenteral nutrition to overcome cachexia and attempt to "build up" patients to better endure the deleterious side effects of radiation or chemotherapy. Although in animal experiments a number of promising approaches have been utilized to restrict tumor growth by nutritional modifications, there have been few attempts to apply these findings for the treatment of human cancer patients.

It has been found that initiators of carcinogenesis are usually molecules that become very reactive as the result of metabolic activation. Most substances identified as carcinogens have a common property of being electron deficient, but otherwise, common structural features are unrelated to their carcinogenic activity.[3] Their carcinogenic potential has often been associated with their degree of mutagenicity as determined by a variety of bioassays.[5,6]

A number of generally accepted facts concerning the mechanisms of carcinogenesis have been listed by Newell:[7]

1.  Carcinogenesis is dose dependent, with dose being a function of both carcinogen concentration and exposure time.
2.  There is a long latent period between carcinogen exposure and tumor development. In humans this period could be 5 to 30 years.
3.  There is no evidence of a threshold dose below which there is no risk of cancer.
4.  Carcinogens can act transplacentally and can cause tumors in offspring.
5.  Transformation (normal to cancer) is a multistep process with initiating agents enhanced by promoting agents, hormonal agents, and other co-factors.
6.  Cells with rapid proliferation (high turnover) are more susceptible.

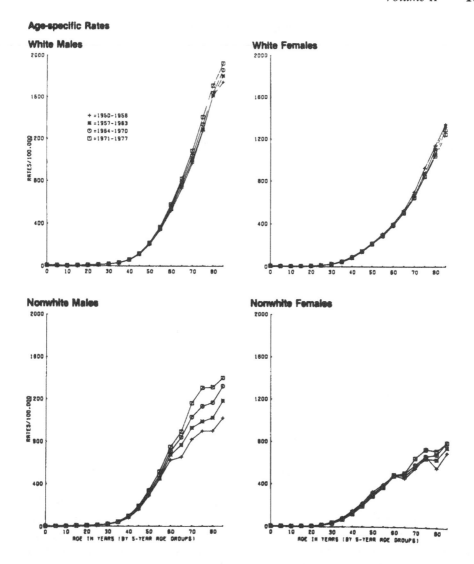

FIGURE 1. Age-specific mortality from all malignant neoplasms in the U.S. from 1950 to 1977. Waterhouse, J., Correa, P., and Muir, C. et al., Eds., *Cancer Incidence in Five Continents*, Vol. 111, IARC Sci. Publ No. 15, IARC, Lyon, 1976, 1. *Directory of On-Going Research in Cancer Epidemiology*, IARC Sci. Publ. No. 35, IARC, Lyon, 1980, 1. Doll, R., The epidemiology of cancer, *Cancer*, 45, 2475, 1980. Pollack, E S. and Horm, J. W., Trends in cancer incidence and mortality in the United States, *J. Natl. Cancer Inst.*, 64, 1091, 1980. (From Cancer Mortality in the United States: 1950—1977, NCI Monogr. 59, U.S. Department of Health and Human Services, Washington, D.C., 1982.)

7. There is antigenic diversity to the same chemical stimulus and immune responses differ.
8. The same chemicals can cause different types of tumors.

Discussion thus far has been limited to chemical carcinogenesis because it is most probable that nutritional factors may be involved at specific stages and thus render them amenable to modification.[8] Our scheme showing these relationships is outlined in Figure 2. However, other initiating causes of cancer have also been identified, such as DNA and RNA viruses[9-11] and radiation.[12,13] These initiators also affect changes in DNA, viruses by "novel juxtaposition of genes from the virus and the host",[14] and radiation by dimer formation between

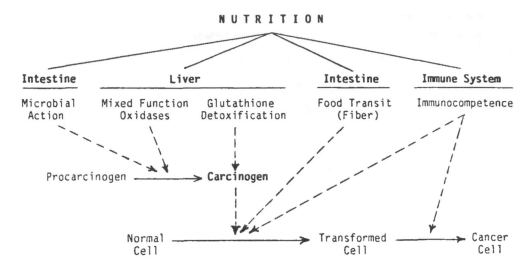

FIGURE 2    Nutritional effects on carcinogenesis

bases.[15] There is no evidence that viruses act as mutagens, and they tend to cause specific types of cancer such as lymphomas, leukemias, or sarcomas rather than acting as nonspecific carcinogens.[14]

## B. Diet and Cancer

The concept that common cancers may be related to diet has derived largely from epidemiological data, particularly studies of migrant populations that have undergone significant changes in lifestyle, especially diet. It has been possible to compare the incidence of a specific type of cancer in one country to the incidence of the same cancer in groups of people who moved to new environments. This kind of comparison is dramatically illustrated in the reversals of stomach and colon cancers in Japanese migrants (Figure 3).[16] Although it has been stated that nutritional factors are involved in the etiology of greater than 50% of cancer cases in the U.S.,[17] data derived from human epidemiological studies are notoriously unreliable. People have poor recall of their food histories in terms of kinds and quantities of food ingested. Moreover, the long-term effects of modes of food storage and preparation cannot be evaluated. Again, the long latent period between exposure and clinical manifestations of cancer renders most information useless. The best insight into changes that are due specifically to nutrition, cancer, or age is derived from animal experiments that are strictly controlled in terms of pathogen exposure, diet, temperature, and other environmental variables.

The recent publication of the National Research Council[14] stated that common cancers are more likely related to certain long-standing features of our lifestyles rather than to industrialization and its attendant effects. In other words, we need to examine the lifestyle factors that we can control, especially diet! An interesting observation may be that the incidence of tumors is related more to *excess* of nutrients than to deficiency. Among the few deficiencies associated with human cancer are those of Fe, I, vitamin $B_2$, vitamin A, and pyridoxine.[18] Also, a high incidence of spontaneous malignant lymphomas have been found in Mg-deficient rats.[19]

## C. Aging and Cancer

Although it is increasingly apparent that diet and cancer are in some way related, the relationship of aging to cancer is more obscure. A primary problem has been lack of understanding of the mechanisms that underlie these relationships. One premise derived from

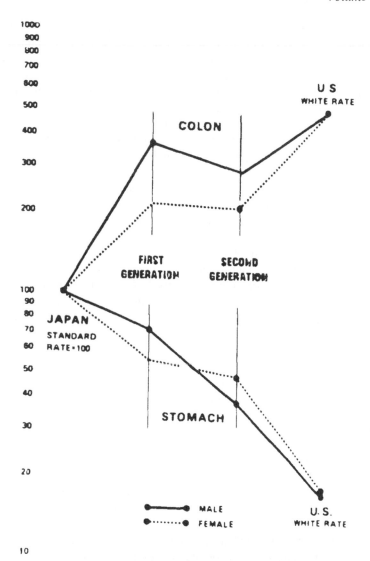

FIGURE 3. Mortality trends of Japanese migrants to the U.S. Note the relatively rapid increase of the risk of colon cancer even in the first generation, and progressive decrease of the risk for stomach cancer (Adapted from Haenszel, W. and Kurihara, M., Studies of Japanese migrants. I. Mortality from cancer and other diseases among Japanese in the United States, *J. Natl. Cancer Inst.*, 40, 43, 1968. (Reprinted with permission from Raven Press, New York.)

animal experiments is that the high incidence of cancer in old age is a function of *duration* of exposure to cancer-producing agents and that there is no intrinsic effect of aging per se, such as decreased immunological surveillance or age-related hormonal changes.[20] Nonetheless, is it possible that there is a common connection and that the same nutritional factors that affect longevity also affect the susceptibility to cancer? A key observation that indicates a common underlying mechanism is that tumor incidence is decreased in animals that have prolonged longevity resulting from dietary restriction.

There is no agreement among gerontologists as to the causes of aging and a variety of so-called "Theories of Aging" have been suggested. Indeed, Moment[21] has pointed out that a major impediment to nutritional research in relation to aging is that "...there is no certain theory, not even an agreed-upon, most likely hypothesis, about the nature of the timing mechanism or mechanisms which set the pace for aging.'' Since this topic has been discussed in depth by others,[22,23] we need only indicate the diversity of opinion by noting that hypotheses include such concepts as ''rate of living'', cross-linkage of macromolecules, accumulation of cellular waste products, loss of immunologic function, and free radical damage.

Carcinogenesis is likely to be related to alterations at the cellular or molecular levels and thus, it may be more productive to discuss the biochemical changes of senescence. The list of biochemical changes that have been reported in aging organisms is extensive and includes alterations that seem to have no possible relationship to cancer, such as increased cross-linkage of collagen and elastin, lipofuscin accumulation, and impaired transmitter function. In contrast, cellular changes that affect information processing, oxidation/reduction status, the capacity for detoxification, and immunologic protection may be highly significant for carcinogenesis and have been observed in aging organisms.[24-28] The relationship of aging changes to carcinogenesis has been reviewed by Pitot.[29] There is some evidence that a number of these changes are affected by nutritional status.

Our hypothesis is that the same basic mechanisms underly both aging and the cancer of old age and that the same chemopreventive nutritional measures can be utilized to retard both processes. In essence, both conditions result from long-term exposure to harmful substances and a decreased capacity for detoxification in senescence. We believe that these conditions are highly dependent upon the oxidation/reduction status of organisms, which can be altered nutritionally to decrease metabolic rate, and the production of free radicals. The cumulative damage to cellular integrity from both exogenous and endogenous reactive molecules could cause both cancer and aging. It is quite feasible that by modifying diet to minimize nutrients that increase metabolic rate and also increase the intake of nutrients that maintain high reducing capacity, we will be able to attain longer cancer-free lives.

The importance of redox balance in the body was discussed by Shapiro,[30] who noted that the redox status of tissues, like the pH, is maintained within a relatively narrow range and that deviations from this range may occur in various pathological states. To extend this concept further, one might add that alterations in the redox state may even be *causative* factors of pathological conditions.

Along this line of thought, it has also been noted that thiols and disulfides are involved in both the normal and pathological growth of tissues — that the SH:SS ratio is a determinant of the oxidation/reduction potential that is critical for cell division.[31]

It becomes apparent that the oxidation/reduction status can be a two-edged sword. A high reducing, antioxidant capacity is favorable for normal growth and protection from free radicals and exogenous electrophiles. On the other hand, an oxidative state that interferes with free SH:SS interchanges, which are essential for malignant proliferation and increases peroxidation, may suppress growth of tumor cells. A practical application of this concept in regard to cancer would be the judicial intake of nutrients that increase the natural defense mechanisms at the stages of initiation and promotion, but to reduce intake of specific nutrients once a neoplasm is established and in progression. The identification of such nutrients is a key area for investigation of the nutritional modulation of tumor growth.

## II. NUTRITIONAL ASPECTS OF CANCER PREVENTION

### A. Macronutrients

A key problem in determining roles that dietary protein, fat, and carbohydrate may have in the genesis of tumors of different kinds is that most foods are complex mixtures that vary in the percentages of these components. Determination of their specific contributions in terms of caloric content and effects on biochemical processes in humans is impossible. Again we must refer to controlled animal experiments for information.

### 1. Calories and Protein

The role of dietary restriction in increasing longevity has been discussed in a previous chapter. However, it became apparent in animal studies that chronic dietary restriction also resulted in a lower incidence of spontaneous or chemically induced tumors.[32-36] Major questions arose concerning this phenomenon. Was it brought about by reduction of total calories or of only the protein component, or by change in the ratios of the macronutrients? Also, is lifetime restriction essential or are there certain sensitive periods during which restriction has maximal effect? Although these questions have not been completely resolved, most recent evidence indicates that restriction of total calories is probably more significant than reduction of protein[37] and that restriction initiated at later stages of the life span can also be effective.[38,39] The mechanisms by which nutritional restriction increases longevity and reduces tumor incidence may not be identical. Although life-long restriction was most effective for increased life span of mice, restriction initiated after 61 weeks resulted in the least number of spontaneous tumors.[40] A key point, however, is to differentiate between *undernutrition*, which limits calories or protein but provides essential minerals and vitamins, and *malnutrition*, which can produce severe adverse effects, particularly for young growing animals.[41]

In a summary of studies of the relationship between underfeeding and tumor formation, White[42] concluded that the basic effect of undernutrition was to increase the latent period of tumorigenesis and thus decrease the incidence of spontaneous tumors. More recently, Masoro et al.[43] and Liepa et al.[44] concluded that food restriction results in *delay* of time of onset and change in the chronologic course of physiologic decline. For example, in caloric-restricted rats with prolonged longevity, there was delayed loss of lipolytic responsiveness of adipocytes to both glucagon and epinephrine and also a delay in the age-related increase in serum cholesterol and a reduction in the magnitude of the increase. Further, caloric restriction delayed the onset of testicular Leydig cell tumors that commonly occurred with 100% frequency in male Fischer 344 rats.[45] Evidence of this kind indicates that delay of physiologic decline may also retard the occurrence of other age-related diseases.

A major difficulty lies in the fact that all tumors are not reduced by underfeeding and may even be increased.[46] There is an obvious need to identify those biochemical and/or physiological changes that occur with food restriction to determine if they have an anticarcinogenesis function. One possible mechanism is a decrease in metabolic rate, resulting in a lower production of harmful by-products of oxygen metabolism, in particular free radicals that may damage DNA irreversibly. Indeed, it was concluded by Totter[47] that most cancer is spontaneous due to endogenous injury from steady univalent oxygen reduction and subsequent production of harmful radicals. This concept is supported in that few cancers occur in tissues where metabolism is largely controlled by physical activity and that there is a lower cancer mortality in women, who also have a metabolic rate about 10% lower than men.

Harman[48] suggested that dietary protection from free-radical reactions might be attained by reducing intake of calories, Cu, and polyunsaturated fats that tend to increase levels of free-radical reactions and by adding antioxidant compounds capable of slowing free-radical damage. Support for this hypothesis comes from experiments in which diets fortified with antioxidants significantly reduced or delayed cancer incidence in mice or rats or inhibited chemical carcinogenesis.[49-53] It has been argued, however, that the observed effects possibly resulted from a self-imposed nutritional restriction since animals fed the antioxidant diets had reduced food intake. There is recent evidence that free radical-mediated cellular damage may be decreased by dietary restriction. In mice fed one half the *ad libitum* consumption for 1 year there was less accumulation of lipofuscin in brain and heart and lower rates in vitro of lipid peroxidation and lysosomal enzyme activities.[54]

There is evidence that dietary restriction, whether of total calories or of protein alone, affects enzyme activities that may play significant roles in modifying carcinogenic agents.

In rats restricted to 50% of the *ad libitum* intake, the in vitro activities of hepatic drug-metabolizing enzymes were increased two- to threefold and there was a significant increase in NADPH-generating enzymes of liver and adipose tissue.[55] Protein deprivation resulted in decreased mixed function oxidase (MFO) activities in rat liver,[56] reduced the binding of aflatoxin $B_1$ to DNA, chromatin, and protein,[57] and inhibited the formation of preneoplastic lesions in rat liver.[58] This series of experiments led to the conclusion that dietary protein was a promoter or modifier of carcinogens.

The increase in longevity and reduced tumor incidence resulting from dietary restriction may be mediated via the immune system. It is generally accepted that immune dysfunction is a major consequence of normal aging as assessed by various parameters.[28,39,59] Concurrently, increases in manifestations of autoimmunity have been demonstrated. Unfortunately, the degree and mode of protection from cancer that is derived from the immune system is still not clarified. Since the immune system involves complex interactions of lymphocytes, phagocytes, the vascular system, immunoglobulins, complement, endogenous protein mediators, lymphokines, and other factors,[60] it is difficult to determine which type of immune response can control tumor development. According to Stutman,[61] the immune functions evolved as a mechanism to prevent the *emergence* of cancer cells rather than to mediate the regression of established tumors. This concept is based on the idea that aberrant cells with proliferative potential carry new antigenic determinants that will initiate a thymus-dependent immune response. Recently it was demonstrated that natural killer (NK) cell activity and tumor cell killing mediated by cytotoxic T lymphocytes were decreased in old mice.[62] However, substantially higher NK activity was elicited in dietarily restricted old mice compared to age-matched controls as well as increased in vitro generation of cytotoxic lymphocytes.

Investigators at Sloan-Kettering Institute postulated that factors which delay, decrease, or reverse immunologic decline of senescence may modify the development of diseases associated with aging, including malignancies.[63-65] In an elaborate series of experiments using mice with varying life spans associated with age-related diseases, they tested the effects of several dietary restrictions. In particular, they found that dietary restriction preserved immune function and that chronic protein deprivation delayed thymic involution and inhibited development of splenomegaly. Tumor immunity and graft-vs.-host reactions are indexes of T cell-mediated immune function. Reduction of total caloric intake doubled the life span of the MRL/1 mouse, which is a model of severe lymphoproliferative diseases, delayed thymic involution, and prevented massive lymphadenopathy. With C3H mice that are predisposed to mammary tumors, reduction of dietary fat resulted in vigorous response to T cell mitogens and the capacity for generation of splenic suppressor cells. An interesting point was their finding that while T cell or thymus-derived immunity was maintained or increased with restricted diets, B cell immunity was decreased. Such findings point out some possible mechanisms by which dietary restriction may better protect aging animals via the immune system.

A caveat stated by this group of investigators in regard to nutritional restrictions was the requirement for dietary Zn to maintain optimal immunological surveillance. Zn is an absolute requirement for lymphocyte transformation and appears to act as a mitogen. The multiple effects of zinc deficiency on immune response such as hypoplasia of lymphoid tissues, changes in lymphocyte numbers, inhibited plaque-forming cell response, deficient cell-mediated immunity, and impaired sensitivity to skin test were reviewed recently by Beisel.[66]

### 2. Fat and Fiber

Epidemiologic data compiled in Table 1 show a strong correlation between dietary fat and the incidence of cancer, in particular hormone-dependent and colon cancers.[67] Similarly, a strong positive correlation between the per capita consumption of dietary fat in numerous countries worldwide and the mortality from breast and prostate cancer has been demon-

## Table 1
## SIGNIFICANT ($p < 0.05$) PEARSON'S CORRELATION COEFFICIENTS: AGE-ADJUSTED MORTALITY RATES[a] VS. PER CAPITA CONSUMPTION OF DIETARY ITEM[b] AND HEIGHT[c]

| | GNP | Wheat | Rice | Maize | Beans | Pork | Cattle meat | Eggs | Milk | Animal oil/fat | Beer | Animal calories | Animal Proteins | Fat | Height (F) |
|---|---|---|---|---|---|---|---|---|---|---|---|---|---|---|---|
| Colon | 0.67 | — | −0.34 | −0.67 | −0.68 | 0.60 | 0.54 | 0.75 | 0.48 | 0.64 | 0.62 | 0.84 | 0.49 | 0.74 | 0.56 |
| Breast (F) | 0.66 | 0.34 | −0.50 | −0.66 | −0.70 | 0.62 | 0.58 | 0.75 | 0.55 | 0.64 | 0.56 | 0.84 | 0.48 | 0.75 | 0.63 |
| Prostate | 0.64 | — | −0.58 | −0.53 | −0.66 | 0.48 | 0.47 | 0.48 | 0.52 | 0.51 | 0.44 | 0.76 | 0.52 | 0.72 | 0.58 |
| Arteriosclerotic heart disease | 0.52 | — | −0.40 | — | — | — | 0.48 | — | 0.59 | 0.48 | — | 0.74 | 0.51 | 0.73 | — |
| Height | 0.70 | — | −0.42 | −0.66 | — | 0.71 | — | 0.64 | 0.48 | 0.58 | 0.47 | 0.69 | — | 0.61 | — |

[a] Segi, M., Age-Adjusted Death Rates for Cancer for Selected Sites (A-Classification) in 52 Countries in 1973, Segi Institute of Cancer Epidemiology, Nagoya, Japan, 1978.

[b] Food and Agricultural Organization. Food Balance Sheets, 1964—1966, Rome, 1971.

[c] Meredith, H. V., Worldwide somatic comparisons among contemporary human groups of adult females, *Am. J. Phys. Anthropol.*, 34, 89, 1971.

From Correa, P., 1981. Reprinted with permission from Raven Press, New York.

strated.[68] However, there is little agreement as to the specific role(s) of fat in carcinogenesis. Although experimental evidence is weak, obesity is associated with increased risk of breast and endometrial cancers, obese women being at two- to fourfold greater risk than nonobese women.[69]

Although the early work of Tannenbaum showed that high-fat diets increased the incidence of spontaneous mammary tumors in mice[70] and promoted the induction of skin tumors by carcinogens,[71] no mechanism was suggested. As noted earlier, restriction of total calories decreased the incidence of spontaneous tumors in rodents, and this was confirmed by Good et al.,[72] who found that mammary adenocarcinomas were reduced from 60 to 70% to 0 to 5% in calorie-restricted mice.[72] However, they also found that restriction of calories without restriction of fat was only minimally effective.

It has now been demonstrated, in addition, that the kind of fat is important, with polyunsaturated fats generally being more effective than saturated or monosaturated fatty acids in promoting tumor growth.[73,74] In mice, essential fatty acids were required for normal ductal growth[75] and dietary polyunsaturated fatty acids enhanced the growth of mammary adenocarcinomas.[76] In other experiments the incidence of tumors was lowest in rats fed cottonseed oil, butter, or beef fat.[77] In these same experiments the life span of rats fed a 20% fat diet was shortest when the diet contained coconut oil and longest with beef fat or a random mixture of medium-chain fatty acids.

Although there is little evidence that dietary fat is in itself carcinogenic, it may play an important role as a promoter or cocarcinogen.[78] Several possibilities have been suggested for the tumor-enhancing effects of dietary fat: (1) host endocrine metabolism is altered to produce high circulating levels of hormones;[69,79] (2) unsaturated fatty acids act directly on mammary gland epithelium to stimulate growth;[80] (3) bile acids derived from cholesterol may provide an important cancer promotion stimulus;[81] (4) tumor growth is mediated via prostaglandins derived from polyunsaturated fatty acids.[76]

It has been noted that serum levels of steroid hormones are increased with a high intake of dietary fat.[69] In contrast, a calorie restricted, low-fat high-protein diet resulted in a marked decrease in serum prolactin and fewer mammary alveolar lesions in older rats.[82] Numerous mechanisms by which sex steroids may be implicated in tumorigenesis have been presented,[83] but a possible sequence may be high fat → high estrogen → high prolactin → increased tumor prolactin receptors. The exact role of prolactin in the stimulation of tumor growth has not been elucidated, but it may act as a hormonal stimulus for the release of fatty acids that enhance cell growth.[84] Prolactin has been implicated in the control of prostatic tissue growth as well as of mammary gland.[69] Strong evidence that estrogen and prolactin are essential for mammary tumor growth was a reversal of tumor growth inhibition in half-fed rats by administration of either estrogen or haloperidol, a drug that increases prolactin secretion.[85]

A provocative summary of the role of sex steroids in the aging-related development of neoplasia was prepared by MacDonald.[86] He noted that an increase in the endogenous production of estrogen was associated with an increased risk of endometrial cancer. Since after cessation of menses, the ovary does not secrete significant amounts of estrogens or the precursor, androstenedione, the most important site quantitatively for extraglandular estrone formation in post-menopausal women is in adipose tissue. Moreover, the conversion of androstenedione to estrone is increased with aging, obesity, and hepatic disease — all features associated with increased risk of endometrial cancer.

The role of dietary fat in colon cancer may be mediated by increasing in the gut the concentrations of secondary bile acids which act as promoters. Persons fed Western diets high in animal fats excrete increased amounts of fecal steroids that represent products of cholesterol metabolism (Table 2).[81] In population samples from six countries there was an almost linear relationship between the fecal concentration of secondary bile acids and car-

**Table 2**
**DAILY FECAL EXCRETION OF BILE ACIDS IN DIFFERENT
POPULATIONS CONSUMING A VARIETY OF DIETS**

| Bile acids | Americans on a mixed Western diet (45)[a] | American vegetarians (18) | Seventh-Day Adventists (22) | Japanese (28) | Chinese (19) |
|---|---|---|---|---|---|
| Cholic acid | 25 | 29 | 13 | 20 | 15 |
| Deoxycholic acid | 180 | 68 | 64 | 60 | 69 |
| Lithocholic acid | 129 | 50 | 46 | 59 | 62 |
| Total | 325 | 182 | 170 | 160 | 171 |

[a]  Number in parentheses is number of subjects.
[b]  Total bile acids include cholic acid, deoxycholic acid, chenodeoxycholic acid (not shown in table), lithocholic acid, and other microbially modified bile acids (not shown in table).

Reprinted with permission from *Clinical Chemistry*.

cinoma of the colon.[87] Sequentially, cholesterol is converted by liver enzymes to primary bile acids that are then dehydroxylated by intestinal bacteria to the secondary bile acids, deoxycholic and lithocholic. The fact that cholic or deoxycholic acids can be chemically converted by pyrolysis to methylcholanthrene, a known carcinogen, has presented the possibility that the reaction may occur in vivo at 37°C if the proper enzymes are present.[88]

Epidemiologic data indicate that dietary fiber plays a significant role in reducing the incidence of colon cancer. Although the exact mechanism of this protective effect is not known, it may be that fiber increases the transit time of carcinogenic agents in food through the gut. Possibly, an increase in fecal bulk dilutes secondary bile acids and also any suspected carcinogens in direct contact with large bowel mucosa.[81] Much work is still needed to identify effective components of fiber since it is composed of various carbohydrates including cellulose, hemicellulose, various starches, noncellulose polysaccharides, pectin, waxes, and also lignin.[89] Cereal grain fibers differ in composition from fruit or vegetable fiber.[90]

In recent years there has been increased interest in the relationship of prostaglandins and cancer.[91] Prostaglandins are a series of 20-carbon unsaturated carboxylic acids containing a 5-carbon ring. Their immediate precursors are *bis*-homo-γ-linolenic and arachidonic acids of the cellular phospholipid pool derived from dietary linoleic and linolenic acids. These compounds have been implicated in many key biological regulatory processes and may also play an important role in carcinogenesis. The possible interactions by which prostaglandins may be involved in the initiation, promotion, or progression of cancer have been extensively reviewed.[92,93] Recent evidence that $PGE_2$ can be angiogenic suggested that it could be a key factor in the neovascularization of tumors.[94] The most convincing evidence that dietary fat enhances tumor growth via prostaglandins is derived from animal experiments in which the growth of either transplantable or chemically induced tumors was reduced by administration of prostaglandin synthetase inhibitors[95–97] and increased by administration of prostaglandins.[98,99] A rapid increase in tumor growth, after cessation of an indomethacin treatment that initially inhibited their development, led to the conclusion that the prostaglandin inhibitor did not actually reject or kill growing tumors.[100]

As noted earlier, changes occur in the lipid metabolism of old rats, such as loss of lipolytic responsiveness to glucagon and epinephrine and increase in serum cholesterol concentration.[43,44] In addition, the lipid and cholesterol content of many tissues accumulate with age.[101,102] The significance of these aging changes for tumorigenesis is not known. However, if lipids per se are involved in tumor promotion in old age, it might be related to a decreased ability to synthesize, degrade, and excrete lipids, so that lipid accumulation results in blood and tissues.[103]

## B. Micronutrients

It may well be that micronutrients (vitamins, minerals, antioxidants, and minor food constituents) comprise the major protection against carcinogenesis, yet it has been shown that some of these nutrients are decreased in the elderly due to lowered food intake. A review of the vitamin status in old persons noted that deficiencies of vitamins C, D, K, folic acid, and of the B group were clinically significant, but symptoms of vitamin A or E deficiency were rare.[104] In general, information on senescent changes in trace element concentrations is relatively sparse. Hsu,[105] in 1979, reviewed the current knowledge on Zn, Cu, and Cr in aging. Several changes in the concentrations of trace elements in plasma and blood have been found in old age. However, there is considerable doubt that plasma levels are valid indicators of the trace element status of the entire organism.

In old human subjects, plasma Zn concentrations were significantly decreased although red cell Zn was unchanged.[106] Several investigators have also found significant decreases in the Se concentrations of whole blood or red cells from old humans.[107-109] Serum Se was also decreased in very old people, but to a lesser degree.[107,109,110] Serum Cu levels were found to be increased only in males in old age.[111] A significantly increased ratio of Cu to ceruloplasmin was interpreted to indicate an increased amount of "free" Cu in old human males, although there were no changes in either serum Cu or ceruloplasmin.[112]

### 1. Mechanisms of Inhibition

Wattenberg[113] classified inhibitors of carcinogenesis into three categories dependent upon the stage at which they act. Thus they include compounds that (1) prevent the formation of carcinogens from precursor substances, (2) inhibit carcinogenic agents from reaching or reacting with target sites, and (3) suppress the expression of neoplasia by exposed cells. This classification should not be considered too rigorously since some compounds may be effective at more than one stage and exert their effects through different mechanisms. These aspects have been discussed in detail by Wattenberg.[114]

### a. Prevention of Carcinogen Formation

The antioxidants, $\alpha$-tocopherol (vitamin E) and ascorbic acid (vitamin C), are examples of compounds that act to prevent the formation of ultimate carcinogens. A putative role for vitamin E is its ability to deactivate free radicals from a wide variety of oxidative reactions induced by both exogenous and endogenous factors. In particular, vitamin E has been ascribed the function of protecting biomembranes against lipid peroxidation. The phenolic group on the chroman ring of $\alpha$-tocopherol permits its reactivity with free radicals, and its fatty chain makes it soluble in membrane phospholipids. Two reactions by which $\alpha$-tocopherol may be involved in the destruction of free radicals may be (1) the formation of $\alpha$-tocopheryl-quinone via an unstable intermediate or (2) the formation of a tocopherol radical and its subsequent detoxification by glutathione disulfide.[115]

Although it is difficult to determine a direct relationship between membrane lipid peroxidation and carcinogenesis, it may be that the structural integrity of cellular membranes is essential for protection. It is known that lipid peroxidation in biological systems increases with age and that deficiency of vitamin E accelerates the process.[116]

At certain critical levels, dietary vitamin E was effective in inhibiting the growth of sarcoma tumor cells injected into mice.[117] The effect was dependent both on the levels of vitamin E and of polyunsaturated fats in the diet. In contrast to the notion that vitamin E acts primarily as a free radical scavenger, these investigators suggested that it inhibited tumor production by strengthening the antigenicity of tumor cells which normally have weak antigens.

In recent experiments, α-tocopherol succinate inhibited the in vitro growth of mouse melanoma and neuroblastoma cells and rat glioma cells.[118] Since the alterations on growth and cellular morphology were similar to those produced by butylated hydroxyanisole (BHA) and butylated hydroxytoluene (BHT), this data indicated that the vitamin E effect could be mediated by antioxidant mechanisms. Alternately, it has been suggested that vitamin E has a "sparing" effect on vitamin A by protecting it from oxidation in the gut, especially when the diet contains large amounts of polyunsaturated fats, and by increasing the absorption, utilization, and storage of vitamin A.[119]

Vitamin C (ascorbate) may also act as a direct blocking agent against *N*-nitroso compounds in the etiology of gastric cancer. *N*-Nitroso compounds are formed by the reaction of nitrite and secondary amines or ureas in the presence of acid, and the majority of them tested were found to be carcinogenic.[120] Nitrites are added to cured meat and fish products and are also found in certain leafy vegetables. Ascorbic acid prevents amine nitrosation by reacting with nitrite. In humans the synthesis of nitrosoproline, a noncarcinogenic indicator of endogenous nitrosation,[121] was effectively decreased by 2 g of vitamin C per day.[122] A comprehensive review of nitrite and nitrite flow in the human upper GI tract has been compiled by Hartman.[123]

There is only sparse evidence that vitamin C is effective against cancer in vivo,[124] but studies in vitro have demonstrated an inhibitory effect on tumor cell growth.[125,126] It has been hypothesized that malignant invasiveness may be retarded by inhibition of a plasma hyaluronidase that requires ascorbic acid for its synthesis.[127]

Protection afforded by endogenous vitamin C is likely to be decreased in old age. In mice, levels of ascorbic acid in serum, liver, thymus, and lung decreased with advancing age.[128] In humans, levels of ascorbate in serum were decreased from 12 to 96 years of age,[129,130] and also in leukocytes of females.[129,131] The administration of 1 g of supplemental vitamin C for 28 days to elderly subjects was able to increase vitamin C levels in their leukocytes and plasma.[132]

### b. Detoxification of Carcinogens

A second group of inhibitors are those that prevent carcinogenic agents from reaching or reacting with target sites by inducing the activities of detoxification systems such as Phase I and Phase II reactions.

Phase I reactions of the mixed function oxidase system (MFO) involve oxidation, reduction, and hydrolysis and may deactivate or activate compounds dependent upon the specific substance. Phase II reactions consist of conjugation and synthesis to form nontoxic products.[133] The protective effects of dietary antioxidants may be produced via either system dependent upon the carcinogenic compound, the specific antioxidant, and the fat composition of the diet. For example, BHT and propyl gallate reduced the incidence of tumors in rats given oral 7,12-dimethylbenz(*a*)anthracene (DMBA), which requires metabolic activation, but 2(3)-*tert*-butyl-4-hydroxyanisole (BHA) and α-tocopherol were ineffective.[134] The differential effects of both antioxidant and carcinogen on the MFO system in mice suggested that the protective action of antioxidants may be produced by a decrease in metabolic activation and/or enhancement of metabolic deactivation.[135] It was found that BHA in the diet involved both Phase I and Phase II reactions by increasing the activities of both epoxide hydratase and glutathione (GSH) *S*-transferase in mice and also increased the tissue levels of nonprotein sulfhydryls.[136]

Alterations in MFO activities have also been brought about by naturally occurring constituents (benzyl isothiocyanate, phenethyl isothiocyanate, benzylthiocyanate, flavones, and indoles) that are found in edible cruciferous plants. These compounds and citrus fruit oils may also exert their protective effect against electrophiles by their ability to induce GSH

*S*-transferase activity in Phase II reactions. Phase II conjugation reactions that convert carcinogenic agents to noncarcinogenic agents provide another mechanism for blocking them from reaching target sites. The group of enzymes known as glutathione *S*-transferases catalyzes the conjugation of hydrophobic electrophilic substances with endogenous glutathione (GSH) in the initial step for the formation of water-soluble mercapturic acids. Novi[137] demonstrated a dramatic regression of aflatoxin $B_1$-induced hepatic carcinomas by oral administration of GSH, although the mechanism responsible was not elucidated. However, it was shown that the intact GSH molecule was recovered in the liver.[138]

The tripeptide, glutathione (L-γ-glutamyl-L-cysteinyl-glycine) is the most abundant antioxidant in mammalian tissues and its concentration is nutritionally controlled in part by dietary protein and sulfur amino acids.[139–143] Also, GSH levels in blood and liver of rats were decreased by dietary depletions of Zn or Mg[144-146] and increased by excess Se.[147] An increased susceptibility to carcinogenesis in old age may be attributable in part to the decreased levels of GSH found in senescent organisms, including mosquito, mouse, and man.[148-152]

It has not been demonstrated that glutathione levels are increased by consumption of the compound itself, but other substances, such as thiazolidine-4-carboxylic acid (TC) and its analogue, L-2-oxothiazolidine-4-carboxylate, are precursors of GSH when administered orally or injected.[153.154]

Thiazolidine-4-carboxylic acid has been used in Europe for a number of clinical conditions including various liver diseases and GI disorders,[155] but the anticarcinogenic potential of TC has not been clarified with both positive and negative results reported.[156.158] However, TC was demonstrated to increase longevity in a number of different animal species.[159,160] Similarly, another antioxidant, nordihydroguaiaretic acid, increased the median life span of mosquitoes.[161] The evidence relating antioxidants to increased life span and carcinogenesis and possible mechanisms have been reviewed elsewhere.[162]

### c. Suppression of Transformation

A third important group of anticarcinogenic compounds are those thought to suppress the expression of neoplasia. Prime candidates in this group are retinoids, protease inhibitors, and prostaglandin synthesis inhibitors. It is already well established that retinoids are essential for normal differentiation of most epithelial cells from which the majority of all cancers are derived. An anticarcinogenic role for vitamin A or its synthetic analogues, the retinoids, is suggested by both epidemiological and experimental data. The risk of developing lung cancer is inversely proportional to vitamin A intake.[163] Plasma vitamin A levels were also decreased in patients with several human cancers,[164–166] but the question exists whether the deficiency was a predisposing factor or a consequence of the disease.

Animal experiments demonstrated that supplemental retinoids were inhibitory against chemical carcinogenesis and conversely, animals deficient in vitamin A had marked increases in carcinogen-induced tumors.[167] In rats, vitamin A or its analogue not only prevented the development of tumor transplants, but also arrested the growth of established tumors.[168] There is belief that retinoids may not only prevent tumor progression, but may actually reverse abnormal differentiation.[169]

It is not yet clear whether these effects on tumor growth are due to vitamin A itself or its precursors, the carotenoids in yellow and green vegetables. Experiments utilizing β-carotene at 90 mg/kg of diet showed that tumor growth was inhibited and the survival time of rats was increased even when the diets were initiated after tumors were well established.[170] A problem with naturally occurring retinoids is that their use may well be limited due to inadequate tissue distribution or excessive toxicity.

Other anticarcinogenic agents that may suppress the expression of neoplasia are protease inhibitors and prostaglandin synthesis inhibitors, such as indomethacin, or several nonsteroidal anti-inflammatory drugs. Evidence for the involvement of prostaglandins in carcinogenesis was previously discussed in relation to dietary fats. Less well known may be a possible role for protease inhibitors which are not strictly micronutrients. The observation that proteolytic activity increased in tumors compared to their normal tissue counterparts suggested that proteinases could be involved in the process of malignant invasion and altered growth control.[171] In early experiments, synthetic protease inhibitors topically applied to areas treated with DMBA and a promoter delayed the onset and reduced the incidence of mouse skin tumors.[172] More recently it was found that a soybean diet lowered the incidence of breast tumors in irradiated rats.[173] The soybean inhibitor survives inactivation by stomach digestion in rodents and seems to be fully active as an inhibitor against trypsin and chymotrypsin in the small intestine.[174] Good dietary sources of protease inhibitors are legumes and cereals.

## 2. Trace Elements

There is no firm evidence that dietary minerals and trace elements are carcinogenic, although there appears to be a relationship between cancer incidence and other kinds of exposure to high concentrations of As, Be, Cd, Ni, and Zn.[175] Schrauzer,[176,177] who compiled comprehensive reviews on the roles of trace elements in carcinogenesis, suggested that even elements essential for the maintenance of life at low concentrations may be mutagenic or carcinogenic in unusual oxidation states or at high concentrations. Further, apparent carcinogenic effects of elements such as As or Zn may be related more to their diminishing the anticarcinogenic effect of Se.

In recent years Se has been the focus of much interest because both epidemiological and experimental data have suggested that it has a strong protective effect against tumorigenesis. The mechanism of action is not known, but Schrauzer[176] outlined the various properties and functions of Se that may be related to an anticarcinogenic function. The role of Se in cancer chemoprevention has also been discussed at length by Griffin.[178]

The significant inverse correlations of dietary Se intakes with the age-adjusted mortalities from the major forms of cancer suggest a strong nutritional relationship.[177] Schrauzer pointed out that the average U.S. diet contains less Se than is optimal for cancer protection and that to achieve adequate intake it would be necessary to increase significantly the amounts of wheat grain cereal products, seafoods, and organ meats.

Careful examination of experimental data shows that adequate Zn status is probably critical for the chemoprevention of tumors. Several investigations have demonstrated that supplemental Zn inhibited the induction and onset of both implanted and chemically induced tumors.[179–181] The extensive series of experiments carried out by the Sloan-Kettering group led to the conclusion that the protective effect was mediated via the immune system and was highly dependent upon Zn.[62–64]

It has been unclear as to whether the altered levels of some elements in the blood of cancer-bearing patients or animals is a cause or consequence of the disease. For example, there are numerous reports of decreased serum concentrations of Zn and Se associated with cancer, and serum Cu levels are often elevated.[109,182-188] The low levels of Zn and Se have been considered indexes of nutritional deficiency by some investigators. However, in our opinion, these changes reflect altered host metabolism as a result of tumor presence. In rats, we found that plasma concentrations of Zn decreased and Cu levels increased within 3 days after tumor implantation, well before tumors were palpable or any changes in food intake or weight occurred.[189] Similar changes in the concentrations of these elements did not occur in sham-implanted rats. In other cases, abnormal levels of Zn or Cu were reversed by treatments that resulted in increased survival and remission of disease.[190-193]

Interpretation of these findings of altered trace element concentrations in plasma must be considered with caution. There is considerable disagreement as to whether plasma or blood levels are truly representative of the trace element status of the whole organism. Further, although it is tempting to suggest that reduced Se and Zn and increased Cu may be factors in carcinogenesis, it is probably more prudent to view, as did Schrauzer[177] in the case of Se, that a low status must not be considered a cause of cancer but rather a condition that enhances the susceptibility to cancer. An enhanced susceptibility to the induction of neoplasms has been demonstrated in both Zn- and Mg-deficient rats.[194-196]

## III. NUTRITIONAL MODULATION OF TUMOR GROWTH

### A. Basic Considerations

Discussion thus far has focused on the nutritional chemoprevention of cancer operating at the stages of initiation and promotion. We now ask if nutritional modalities are also effective during the stage of tumor *progression*. It is only at this point that the physician identifies that a malignancy is present and must determine appropriate treatment. It has been assumed that older persons survive longer after a cancer has been diagnosed because of a decreased capacity for cell turnover. However, in a long-term follow-up of 3558 women with breast cancer, patients over 71 years of age had a 5-year 50% survival rate; those 51 to 70 years had a 8-year rate; and those less than 50 years had a 13-year survival rate.[197] These data clearly show that the prognosis for survival after diagnosis is substantially poorer in older women.

There is no evidence that cancer cells can be killed by nutritional means. However, a substantial body of data has demonstrated that various kinds of dietary manipulations can alter the rate of tumor growth. Along these lines, inducers of differentiation in vivo and growth modulators which may arrest cancer have been discussed by Howell[198] and Tisdale.[199] To evaluate the results of nutritional interventions on the rate of tumor progression we shall focus specifically on those experiments in which the dietary regimens were initiated at day 0 or after tumors were positively identified. Although experiments in which animals were dietarily preconditioned before tumor implantation can yield valuable information, they do not simulate the clinical situation in which patients present with established tumors.

Some intriguing findings in rats may shed some light on basic mechanisms that should be considered in relation to the regulation of tumor growth by nutritional means. Comparison of growth of two different neoplasms showed significantly lower growth rates in old compared to young hosts.[200] However, the finding of highest mitotic indexes in tumors from old rats suggested that the slower growth was not due to failure to maintain the capacity for cell proliferation. Thus, a key issue is to identify and deplete those *specific* nutrients that are essential for rapid cellular growth and proliferation. There is no sound basis for assuming that nonspecific deprivation of nutrients is effective in limiting tumor growth. No differences in tumor growth rates have been demonstrated between fed and starved animals,[201-203] and it was concluded that tumors have a biochemical advantage in nutrient acquisition in situations of nutrient depletion.[203] Therefore, any successful control of tumor growth must selectively inhibit rapidly multiplying tumor cells to a greater extent than host cell populations with normal turnover.

A major point of controversy regarding nutrition of the cancer patient has been the question, "By feeding the host, do we also feed the tumor and increase its growth?" The answer derived from experimental animal data is probably, yes. Tumor growth generally has been stimulated in animals that have been force-fed.[204,205] The mere presence of a tumor will produce a decrease in carcass mass,[206] and in cancer patients the metabolic disturbances that interfere with normal nutrient utilization are not necessarily resolved by hyperalimentation. In patients with small cell lung carcinoma, a group receiving total parenteral nutrition (TPN) did gain in body weight, fat, and total body K compared to a patient group that received a normal diet orally, but both groups had a similar loss of total body nitrogen and there was no difference in survival.[207] Such results are not necessarily contraindications to all use of hyperalimentation, but long-term support is still problematical and it may best be utilized to counteract acute metabolic insults such as surgery, radiation, or chemotherapy.

## B. Protein and Amino Acids

Attention has centered in large part on the contribution of proteins or amino acids to tumor growth since they provide the major building blocks of tissues. In general, these results have been disappointing. Deletion or reduction in dietary protein has resulted on occasion in somewhat smaller tumors, but with few differences in the ratio of tumor weight to carcass weight.[208-210] Rats repleted orally or intravenously after protein depletion rapidly gained in both body and tumor weights.[211,212]

The concept that deprivation of amino acids would inhibit tumor growth has not met with significant success, apparently because of the remarkable ability of tumors to derive nitrogen from host tissues and utilize whatever substrate is in abundant supply.[213] The requirement for amino acids for tumor growth was demonstrated in a study comparing three nutrient solutions fed intravenously: (1) 30% glucose + 5% amino acids, (2) 5% amino acids, and (3) 5% glucose. Tumor growth was inhibited only in rats not supplied with amino acids.[214]

Early work suggested that implanted sarcomas grew less rapidly in mice fed synthetic diets deficient in valine (Val), leucine (Leu), threonine (Thre), phenylalanine (Phe), histidine (His), or methionine (Met).[215] Later investigations found growth restriction of adenocarcinomas in mice fed reduced levels of tryptophan (Trp), Leu, or Met, but the animals suffered considerable weight loss.[216] On the other hand, dietary restriction of Phe, Val, or isoleucine (IsoLeu) significantly inhibited tumor growth without affecting host weight.

A review of the data seems to indicate a specific requirement for Met for tumor cell growth. The growth rate of mammary tumors significantly increased after transfer from a Met-deficient to a Met-supplemented diet.[217] Also, both i.p. administration of a Met-degrading enzyme, methionase, or a Met-free diet inhibited the growth of Walker 256 tumors in rats.[218] Treatment with the enzyme, however, depressed host weight less than the Met-free diet.

Effective inhibition of tumor growth by amino acid deficiency seems highly dependent upon tumor type. The unique nature of melanomas has enabled retardation of their growth by dietary deprivation of Phe and Tyr, but the same diets had no effect on sarcoma growth.[219,220] In other work, hepatoma growth in mice was also reduced by Phe-Tyr deficiency, but two different mammary adenocarcinomas responded in the opposite manner.[221] Reduced intake of Phe and Tyr by patients with advanced melanomas halted the development of new masses and resulted in regression of existing lesions.[222] Unfortunately, the dietary regimens were difficult to maintain and tumor growth recurred when patients resumed their regular food intake.

## C. Carbohydrates

High rates of anaerobic glycolysis, lactic acid production, and gluconeogenesis are characteristics of cancerous animals and humans. If these unique metabolic aberrations favor growth of tumors, the possibility exists that reversals of these conditions may inhibit their growth. Two different approaches have been used to test this possibility. In one, selective inhibition of gluconeogenesis, without inhibition of glycolysis that is necessary for host tissues, has been attempted by blocking the activity of the enzyme, phosphoenolpyruvate carboxykinase.[223] Administration of the enzyme inhibitors, hydrazine sulfate, pyridine-2,3-dihydrazide, or L-tryptophan produced significant reduction in the growth of several different tumors in rats and also increased survival times. Toxicity was manifested in weight loss alone with no mortality due to treatment.

A contrasting approach has been to induce hyperglycemia by high oral intake or by intravenous infusions of glucose. The idea underlying this approach is that by supplying exogenous glucose for the energy needs of the tumor, it reduces the pressure to convert host glucogenic amino acids to glucose.[224] A combined treatment of glucose infusions along with other therapeutic measures used on 80 patients with advanced forms of cancer resulted in tumor regressions and increased survival. An interesting sidelight on this aspect was from earlier studies in which tumors implanted into alloxanized diabetic rats had significant reduction in growth that was inversely related to the degree of hyperglycemia.[225]

## D. Vitamins

Except for the numerous studies of vitamins A and E, which are still inconclusive in regard to their ability to inhibit the growth of established neoplasms, there has been limited research on other vitamins. In contrast to the notion that vitamin supplements are nutritionally beneficial to the cancer patient, a few animals studies indicate that deficiencies of specific vitamins may retard tumor growth. There is some data demonstrating that a deficiency of either pantothenic acid or vitamin C decreased tumor growth rates.[226,227] On the other hand, thiamin deficiency slightly accelerated the growth of mammary adenocarcinomas in mice while pyridoxine had no effect at all.[226]

The most convincing evidence, however, related to riboflavin. In mice fed riboflavin-deficient diets, complete regression of implanted lymphosarcomas occurred as did a marked increase in survival.[228] Similar results were obtained when the diets included riboflavin and also the riboflavin antagonists, isoriboflavin or galactoflavin. Rivlin[229] reviewed the relationship of riboflavin to cancer and suggested that neoplastic tissue utilizes more riboflavin than normal tissue.

## E. Trace Elements

The essential roles of trace elements for normal growth and health maintenance have been well established. We have focused on the specific elements, Zn and Mg, because of the requirement for Zn for DNA, RNA, and protein synthesis, and for Mg in all enzyme reactions catalyzed by ATP and involving the transfer of phosphate. Their essentiality for tumor growth has also been demonstrated as the growth of tumors implanted into animals with preconditioned Zn or Mg deficiencies was significantly inhibited.[230-234] Moreover, in these deficient animals tumor growth was stimulated by mineral repletion.[232,233]

Recently we demonstrated in the rat that Zn or Mg depletion initiated after implanted mammary adenocarcinomas were well established also resulted in significant retardation of tumor growth.[235,236] Although there was carcass weight loss in Zn-depleted rats, the weight of Mg-depleted rats actually increased 14%, indicating that the Mg depletion had a greater differential effect on growth of tumor and host tissue.

There have also been encouraging reports on the use of Mg depletion in human cancer patients. Parsons and co-workers[237,238] achieved Mg and K deficiency in patients with advanced malignancies with resultant regression of tumors, weight gain, and relief of symptoms.[237,238] An important finding was an apparent absence of bone marrow depression so that resistance to infection was not impaired. Although their aim was to induce deficiencies by diet alone, hemodialysis was necessary to accelerate the process.

Although it appears that Se may have an anticarcinogenic function, its effectiveness in suppressing the growth of established tumors has not been established.

## IV. CONCLUDING REMARKS

All evidence points to the conclusion that cancers are of multiple origins, are heterogeneous in both tissue type and metabolism, and produce metabolic aberrations that affect the whole organism. A scheme diagrammed by Apple[239] showed that cancer cells are distinguished by a shift of catabolism of amino acids and nucleotides into anabolic pathways, building up polyamino acids and polynucleotides by utilizing the breakdown of carbohydrate energy for their creation.

The high incidence of cancer in later years is not questioned, and although similarities between aging and carcinogenesis suggest common underlying mechanisms, the actual causative factors remain in dispute. The somatic mutation hypothesis as a common molecular mechanism is attractive, but must be viewed in relation to genetic and environmental factors that alter the cellular homeostatic ability to adapt.[240] We believe that nutrition as an environmental factor can be utilized to enhance protective mechanisms against insults that induce cancer and accelerate aging. A review of the literature indicates that the most likely biochemical areas for chemoprevention against cancer are those that maintain immunologic competence and detoxification capacity.

There is an enormous amount of investigation still needed to elucidate the appropriate dietary modifications to reduce cancer incidence. The recommendations of the National Research Council,[14] as reviewed by Mendeloff,[241] are those which apply for the general maintenance of overall good health. Caution is in order since possible roles of food components in preventing carcinogenesis have been identified and partially informed persons may modify their diets unwisely by either avoidance or excessive intake of specific foods. These practices may be detrimental as even common foods contain compounds that are mutagens, carcinogens, or promoters of cancer, as well as anticarcinogens.[242,243] Therefore, with our current knowledge, moderation and balance in the diet are still recommended. A major point of agreement is that as a population, Americans consume too many calories. The basal metabolic rate declines between 20 and 90 years of age, a period during which activity levels also usually decrease, but often persons do not adjust their intake to the decreased need.

It was recognized as early as 1953 by Tannenbaum and Silverstone[244] that the aspects of the genesis of tumors and their continued growth must be considered separately and that nutritional needs may differ dependent upon the stage. It is this very point that demands utmost attention and yet has seemingly been dismissed or ignored. Conflicting reports that either an excess or a deficiency of a nutrient can inhibit tumor growth can be resolved if one distinguishes between the processes of initiation or promotion of tumorigenesis and the progression of tumor growth which may be influenced by different mechanisms. In the literature there are many examples from animal studies showing that nutritional supplements or deficiencies having an inhibitory effect initially, later accelerated the process of cancer growth. This point has been demonstrated in the cases of Zn, Mg, riboflavin, protein, and starvation, to cite a few examples.

The need to discriminate differing nutritional needs dependent upon the stage of carcinogenesis has particular implications for the treatment of cancer patients. Thus, it is important to determine the unique nutritional requirements for tumor growth in relation to the requirements for basal maintenance of host tissues. Knowledge of nutrients that enhance tumor growth can be exploited for use in combination with other various kinds of therapy. Depletion, followed by repletion, has been shown to stimulate tumor growth, acting as a pseudo-cell synchronization mechanism. Several investigators have suggested that nutritional repletion in combination with cell-cycle specific chemotherapeutic agents increase their effectiveness and improve host tolerance.[214,245-247]

A comprehensive and interdisciplinary review of three distinct areas, nutrition, cancer and aging, by its very nature is extensive, but because of the abundance of available information the author must also exercise some selectivity. Therefore, it will be obvious to the discerning reader that certain areas of interest have been omitted. These aspects, which include the effects of alcohol, artificial sweetening agents, and caffeine, have been reviewed by others.[248-252] Also, a specific role of carbohydrate in carcinogenesis has not been elucidated although its contribution to total calories and conversion to fat may be a factor. Another omission has been a discussion of the effects of cooking methods and the possible conversion of food components to carcinogens.[253,254] It is our hope that the reader will be indulgent and utilize the information included herein as a springboard from which to launch a more in-depth research of specific topics of interest.

# REFERENCES

1. Cancer Mortality in the United States. 1950—1977, NCI Monogr 59, U.S Department of Health and Human Services, Washington, D C., 1982, 3
2. **Pitot, H. C.**, Interactions in the natural history of aging and carcinogenesis, *Fed. Proc Fed. Am Soc. Exp. Biol.*, 37, 2841, 1978.
3. **Miller, J. A. and Miller, E. C.**, Ultimate chemical carcinogens as reactive mutagenic electrophiles, in *Origins of Human Cancer*, Hiatt, H. H , Watson, J D , and Winsten, J A., Eds , Cold Spring Harbor Laboratory, Cold Spring Harbor, N Y., 1977, 605.
4 **Sugimura, T., Kawachi, T., Nagao, M., and Yakagi, T.**, Mutagens in food as causes of cancer, in *Nutrition and Cancer. Etiology and Treatment, Progress in Cancer Research and Therapy*, Vol. 17, Newell, G. R. and Ellison, N. M , Eds., Raven Press, New York, 1981, 59.
5. **McCann, J. and Ames, B. N.**, Detection of carcinogens as mutagens in the *Salmonella* microsome test: assay of 300 chemicals, *Proc. Natl. Acad. Sci. U.S A.*, 73, 950, 1976.
6. **Sugimura, T., Sato, S., Nagao, M., Yakagi, T., Matsushima, T., Seino, Y., Takeuchi, M., and Kawachi, T.**, Overlapping of carcinogens and mutagens, in *Fundamentals in Cancer Prevention*. Magee, P. N., Takayama, S., Sugimura, T , and Matsushima, T., Eds , University Park Press, Baltimore, 1976, 191.
7 **Newell, G. R.**, General mechanisms of carcinogenesis, in *Nutrition and Cancer: Etiology and Treatment, Progress in Cancer Research and Therapy*, Vol. 17, Newell, G. R and Ellison, N M , Eds., Raven Press, New York, 1981, 49.
8 **Petering, H. G.**, Diet, nutrition and cancer, in *Advances in Experimental Medicine and Biology*, Vol 91, Schrauzer, G. N.. Ed.. Plenum Press, New York, 1978, 207
9. **Gross, L.**, *Oncogenic Viruses*, 2nd ed., Pergamon Press, Oxford, 1970.
10 **Becker, F. F., Ed.**, *Cancer 2, A Comprehensive Treatise, Etiology. Viral Carcinogenesis*, Plenum Press, New York, 1975.
11. **de-Thé, G.**, Viruses as causes of some human tumors? Results and prospectives of the epidemiologic approach, in *Origins of Human Cancer*, Hiatt, H. H., Watson, J D., and Winsten, J. A , Eds , Cold Spring Harbor Laboratory, Cold Spring Harbor, N.Y , 1977, 1113.
12. **Stewart, A.**, Low dose radiation cancers in man, *Adv. Cancer Res.*, 14, 359, 1971
13. **Little, J. B.**, Radiation carcinogenesis in vitro: implications for mechanisms, in *Origins of Human Cancer*, Hiatt, H H., Watson, J D., and Winsten, J. A., Eds., Cold Spring Harbor Laboratory, Cold Spring Harbor, N.Y., 1977, 923.
14 Committee on Diet, Nutrition and Cancer, *Diet, Nutrition, and Cancer*, Assembly of Life Sciences, National Research Council, National Academy Press, Washington, D.C., 1982
15. **Setlow, R. B., Ahmed, F. E., and Grist, E.**, Xeroderma pigmentosum. Damage to DNA is involved in carcinogenesis, in *Origins of Human Cancer*, Hiatt, H H., Watson, J. D., and Winsten, J. A , Eds., Cold Spring Harbor Laboratory, Cold Spring Harbor, N.Y., 1977, 889.
16 **Wynder, E. L., McCoy, G. D., Reddy, B. S., Cohen, L., Hill, P., Spingarn, N. E., and Weisburger, J. H.**, Nutrition and metabolic epidemiology of cancers of the oral cavity, esophagus, colon, breast, prostate, and stomach, in *Nutrition and Cancer· Etiology and Treatment, Progress in Cancer Research and Therapy*, Vol. 17, Newell, G. R and Ellison, N. M., Eds., Raven Press, New York, 1981, 11.
17. National Dairy Council, Update on nutrition, diet and cancer, *Dairy Council Digest*, 51, 25, 1980.
18. **Wynder, E. L.**, Nutrition and cancer, *Fed. Proc Fed. Am Soc. Exp. Biol*, 35, 1309, 1976.
19 **McCreary, P. A., Laing, G. H., Coogan, P. S., and Hass, G. M.**, Magnesium versus lead in dietary induction of rat neoplasms, *Am J. Pathol*, 86, 26a, 1977.
20. **Peto, R., Roe, F. J. C., Lee, P. N., Levy, L., and Clack, J.**, Cancer and ageing in mice and men, *Br. J. Cancer*, 32, 411, 1975.
21 **Moment, G. B.**, Introduction, in *Nutritional Approaches to Aging Research*, Moment, G. B., Ed , CRC Press, Boca Raton, Fla., 1982, 1.
22. **Strehler, B. L.**, *Time, Cells, and Aging*, 2nd ed Academic Press, New York, 1977.

23. **Comfort, A.,** *The Biology of Senescence,* 3rd ed., Elsevier, New York, 1979

24. **Cutler, R. G.,** Evolutionary biology of senescence, in *The Biology of Aging,* Behnke, J. A., Finch, C E., and Moment, G. B., Eds., Plenum Press, New York, 1978, chap 20.

25. **Price, G. B. and Makinodan, T.,** Aging alteration of DNA-protein information, *Gerontologia,* 19, 58, 1973.

26 **Hazelton, G. A. and Lang, C. A.,** Glutathione S-transferase activities in the yellow-fever mosquito *[Aedes aegypti]* (Louisville) during growth and aging, *Biochem. J.,* 210, 281, 1983.

27 **Leibovitz, B. E. and Siegel, B. V.,** Aspects of free radical reactions in biological systems aging, *J Gerontol.,* 35, 45, 1980

28. **Weindruch, R. H. and Walford, R. L.,** Aging and functions of the RES, in *The Reticuloendothelial System* Vol. 3, Cohen, N. and Sigel, M. M., Eds., Plenum Press, New York, 1982, 713.

29 **Pitot, H. C.,** Carcinogenesis and aging — two related phenomena?, *Am. J. Pathol.,* 87, 444, 1977.

30. **Shapiro, H. M.,** Redox balance in the body: an approach to quantitation, *J. Surg. Res.,* 13, 138, 1972

31. **Apffel, C. A.,** Nonimmunological host defenses: a review, *Cancer Res.,* 36, 1527, 1976.

32. **Tannenbaum, A.,** The initiation and growth of tumors. Introduction. I. Effects of underfeeding, *Am. J. Cancer,* 38, 335, 1940.

33. **Tannenbaum, A.,** The genesis and growth of tumors II. Effects of caloric restriction per se, *Cancer Res.,* 2, 460, 1942.

34. **Visscher, M. B., Ball, Z. B., Barnes, R. H., and Sivertsen, I.,** The influence of caloric restriction upon the incidence of spontaneous mammary carcinoma in mice, *Surgery,* 11, 48, 1942.

35 **Saxton, J. A., Jr., Boon, M. C., and Furth, J.,** Observations on the inhibition of development of spontaneous leukemia in mice by underfeeding, *Cancer Res.,* 4, 401, 1944.

36. **Zamenhof, S. and van Marthens, E.,** Effects of prenatal and chronic undernutrition on aging and survival in rats, *J. Nutr,* 112, 972, 1982.

37. **Fernandes, G., Yunis, E. J., and Good, R. A.,** Influence of diet on survival of mice, *Proc. Natl. Acad. Sci. U.S.A.,* 73, 1279, 1976.

38. **Ross, M. H. and Bras, G.,** Lasting influence of early caloric restriction on prevalence of neoplasms in the rat, *J. Natl. Cancer Inst.,* 47, 1095, 1971

39. **Weindruch, R. and Walford, R. L.,** Dietary restriction in mice beginning at 1 year of age effect on life span and spontaneous cancer incidence, *Science,* 215, 1415, 1982.

40. **Cheney, K. E., Liu, R. K., Smith, G. S., Meredith, P. J., Mickey, M. R., and Walford, R. L.,** The effect of dietary restriction of varying duration on survival, tumor patterns, immune function, and body temperature in B10C3F, female mice, *J. Gerontol.,* 38, 420, 1983.

41. **Weindruch, R. H., Kristie, J. A., Cheney, K. E., and Walford, R. L.,** Influence of controlled dietary restriction on immunologic function and aging, *Fed. Proc. Fed. Am Soc. Exp. Biol.,* 38, 2007, 1979.

42 **White, F. R.,** The relationship between underfeeding and tumor formation. Transplantation and growth in rats and mice, *Cancer Res,* 21, 281, 1961.

43 **Masoro, E. J., Bertrand, H., Liepa, G., and Yu, B. P.,** Analysis and exploration of age-related changes in mammalian structure and function, *Fed Proc. Fed. Am. Soc. Exp. Biol.,* 38, 1956, 1979.

44. **Liepa, G. U., Masoro, E. J., Bertrand, H. A., and Yu, B. P.,** Food restriction as a modulator of age-related changes in serum lipids, *Am. J. Physiol,* 238, E253, 1980.

45 **Yu, B. P., Masoro, E. J., Murata, I., Bertrand, H. A., and Lynd, F. T.,** Life span study of SPF Fischer 344 male rats fed *ad libitum* or restricted diets. longevity, growth, lean body mass and disease, *J Gerontol.,* 37, 130, 1982

46. **Ross, M. H.,** Nutrition and longevity in experimental animals, in *Nutrition and Aging,* Winick, M., Ed., John Wiley & Sons, New York, 1976, 43.

47. **Totter, J. R.,** Spontaneous cancer and its possible relationship to oxygen metabolism, *Proc. Natl. Acad. Sci. U.S.A.,* 77, 1763, 1980.

48. **Harman, D.,** Nutritional implications of the free-radical theory of aging, *J. Am. Coll. Nutr.,* 1, 27, 1982.

49. **Harman, D.,** Prolongation of the normal life span and inhibition of spontaneous cancer by antioxidants, *J. Gerontol.,* 16, 247, 1961.

50. **Apffel, C. A., Walker, J. E., and Issarescu, S.,** Tumor rejection in experimental animals treated with radioprotective thiols, *Cancer Res.,* 35, 429, 1975.

51. **Slaga, T. J.,** Food additives and contaminants as modifying factors in cancer induction, in *Nutrition and Cancer: Etiology and Treatment, Progress in Cancer Research and Therapy,* Vol. 17, Newell, G. R and Ellison, N. M., Eds., Raven Press, New York, 1981, 279.

52. **Ellison, N. M. and Londer, H.,** Vitamins E and C and their relationship to cancer, in *Nutrition and Cancer Etiology and Treatment, Progress in Cancer Research and Therapy,* Vol. 17, Newell, G. R. and Ellison, N. M, Eds., Raven Press, New York, 1981.

53 **Wattenberg, L. W., Loub, W. D., Lam, L. K., and Speier, J. L.,** Dietary constituents altering the responses to chemical carcinogens, *Fed. Proc. Fed Am. Soc. Exp. Biol.,* 35, 1327, 1976.

54. **Chipalkatti, S., De, A. K., and Aigar, A. S.,** Effect of diet restriction on some biochemical parameters related to aging in mice, *J. Nutr.,* 113, 944, 1983

55. **Sachan, D. S. and Das, S. K.,** Alterations of NADPH-generating and drug-metabolizing enzymes by feed restriction in male rats, *J. Nutr.,* 112, 2301, 1982.

56 **Mgbodile, M. V. K. and Campbell, T. C.,** Effect of protein deprivation of male weanling rats on the kinetics of hepatic microsomal enzyme activity, *J Nutr.,* 102, 53, 1972.

57. **Campbell, T. C.,** Nutritional modulation of carcinogenesis, in *Molecular Interrelations of Nutrition and Cancer,* Arnott, M. S., van Eys, J., and Wang, Y.-M., Eds., Raven Press, New York, 1982, 359.

58. **Appleton, B. S. and Campbell, T. C.,** Effect of high and low dietary protein on the dosing and postdosing periods of aflatoxin B$_1$-induced hepatic preneoplastic lesion development in the rat, *Cancer Res ,* 43, 2150, 1983.

59. **Walford, R. L.,** Studies in immunogerontology, *J. Am. Geriatr. Soc.,* 30, 617, 1982.

60. **Hoffman-Goetz, L. and Blackburn, G. L.,** Relationship of nutrition to immunology and cancer, in *Nutrition and Cancer. Etiology and Treatment, Progress in Cancer Research and Therapy,* Vol. 17, Newell, G. R. and Ellison, N. M., Eds., Raven Press, New York, 1981, 73

61. **Stutman, O.,** Immunological surveillance, in *Origins of Human Cancer,* Hiatt, H. H., Watson, J. D., and Winsten, J. A., Eds , Cold Spring Harbor Laboratory, Cold Spring Harbor, N.Y., 1977, 729.

62. **Weindruch, R., Devens, B. H., Raff, H. V., and Walford, R. L.,** Influence of dietary restriction and aging on natural killer cell activity in mice, *J. Immunol.,* 130, 993, 1983.

63. **Good, R. A., Fernandes, G., and West, A.,** Nutrition, immunity and cancer — a review. I Influence of protein or protein-calorie malnutrition and zinc deficiency on immunity, *Clin. Bull.,* 9, 3, 1979.

64. **Schloen, L. H., Fernandes, G., Garofalo, J. A., and Good, R. A.,** Nutrition, immunity and cancer — a review. II. Zinc, immune function and cancer, *Clin. Bull.,* 9, 63, 1979.

65 **Fernandes, G., West, A., and Good, R. A.,** Nutrition, immunity and cancer — a review III Effects of diet on the diseases of aging, *Clin Bull.,* 9, 91, 1979

66 **Beisel, W. R.,** Single nutrients and immunity, *Am. J. Clin. Nutr.,* 35, 417, 1982.

67. **Correa, P.,** Nutrition and cancer: epidemiologic correlations, in *Nutrition and Cancer: Etiology and Treatment, Progress in Cancer Research and Therapy,* Vol. 17, Newell, G. R and Ellison, N. M., Eds., Raven Press, New York, 1981, 1.

68 **Carroll, K. K. and Khor, H. T.,** Dietary fat in relation to tumorigenesis, in *Lipids and Tumors, Progress in Biochemical Pharmacology,* Vol. 10, Carroll, K. K., Ed., S. Karger, Basel, 1975, 308.

69. **Mohla, S. and Criss, W. E.,** The relationship of diet to cancer and hormones in *Nutrition and Cancer: Etiology and Treatment, Progress in Cancer Research and Therapy,* Vol. 17, Newell, G. R. and Ellison, N. M., Eds., Raven Press, New York, 1981, 93.

70. **Tannenbaum, A.,** The genesis and growth of tumors. III Effects of high fat diet, *Cancer Res.,* 2, 49, 1942.

71. **Tannenbaum, A.,** The dependence of the genesis of induced skin tumors on the fat content of the diet during different stages of carcinogenesis, *Cancer Res.,* 4, 683, 1944.

72 **Good, R. A., Fernandes, G., and Day, N. K.,** The influence of nutrition on development of cancer immunity and resistance to mesenchymal diseases, in *Molecular Interrelations of Nutrition and Cancer,* Arnott, M. S., van Eys, J., and Wang, Y -H., Eds., Raven Press, New York, 1982, 73.

73. **Carroll, K. K. and Davidson, M. B.,** The role of lipids in tumorigenesis, in *Molecular Interrelations of Nutrition and Cancer,* Arnott, M. S., van Eys, J., and Wang, Y.-M., Eds., Raven Press, New York, 1982, 237.

74. **Chan, P.-C., Ferguson, K. A., and Dao, T. L.,** Effects of different dietary fats on mammary carcinogenesis, *Cancer Res.,* 43, 1079, 1983

75 **Miyamoto-Tiaven, M. J., Hillyard, L. A., and Abraham, S.,** Influence of dietary fat on the growth of mammary ducts in BALB/c mice, *J. Natl. Cancer Inst.,* 67, 179, 1981.

76. **Hillyard, L. A. and Abraham, S.,** Effect of dietary polyunsaturated fatty acids on growth of mammary adenocarcinomas in mice and rats, *Cancer Res.,* 39, 4430, 1979.

77 **Kaunitz, H. and Johnson, R. E.,** Influence of dietary fats on disease and longevity, *Proceedings of the 9th International Congress on Nutrition,* Vol 1, S. Karger, Basel, 1972, 362.

78. **Dao, T. L. and Chan, P.-C.,** Effect of duration of high fat intake on enhancement of mammary carcinogenesis in rats, *J. Natl. Cancer Inst.,* 71, 201, 1983.

79 **Lipsett, M. B.,** Carcinogenesis of endocrine-related human cancers interactions of diet and nutrition with the endocrine system, *Molecular Interrelations of Nutrition and Cancer,* Arnott, M S., van Eys, J , and Wang, Y -M., Eds , Raven Press, New York, 1982, 61

80. **Kidwell, W. R., Knazek, R. A., Vonderhaar, B. K., and Losonczy, I.,** Effects of unsaturated fatty acids on the development and proliferation of normal and neoplastic breast epithelium, in *Molecular Interrelations of Nutrition and Cancer,* Arnott, M S., van Eys, J , and Wang, Y.-M , Eds , Raven Press, New York, 1982, 219.

81. **Reddy, B. S., McCoy, G. D., and Wynder, E. L.,** Nutritional and environmental factors in carcinogenesis, in *The Clinical Biochemistry of Cancer,* Fleisher, M., Ed., American Association for Clinical Chemistry, Washington, D.C., 1979, 239.

82. **Sarkar, N. H., Fernandes, G., Telang, N. T., Kourides, I. A., and Good, R. A.,** Low-calorie diet prevents the development of mammary tumors in $C_3H$ mice and reduces circulating prolactin level, murine mammary tumor virus expression, and proliferation of mammary alveolar cells, *Proc. Natl. Acad. Sci. U.S.A.,* 79, 7758, 1982.

83. **Bakshi, K., Brusick, D., Bullock, L., and Bardin, C. W.,** Hormonal regulation of carcinogen metabolism in mouse kidney, in *Origins of Human Cancer,* Vol. 4, Hiatt, H. H., Watson, J. D., and Winsten, J. A., Eds., Cold Spring Harbor Laboratory, Cold Spring Harbor, N.Y., 1977, 683.

84. **Kelly, P. A., Labrie, F., and Asselin, J.,** The role of prolactin in tumor development, in *Influence of Hormones in Tumor Development,* Vol. 2, Kellen, J. A. and Hilf, R., Eds., CRC Press, Boca Raton, Fla., 1979, chap 7.

85. **Leung, F. C., Aylsworth, C. F., and Meites, J.,** Counteraction of underfeeding-induced inhibition of mammary tumor growth in rats by prolactin and estrogen administration, *Proc. Soc. Exp. Biol. Med.,* 173, 159, 1983.

86. **MacDonald, P. C.,** Age-related aberrations in proliferative homeostasis. The role of endocrines in age-related hyperplasias and neoplasias of the endometrium and prostate: a summary statement, in *Biological Mechanisms in Aging Conference Proceedings,* Schmike, R. T., Ed., U.S. Department of Health and Human Services, Bethesda, Md., 1981, 83.

87. **Hill, M. J., Crowther, J. S., Drasar, B. S., Hawksworth, G., Aries, V., and Williams, R. E. O.,** Bacteria and aetiology of cancer of large bowel, *Lancet,* 1, 95, 1971.

88. **Kritchevsky, D. and Klurfeld, D. M.,** Fat and cancer, in *Nutrition and Cancer,* Vol. 17, Newell, G. R. and Ellison, N. M., Eds., Raven Press, New York, 1981, 173.

89. **Graham, S. and Mettlin, C.,** Fiber and constituents of vegetables in cancer epidemiology, in *Nutrition and Cancer: Etiology and Treatment, Progress in Cancer Research and Therapy,* Vol. 17, Newell, G. R. and Ellison, N. M., Eds., Raven Press, New York, 1981, 189.

90. **Pomare, E. W. and Heaton, K. W.,** Alteration of bile salt metabolism by dietary fiber (Bran), *Br. Med. J.,* 4, 262, 1973.

91. **Jaffe, B. M. and Santoro, M. G.,** Prostaglandins and cancer, in *The Prostaglandins,* Vol. 3, Ramwell, P. W., Ed., Plenum Press, New York, 1977, 329.

92. **Honn, K. V., Bockman, R. S., and Marnett, L. J.,** Prostaglandins and cancer: a review of tumor initiation through tumor metastases, *Prostaglandins,* 21, 833, 1981.

93. **Levine, L.,** Arachidonic acid transformation and tumor production, *Adv. Cancer Res.,* 35, 49, 1981.

94. **Form, D. M. and Auerbach, R.,** $PGE_2$ and angiogenesis, *Proc. Soc. Exp. Biol. Med.,* 172, 214, 1983.

95. **Leaper, D. J., French, B. T., and Bennett, A.,** Breast cancer and prostaglandins: a new approach to treatment, *Br. J. Surg.,* 66, 683, 1979.

96. **Bennett, A., Berstock, D. A. and Carroll, M. A.,** Increased survival of cancer-bearing mice treated with inhibitors of prostaglandin synthesis alone or with chemotherapy, *Br. J. Cancer,* 45, 762, 1982.

97. **Carter, C. A., Milholland, R. J., Shea, W., and Ip, M. M.,** Effects of the prostaglandin synthetase inhibitor Indomethacin on 7,12-dimethylbenz(a)anthracene-induced mammary tumorigenesis in rats fed different levels of fat, *Cancer Res.,* 43, 3559, 1983.

98. **Lupulescu, A.,** Enhancement of carcinogenesis by prostaglandins in male albino Swiss mice, *J. Natl. Cancer Inst.,* 61, 97, 1978.

99. **Lupulescu, A.,** Effect of prostaglandins on tumor-transplantation, *Oncology,* 37, 418, 1980.

100. **Narisawa, T., Sato, M., Tani, M., Kudo, T., Takahashi, T., and Goto, A.,** Inhibition of development of methylnitrosurea-induced rat colon tumors by Indomethacin treatment, *Cancer Res.,* 41, 1954, 1981

101. **Hrachovec, J. P. and Rockstein, M.,** Age changes in lipid metabolism and their medical implications, *Gerontologica,* 3, 305, 1959.

102. **Crouse, J. R., Grundy, S. M., and Ahrens, E. H., Jr.,** Cholesterol distribution in the bulk tissues of man: variations with age, *J. Clin. Invest.,* 51, 1292, 1972.

103. **Kritchevsky, D.,** Diet, lipid metabolism and aging, *Fed. Proc. Fed. Am. Soc. Exp. Biol.,* 38, 2001, 1979.

104. **Exton-Smith, A. N.,** Vitamins, in *Metabolic and Nutritional Disorders in the Elderly,* Exton-Smith, A. N. and Caird, F. I., Eds., John Wright & Sons, Bristol, England, 1980, 26.

105. **Hsu, J. M.,** Current knowledge on zinc, copper and chromium in aging, in *World Review of Nutrition and Dietetics,* Vol. 33, Bourne, G H., Ed , S Karger, Basel, 1979, 42

106. **Lindeman, R. D., Clark, M. L., and Colmore, J. P.,** Influence of age and sex in plasma and red-cell zinc concentrations, *J. Gerontol.,* 26, 358, 1971.

107. **Dickson, R. C. and Tomlinson, R. H.,** Selenium in blood and human tissues, *Clin. Chim. Acta,* 16, 311, 1967.

108. **Thomson, C. D., Rea, H. H., and Robinson, M. F.,** Low blood selenium concentrations and glutathione peroxidase activities in elderly people, *Proc. Univ. Otago Med. School,* 55, 18, 1977.

109. **Miller, L., Mills, B. J., Blotcky, A. J., and Lindeman, R. D.,** Red blood cell and serum selenium concentrations as influenced by age and selected diseases, *J. Am. Coll. Nutr.*, 2(4), 331, 1983

110 **Kasperek, K., Schicha, H., Seller, V., and Feinendegen, L. E.,** Normalwerte von Spurenelementen in menschlichen Serum und Korrelation zum Lebensalter und zur serum-Eiweiss-Konzentration, *Strahlentherapie*, 143, 468, 1972.

111. **Yunice, A. A., Lindeman, R. D., Czerwinski, A. W., and Clark, M.,** Influence of age and sex on serum copper and ceruloplasmin levels, *J. Gerontol.*, 29, 277, 1974

112. **Massie, H. R., Colacicco, J. R., and Riello, V. R.,** Changes with age in copper and ceruloplasmin in serum from humans and C57BL/6J mice, *Age*, 2, 97, 1979.

113. **Wattenberg, L. W.,** Inhibition of neoplasia by minor dietary constituents, *Cancer Res. (Suppl.)*, 43, 2448S, 1983.

114 **Wattenberg, L. W.,** Inhibition of chemical carcinogens by minor dietary components, in *Molecular Interrelation of Nutrition and Cancer*, Arnott, M. S., van Eys, J., and Wang, Y.-M., Eds., Raven Press, New York, 1982, 43.

115 **Wang, Y.-M., Howell, S. K., Kimball, J. C., Tsai, C. C., Sato, J., and Gleiser, C. A.,** Alpha-tocopherol as a potential modifier of daunomycin carcinogenicity in Sprague-Dawley rats, in *Molecular Interrelations of Nutrition and Cancer*, Arnott, M. S., van Eys, J., and Wang, Y.-M., Eds., Raven Press, New York, 1982, 369

116 **Sun, A. Y. and Sun, G. Y.,** Dietary antioxidants and aging on membrane functions, in *Nutritional Approaches to Aging Research*, Moment, G. B., Ed., CRC Press, Boca Raton, Fla., 1982, 135

117. **Kurek, M. P. and Corwin, L. M.,** Vitamin E protection against tumor formation by transplanted murine sarcoma cells, *Nutr. Cancer*, 4, 128, 1982.

118. **Rama, B. N. and Prasad, K. N.,** Study on the specificity of $\alpha$-tocopherol (vitamin E) acid succinate effects on melanoma, glioma and neuroblastoma cells in culture, *Proc. Soc. Exp. Biol. Med.*, 174, 302, 1983.

119 **Green, J.,** Vitamin E and the biological antioxidant theory, *Ann. N.Y. Acad. Sci.*, 203, 29, 1972.

120 **Issenberg, P.,** Nitrite, nitrosamines, and cancer, *Fed. Proc. Fed. Am. Soc. Exp. Biol.*, 35, 1322, 1976.

121. **Magee, P. N., Ed.,** *Nitrosamines and Human Cancer*, Banbury Rep. 12, Cold Spring Harbor Laboratory, Cold Spring Harbor, N.Y., 1982

122 **Oshima, H. and Bartsch, H.,** Quantitative estimation of endogenous nitrosation in humans by monitoring *N*-nitrosoproline excreted in the urine, *Cancer Res.*, 41, 3658, 1981.

123. **Hartman, P. E.,** Nitrite load in the upper gastrointestinal tract — past, present, and future, in *Nitrosamines and Human Cancer*, Banbury Rep. 12, Cold Spring Harbor Laboratory, Cold Spring Harbor, N.Y., 1982, 415.

124. **Varga, J. M. and Airoldi, L.,** Inhibition of transplantable melanoma tumor development in mice by prophylactic administration of Ca-ascorbate, *Life Sci.*, 32, 1559, 1983.

125 **Bishun, N., Basu, T. K., Metcalfe, S., and Williams, D. C.,** The effect of ascorbic acid (vitamin C) on two tumor cell lines in culture, *Oncology*, 35, 160, 1978

126 **Bram, S., Froussard, P., Guichard, M., Jasmin, C., Augery, Y., Sinoussi-Barre, F., and Wray, W.,** Vitamin C preferential toxicity for malignant melanoma cells, *Nature (London)*, 284, 629, 1980

127. **Cameron, E., Pauling, L., and Leibovitz, B.,** Ascorbic acid and cancer: a review, *Cancer Res.*, 39, 663, 1979

128. **Siegel, B. V. and Leibovitz, B. E.,** Vitamin C in aging and cancer, *Int. J. Vitam. Nutr. Res.*, 49(Suppl. 19), 9, 1979.

129. **Burr, M. L., Elwood, P. C., Hole, D. J., Hurley, R. J., and Hughes, R. E.,** Plasma and leukocyte ascorbic acid levels in the elderly, *Am. J. Clin. Nutr.*, 27, 144, 1974.

130 **Sasaki, R., Kurokawa, T., and Tero-Kubota, S.,** Ascorbate radical and ascorbic acid level in human serum and age, *J. Gerontol.*, 38, 26, 1983

131 **Milne, J. S., Lonergan, M. E., Williamson, J., Moore, F. M. L., McMaster, R., and Percy, N.,** Leucocyte ascorbic acid levels and vitamin C intake in older people, *Br. Med. J.*, 4, 383, 1971.

132 **Schorah, C. J., Newell, A., Scott, D. L., and Morgan, D. B.,** Clinical effects of vitamin C in elderly inpatients with low blood-vitamin-C levels, *Lancet*, 1, 403, 1979

133. **Neal, R. A.,** Metabolism of toxic substances, in *Casarett and Doull's Toxicology*, 2nd ed., Doull, J., Klaasen, C. D., and Amdur, M. O., Eds., Macmillan, New York, 1980, 56.

134. **King, M. M. and McCay, P. B.,** Modulation of tumor incidence and possible mechanisms of inhibition of mammary carcinogenesis by dietary antioxidants, *Cancer Res.*, 43, 2485S, 1983.

135. **Cha, Y.-N., Heine, H., Ansher, S., and Bueding, E.,** Comparative effects of dietary anticarcinogens and other inducers on hepatic enzymes of mice, *Fed. Proc. Fed. Am. Soc. Exp. Biol.*, 40, 744, 1981.

136. **Benson, A. M., Cha, Y.-N., Bueding, E., Heine, H. S., and Talalay, P.,** Elevation of extrahepatic glutathione S-transferase and epoxide hydratase activities by 2(3)-*tert*-Butyl-4-hydroxyanisole, *Cancer Res.*, 39, 2971, 1979.

137. **Novi, A. M.,** Regression of aflatoxin $B_1$-induced hepatocellular carcinomas by reduced glutathione, *Science,* 212, 541, 1981.

138. **Novi, A. M., Flörke, R., and Stukenkemper, M.,** Glutathione and aflatoxin-$B_1$-induced liver tumors requirement for an intact glutathione molecule for regression of malignancy in neoplastic tissue, special edition: Cell Proliferation, Cancer, and Cancer Therapy, *Ann. N Y Acad Sci ,* 397, 62, 1982

139. **Tateishi, N. and Higashi, T.,** Turnover of glutathione in rat liver, in *Functions of Glutathione in Liver and Kidney,* Sies, H and Wendel, A , Eds., Springer-Verlag, New York, 1978, 3.

140. **Mainigi, K. D. and Campbell, T. C.,** Effects of low dietary protein and dietary aflatoxin on hepatic glutathione levels in F-344 rats, *Toxicol. Appl. Pharmacol.,* 59, 196, 1981.

141. **Finkelstein, J. D., Kyle, W. E., Harris, B. J., and Martin, J. J.,** Methionine metabolism in mammals: concentration of metabolites in rat tissues, *J. Nutr.,* 112, 1011, 1982.

142. **Fau, D., Chanez, M., Bois-Joyeux, B., Delhomme, B., and Peret, J.,** Effects of excess methionine ingestion on hepatic phosphate, adenine nucleotides and free amino acids in the rat, *J. Nutr ,* 112, 833, 1982.

143 **Glazenburg, E. J., Jekel-Halsema, I. M. C., Scholtens, E., Baars, A. J., and Mulder, G. J.,** Effects of variation in the dietary supply of cysteine and methionine on liver concentration of glutathione and "active sulfate" (PAPS) and serum levels of cystine, methionine and taurine: relation to the metabolism of acetaminophen, *J. Nutr.,* 113, 1363, 1983.

144 **Hsu, J. M.,** Zinc as related to cystine metabolism, in *Trace Elements in Human Health and Disease,* Prasad, A. S., Ed., Academic Press, New York, 1976, 295

145. **Mills, B. J., Lindeman, R. D., and Lang, C. A.,** Effect of zinc deficiency on blood glutathione levels, *J. Nutr ,* 111, 1098, 1981.

146 **Hsu, J. M., Rubenstein, B., and Paleker, A. G.,** Role of magnesium in glutathione metabolism of rat erythrocytes, *J Nutr ,* 112, 488, 1982.

147. **LeBoeuf, R. A. and Hoekstra, W. G.,** Adaptive changes in hepatic glutathione metabolism in response to excess selenium in rats, *J. Nutr.,* 113, 845, 1983.

148. **Abraham, E. C., Taylor, J. F., and Lang, C. A.,** Influence of mouse age and erythrocyte age on glutathione metabolism, *Biochem. J.,* 174, 819, 1978.

149. **Hazelton, G. A. and Lang, C. A.,** Glutathione contents of tissues in the aging mouse, *Biochem J.,* 188, 25, 1980.

150. **Naryshkin, S., Miller, L., Lindeman, R. D., and Lang, C. A.,** Blood glutathione: a biochemical index of human aging, *Fed Proc. Fed. Am. Soc. Exp. Biol ,* 40, 3179, 1981.

151. **Lang, C. A., Schneider, D., and Naryshkin, S.,** Further evidence on blood glutathione as an index of aging women, *Gerontologist,* 22, 54, 1982.

152. **Hazelton, G. A. and Lang, C. A.,** Glutathione levels during the mosquito life span with emphasis on senescence, *Proc. Soc. Exp. Biol. Med.,* 176, 249, 1984

153. **Richie, J. R., Jr., Mills, B. J., and Lang, C. A.,** Magnesium thiazolidine-4-carboxylic acid increases both glutathione levels and longevity, *Gerontologist,* 23, 24, 1983.

154 **Williamson, J. M. and Meister, A.,** Stimulation of hepatic glutathione formation by administration of L–2-oxothiazolidine-4-carboxylate, a 5-oxo-L-prolinase substrate, *Proc Natl Acad. Sci. U.S.A.,* 78, 936, 1981

155. **Weber, H. U., Fleming, J. F., and Miquel, J.,** Thiazolidine-4-carboxylic acid, a physiologic sulfhydryl antioxidant with potential value in geriatric medicine, *Arch. Gerontol. Geriatr ,* 1, 299, 1982.

156 **Armand, J. P., Neulat, J., Marty, C., and Delsol, G.,** Effect of thioproline on transplanted tumors in nude mice, *Biomed. Express (Paris),* 33, 171, 1980.

157. **Brugarolas, A. and Gosalvez, M.,** Treatment of cancer by an inducer of reverse transformation, *Lancet,* 2, 68, 1980

158 **Sappino, A. P. and Smith, I. E.,** Thioproline in squamous cell cancer, *Lancet,* 2, 417, 1980

159. **Miquel, J. and Economos, A. C.,** Favorable effects of the antioxidants sodium and magnesium thiazolidine carboxylate on the vitality and life span of *Drosophila* and mice, *Exp. Gerontol.,* 14, 279, 1979.

160. **Miquel, J., Fleming, J., and Economos, A. C.,** Antioxidants, metabolic rate and aging in *Drosophila, Arch. Gerontol. Geriatr ,* 1, 159, 1982.

161. **Richie, J. P., Jr., Mills, B. J., and Lang, C. A.,** On the mechanism of nordihydroguairetic acid-enhanced longevity, *Gerontologist,* 21, 301, 1981.

162. **Cutler, R. G.,** Life-span extension, in *Aging· Biology and Behavior,* McGaugh, J. L., Kiesler, S. B., and March, J. G., Eds., Academic Press, New York, 1981, 31.

163. **Newberne, P. M. and Rogers, A. E.,** Vitamin A, retinoids and cancer, in *Nutrition and Cancer. Etiology and Treatment,* Vol. 17, Newell, G. R. and Ellison, N. M., Eds., Raven Press, New York, 1981, 217.

164. **Ibrahim, K., Jafarcy, N., and Zuberi, S.,** Plasma vitamin A and carotene levels in squamous cell carcinoma of the oral cavity and oropharynx, *Clin. Oncol ,* 3, 203, 1977.

165. **Atukorala, I., Basu, T. K., Dickerson, J. W. T., Donaldson, D., and Sakula, A.,** Vitamin A, zinc and lung cancer, *Br. J. Cancer,* 40, 927, 1979

166. **Mellow. M. H., Layne, E. A., Lipman, T. O., Kaushik, M., Hostetler, C., and Smith, J. C., Jr.,** Plasma zinc and vitamin A in human squamous carcinoma of the esophagus, *Cancer*, 51, 1615, 1983.

167 **Sporn. M. B., Dunlop, N. M., Newton, D. L., and Smith, J. M.,** Prevention of chemical carcinogenesis by vitamin A and its synthetic analogs (retinoids), *Fed. Proc. Fed Am. Soc Exp Biol*, 35, 1332, 1976

168. **Lotan, R.,** Effects of vitamin A and its analogs (retinoids) in normal and neoplastic cells, *Biochim. Biophys Acta*, 605, 33, 1980.

169. **Sporn, M. B.,** Prevention of epithelial cancer by vitamin A and its synthetic analogs (retinoids), in *Origins of Human Cancer*, Hiatt, H. H , Watson, J. D., and Winsten, J A , Eds , Cold Spring Harbor Laboratory, Cold Spring Harbor, N.Y , 1977, 801.

170. **Seifter, E., Rettura, G., Stratford, F., and Levenson, S. M.,** C₃HBA tumor prevention and treatment with beta carotene, *Fed. Proc. Fed. Am. Soc. Exp. Biol.*, 40, 2413, 1981.

171. **Troll, W., Rossman, T., Katz, J., Levitz, M., and Sugimura, T.,** Proteinases in tumor promotion and hormone action, in *Proteases and Biological Control*, Vol. 2, Reich, E., Rifkin, D. B , and Shaw, E., Eds., Cold Spring Harbor Laboratory, Cold Spring Harbor, N.Y., 1975, 977

172. **Troll, W., Klassen, A., and Janoff, A.,** Tumorigenesis in mouse skin: inhibition by synthetic inhibitors of proteases, *Science*, 169, 1211, 1970.

173. **Troll, W., Wiesner, R., Shellabarger, C. J., Holtzman, S., and Stone, J. P.,** Soybean diet lowers breast tumor incidence in irradiated rats, *Carcinogenesis*, 1, 469, 1980.

174. **Yavelow, J., Finlay, T. H., Kennedy, A. R., and Troll, W.,** Bowman-Birk soybean protease inhibitor as an anticarcinogen, *Cancer Res.*, 43, 2454S, 1983.

175. **Stout, M. G. and Rawson, R. W.,** Minerals, trace elements, and cancer, in *Nutrition and Cancer: Etiology and Treatment*, Vol. 17, Newell, G. R. and Ellison, N. M., Eds., Raven Press, New York, 1981, 243.

176. **Schrauzer, G. N.,** Trace elements, nutrition and cancer: perspectives of prevention, in *Advances in Experimental Medicine and Biology: Inorganic and Nutritional Aspects of Cancer*, Vol. 91, Schrauzer, G. N., Ed., Plenum Press, New York, 1978, 323.

177 **Schrauzer, G. N.,** Trace elements in carcinogenesis, in *Advances in Nutritional Research*, Vol. 1, Draper, H. H., Ed., Plenum Press, New York, 1979, 219.

178. **Griffin, A. C.,** Role of selenium in the chemoprevention of cancer, *Adv. Cancer Res.*, 29, 419, 1979.

179. **Poswillo, D. E. and Cohen, B.,** Inhibition of carcinogenesis by dietary zinc, *Nature (London)*, 231, 447, 1971.

180. **Woster, A. D., Failla, M. L., Taylor, M. W., and Weinberg, E. D.,** Zinc suppression of initiation of sarcoma 180 growth, *J. Natl. Cancer Inst.*, 54, 1001, 1975.

181. **Mathur, A., Wallenius, K., and Abdulla, M.,** Influence of zinc on onset and progression of oral carcinogenesis in rats, *Acta Odontol. Scand.*, 37, 277, 1979.

182. **Shamberger, R. J., Rukovena, E., Longfield, A. K., Tytko, S. A., Deodhar, S., and Willis, C. E.,** Antioxidants and cancer. I. Selenium in the blood of normal and cancer patients, *J. Natl. Cancer Inst ,* 50, 863, 1973.

183. **McConnell, K. P., Broghamer, W. L., Jr., Blotcky, A. J., and Hurt, O. J.,** Selenium levels in human blood and tissues in health and disease, *J. Nutr.*, 105, 1026, 1975.

184. **Broghamer, W. L., Jr., McConnell, K. P., and Blotcky, A. J.,** Relationship between serum selenium levels and patients with carcinoma, *Cancer*, 37, 1384, 1976

185. **Delves, H. T., Alexander, F. W., and Lay, H.,** Copper and zinc concentrations in the plasma of leukaemic children, *Br. J. Haematol.*, 24, 525, 1973.

186. **Inutsuka, S. and Araki, S.,** Plasma copper and zinc levels in patients with malignant tumors of digestive organs, *Cancer*, 42, 626, 1978.

187 **Andrews, G. S.,** Studies of plasma zinc, copper, caeruloplasmin, and growth hormone, *J. Clin Pathol ,* 32, 325, 1979.

188. **Garofalo, J. A., Erlandson, E., Strong, E. W., Lesser, M., Gerold, F., Spiro, R., Schwartz, M., and Good, R. A.,** Serum zinc, serum copper and the Cu/Zn ratio in patients with epidermoid cancers of the head and neck, *J. Surg. Oncol.*, 15, 381, 1980.

189. **Mills, B. J. and Lindeman, R. D.,** Unpublished data, 1981.

190. **Hrgovcic, M., Tessmer, C. F., Thomas, F. B., Ong, P. S., Gamble, J. F., and Shullenberger, C. C.,** Serum copper observations in patients with malignant lymphoma, *Cancer*, 32, 1512, 1973

191. **Flynn, A.,** Copper metabolism in an animal cancer model, in *Trace Substances in Environmental Health*, Hemphill, D. D., Ed., University of Missouri, Columbia, 1977, 179.

192. **Saito, T., Wakui, A., Himori, T., Uiue, S., Sugawara, N., and Sugiyama, Z.,** Serum Zn content in tumor-bearing rats treated with anticancer drugs, *Tohoku J. Exp. Med.*, 129, 111, 1979.

193. **Çavdar, A. O., Babacan, E., Arcasoy, A., Erten, J., and Ertem, U.,** Zinc deficiency in Hodgkin's disease, *Eur. J. Cancer,* 16, 317, 1980.

194. **Fong, L. Y. Y., Swak, A., and Newberne, P. M.,** Zinc deficiency and methylbenzyl-nitrosamine-induced esophageal cancer in rats, *J. Natl Cancer Inst.*, 61, 145, 1978.

195. **McCreary, P., Laing, G., and Hass, G.,** Relation of high incidence of spontaneous malignant lymphoma in magnesium-deficient rats to susceptibility to induction of lymphoma with live cells, *Am. J. Pathol.*, 62, 123A, 1971.

196. **Hass, G. M., Laing, G. H., McCreary, P. A., and Galt, R. M.,** Rat lymphomas due to magnesium deprivation are prevented by dietary supplement of liver powder, *Clin. Res.*, 27, 688A, 1979.

197. **Mueller, C. B., Ames, F., and Anderson, G. D.,** Breast cancer in 3,558 women: age as a significant determinant in the rate of dying and causes of death, *Surgery*, 83, 123, 1978.

198. **Howell, A.,** Induction of differentiation and regression in mammary and other tumors, in *Prolonged Arrest of Cancer*, Stoll, B. A., Ed., John Wiley & Sons, New York, 1982, 349.

199. **Tisdale, M. J.,** Prospects for cancer arrest by growth modulators, in *Prolonged Arrest of Cancer*, Stoll, B A., Ed., John Wiley & Sons, New York, 1982, 407.

200. **Bellamy, D. and Hinsull, S. M.,** Effects of host age on the growth and mitotic index of experimental tumors in laboratory rats, *J Clin. Exp. Gerontol.*, 1, 57, 1979.

201 **LePage, G. A., Potter, V. R., Busch, H., Heidelberger, C., and Hurlbert, R. B.,** Growth of carcinoma implants in fed and fasting rats, *Cancer Res.*, 12, 153, 1951.

202. **Reilly, J. J., Goodgame, J. T., Jones, D. C., and Brennan, M. F.,** DNA synthesis in rat sarcoma and liver, *J. Surg. Res.*, 22, 281, 1977.

203. **Goodgame, J. T., Jr., Lowry, S. F., Reilly, J. J., Jones, D. C., and Brennan, M. F.,** Nutritional manipulations and tumor growth. I. The effects of starvation, *Am. J. Clin Nutr.*, 32, 2277, 1979.

204. **Cameron, I. L., Ackley, W. J., and Rogers, W.,** Responses of hepatoma-bearing rats to total parenteral hyperalimentation and to *ad libitum* feeding, *J. Surg. Res.*, 23, 189, 1977.

205 **Grubbs, B., Rogers, W., and Cameron, I.,** Total parenteral nutrition and inhibition of gluconeogenesis on tumor-host responses, *Oncology*, 36, 216, 1979.

206. **Cameron, I. L., Pavlat, W. A., Stevens, M. D., and Rogers, W.,** Tumor-host responses to various nutritional feeding procedures in rats, *J. Nutr.*, 109, 671, 1979

207. **Shike, M.,** Energy expenditure and body composition in cancer patients; the effect of parenteral nutrition and chemotherapy, *Diet, Nutrition and Cancer, Etiologic and Treatment Issues*, 12th Annual Cancer Course, Harvard Medical School, Department of Continuing Education, Boston, 1983

208. **Daly, J. M., Copeland, E. M., III, Guinn, E., and Dudrick, S. J.,** Relationship of protein nutrition to tumor growth and host immunocompetence, *Surg Forum*, 27, 113, 1977.

209. **Lowry, S. F., Goodgame, J. T., Norton, J. A., Jones, D. C., and Brennan, M. F.,** The effects of protein malnutrition on host-tumor composition and growth, *Surg. Forum*, 28, 143, 1977.

210. **Erickson, K. L., Gershwin, M. E., Canolty, N. L., and Eckels, D. D.,** The influence of dietary protein concentration and energy intake on mitogen response and tumor growth in melanoma-bearing mice, *J. Nutr.*, 109, 353, 1979.

211. **Ota, D. M., Copeland, E. M. III, Strobel, H,. W., Jr., Daly, G., Gum, E. T., Guinn, E., and Dudrick, S. J.,** The effect of protein nutrition on host and tumor metabolism, *J Surg. Res.*, 22, 181, 1977.

212. **Daly, J. M., Reynolds, H. M., Rowlands, B. J., Baquero, G. E., Dudrick S. J., and Copeland, E. M.,** Nutritional manipulation of tumor-bearing animals effects on body weight, serum protein levels and tumor growth, *Surg. Forum*. 29, 143, 1978.

213. **Sauer, L. A., Stayman, J. W., III, and Dauchy, R. T.,** Amino acid, glucose, and lactic acid utilization *in vivo* by rat tumors, *Cancer Res.*, 42, 4090, 1982.

214 **Steiger, E., Oram-Smith, J., Miller, E., Kuo, L., and Vars, H. M.,** Effects of nutrition on tumor growth and tolerance to chemotherapy, *J. Surg. Res.*, 18, 455, 1975.

215. **Skipper, H. E. and Thomson, J. R.,** A preliminary study of the influence of amino acid deficiencies on experimental cancer chemotherapy, in *Ciba Foundation Symposium, on Amino Acids and Peptides with Antimetabolic Activity*, Wolstonholme, G. E. W., and O'Connor, O. M., Eds., Little, Brown, Boston, 1958, 38.

216. **Theuer, R. C.,** Effect of essential amino acid restriction on the growth of female C57B1 mice and their implanted BW 10232 adenocarcinomas, *J. Nutr.*, 101, 223, 1971.

217. **Morris, H. P. and Voegtlin, C.,** The effect of methionine on normal and tumor growth, *J. Biol. Bhem.*, 133, lxix, 1940.

218. **Kreis, W. and Hession, C.,** Biological effects of enzymatic deprivation of L-methionine in cell culture and an experimental tumor, *Cancer Res* , 33, 1866, 1973.

219. **Demopoulos, H. B.,** Effects of low phenylalanine-tyrosine diets on S91 mouse melanomas, *J. Natl. Cancer Inst.*, 37, 185, 1966.

220. **Pine, M. J.,** Effect of low phenylalanine diet on murine leukemia L1210, *J. Natl. Cancer Inst.*, 60, 633, 1978.

221. **Lorincz, A. B., Kuttner, R. E., and Brandt, M. B.,** Tumor response to phenylalanine-tyrosine-limited diets, *J. Am. Dietet. Assoc.*, 54, 198, 1969.

222. **Demopoulos, H. B.,** Effects of reducing the phenylalanine-tyrosine intake of patients with advanced melanoma, *Cancer,* 19, 657, 1966.

223 **Gold, J.,** Cancer cachexia and gluconeogenesis, *Ann. N.Y. Acad. Sci.,* 230, 103, 1974.

224. **Shapot, V. S.,** On the multiform relationships between the tumor and the host, *Adv Cancer Res.,* 30, 89, 1979.

225. **Goranson, E. S. and Tilser, G. J.,** Studies on the relationship of alloxan-diabetes and tumor growth, *Cancer Res.,* 15, 626, 1955

226. **Morris, H. P.,** Effects on the genesis and growth of tumors associated with vitamin intake, *Ann. N.Y. Acad Sci.,* 49, 119, 1947.

227. **Robertson, W. V. B., Dalton, A. J., and Heston, W. E.,** Changes in a transplanted fibrosarcoma associated with ascorbic acid deficiency, *J Natl. Cancer Inst.,* 10, 53, 1949.

228. **Stoerk, H. C. and Emerson, G. A.,** Complete regression of lymphosarcoma implants following temporary induction of riboflavin deficiency in mice, *Proc. Soc. Exp. Biol. Med.,* 70, 703, 1949.

229. **Rivlin, R. S.,** Riboflavin and cancer: a review, *Cancer Res.,* 33, 1977, 1973.

230. **DeWys, W. and Pories, W.,** Inhibition of a spectrum of animal tumors by dietary zinc deficiency, *J. Natl. Cancer Inst.,* 48, 375, 1972

231. **Minkel, D. T., Dolhun, P. J., Calhoun, B. L., Saryan, L. A., and Petering, D. H.,** Zinc deficiency and growth of Ehrlich ascites tumor, *Cancer Res.,* 39, 2451, 1979.

232. **Mills, B. J., Broghamer, W. L., Higgins, P. J., and Lindeman, R. D.,** A specific dietary zinc requirement for the growth of Walker 256/M1 tumor in the rat, *Am. J. Clin Nutr.,* 34, 1661, 1981.

233. **Mills, F. H.,** Behaviour of tumours in conditions of experimental magnesium deficiency, *Lancet,* 2, 781, 1974.

234. **Young, G. A. and Parsons, M. F.,** The effects of dietary deficiencies of magnesium and potassium on the growth and chemistry of transplanted tumours and host tissues in the rat, *Eur. J. Cancer,* 13, 103, 1977.

235 **Mills, B. J., Broghamer, W. L., Higgins, P. J., and Lindeman, R. D.,** Inhibition of tumor growth by zinc depletion of rats, *J. Nutr.,* 114, 746, 1984.

236. **Mills, B. J., Broghamer, W. L., Higgins, P. J., and Lindeman, R. D.,** Inhibition of tumor growth by magnesium depletion of rats, *J. Nutr.,* 114, 739, 1984.

237. **Parsons, F. M., Anderson, C. K., Clark, P. B., Edwards, G. F., Ahmad, S., Hetherington, C., and Young, G. A.,** Regression of malignant tumours in magnesium and potassium depletion induced by diet and haemodialysis, *Lancet,* 1, 243, 1974.

238. **Burnam, N. D. and Parsons, F. M.,** Hyperalimentation in the treatment of advanced carcinoma with induced magnesium and potassium depletion, *S. Afr. Med. J.,* 50, 1695, 1976.

239. **Apple, M. A.,** New anticancer drug design past and future strategies, in *Cancer 5: Chemotherapy,* Becker, F. F., Ed., Plenum Press, New York, 1977, 599.

240. **Trosko, J. E. and Chang, C.,** The role of DNA repair capacity and somatic mutations in carcinogenesis and aging, in *Handbook of Diseases of Aging,* Blumenthal, H T., Ed., Van Nostrand Reinhold, New York, 1983, 252.

241. **Mendeloff, A. L.,** Appraisal of "Diet, Nutrition, and Cancer", *Am. J. Clin. Nutr.,* 37, 495, 1983.

242. **Sugimura, T.,** Tumor initiators and promoters associated with ordinary foods, in *Molecular Interrelations of Nutrition and Cancer,* Arnott, M. S., van Eys, J, and Wang, Y -M, Eds, Raven Press, New York, 1982, 3

243. **Ames, B. N.,** Dietary carcinogens and anticarcinogens, *Science,* 221, 1256, 1983.

244 **Tannenbaum, S. and Silverstone, H.,** Nutrition in relation to cancer, *Adv. Cancer Res.,* 1, 451, 1953.

245. **Meyer, J. A.,** Potentiation of solid-tumor chemotherapy by metabolic alteration, *Ann Surg.,* 179, 88, 1974.

246. **Reynolds, H. M., Daly, J. M., Rowlands, B. J., Dudrick S. J., and Copeland, E. M.,** Effects of nutritional repletion on host and tumor response to chemotherapy, *Cancer,* 45, 3069, 1980.

247. **Buzby, G. P., Mullen, J. L., Stein, T. P., Miller, E. E., Hobbs, C. L., and Rosato, E. F.,** Host-tumor interaction and nutrient supply, *Cancer,* 45, 2940, 1980.

248. **Vitale, J. J., Broitman, S. A., and Gottlieb, L. S.,** Alcohol and carcinogenesis, in *Nutrition and Cancer: Etiology and Treatment,* Newell, G. R. and Ellison, N. M., Eds., Raven Press, New York, 1981, 291.

249. Some non-nutritive sweetening agents, *IARC Monographs on the Evaluation of the Carcinogenic Risk of Chemicals to Humans,* Vol 22, World Health Organization, Geneva, 1980

250. **Newell, G. R.,** Artificial sweeteners and cancer, in *Nutrition and Cancer: Etiology and Treatment,* Newell, G. R. and Ellison, N. M., Eds., Raven Press, New York, 1981, 273.

251 **Minton, J. P., Abou-Issa, H., Foecking, M. K., and Sriram, M. G.,** Caffeine and unsaturated fat diet significantly promotes DMBA-induced breast cancer in rats, *Cancer,* 51, 1249, 1983

252. **Wynder, E. L., Hall, N. E. L., and Polansky, M.,** Epidemiology of coffee and pancreatic cancer, *Cancer Res.,* 43, 3900, 1983.

253. **Matsushima, T.,** Mechanisms of conversions of food components to mutagens and carcinogens, in *Molecular Interrelations of Nutrition and Cancer,* Arnott, M  S., van Eys, J , and Wang, Y.-M., Eds., Raven Press, New York, 1982, 35

254. **Sugimura, T. and Sato, S.,** Mutagens-carcinogens in food, *Cancer Res.,* 43, 2415S, 1983.

Chapter 4

CANCER

Part II

CLINICAL ASPECTS OF NUTRITION AND CANCER

**Nancy L. Keim and Earl S. Shrago**

TABLE OF CONTENTS

# I. INTRODUCTION

Cancer patients have numerous metabolic and nutritional abnormalities. The malignant process itself alters host metabolism adversely, and this problem is compounded by the nutritionally debilitating effects of antitumor treatments. As a result, weight loss, muscle wasting, anorexia, and loss of immunocompetence are commonplace. This deterioration of nutritional status may limit the patient's tolerance to therapy, and subsequently, tumor growth continues. Under these circumstances aggressive nutritional intervention appears warranted to offset this vicious cycle of progression of malnutrition and disease. Nutritional support, both enteral and parenteral, is now a well-established therapeutic modality, particularly in the critically ill patient. Whereas the rationale for total parenteral nutrition in major injury or sepsis is clearly defined, its use and ultimate value in the undernourished cancer patient is less well defined. The crucial question is the therapeutic benefit of nutritional support in terms of the overall prognosis of the disease and the positive response to chemotherapy, radiation treatment, or surgical procedure.

# II. THE EFFECT OF CANCER AND ITS TREATMENT ON NUTRITIONAL STATUS

## A. Systemic Effects of the Tumor

A critical problem of the cancer patient is cachexia, which can be due to anorexia, increased energy expenditure, or a combination of both processes.[1] The underlying cause of cancer cachexia is unknown. Besides obvious mechanical problems due to tumor size, it is apparent that in many cases there is no evident cause for the loss of appetite other than the presence of the tumor. An intriguing hypothesis yet unproved is that there are unknown peptides or other small molecules produced by the tumor which may modify activity of the host tissue to produce anorexia.[2] Another suggestion, again not completely substantiated, is that an increase in certain plasma amino acids, particularly tryptophan, may in turn increase the serotonin content of the brain, producing a satiety or anorectic effect.[3]

Cancer cachexia does not resemble uncomplicated starvation. It has a closer similarity to the hypercatabolic state seen in major injury or sepsis.[4] However, it remains unclear if the mechanism involved in production, trapping, and utilization of energy is altered in the tumor-bearing host. Disturbed glucose metabolism and insulin resistance, reduced peripheral muscle protein synthesis, and increased protein catabolic activity are characteristic features of the altered metabolism.[5,6] The resistance to insulin is associated with the normal number of receptors rather than a decreased number, which typifies the insulin resistance of obesity.[7] An increase in glucose recycling or Cori cycle activity commonly occurs in the cancer patient and may be due to excess lactate produced by the tumor which is subsequently converted to glucose in the liver.[8] Theoretically, this could lead to an energy drain since less lactate would be oxidized through the Krebs cycle for ATP production, and considerable ATP would be utilized by the liver for glucose production. This question of whether the tumor alters host metabolism in ways that preclude efficient utilization of an exogenous nutrient, i.e., obligate gluconeogenesis, has not been answered satisfactorily. Experimentally, an extremely high rate of glucose uptake by cancer cells as compared to the host cells has been observed.[9] The $K_m$ for tumor hexokinase has been found to be ten times lower than that of homologous tissue.[10] Thus, there is an enormous concentration gradient between arterial blood and tissue glucose, which may in turn, lead to excess lactate production and persistent gluconeogenesis.

Protein metabolism of the host is shifted toward catabolism, although the relative importance of depressed protein synthesis vs. increased protein degradation is unclear.[6] Skeletal muscle tissue obtained by biopsy from cancer patients has decreased protein synthesis capacity, increased fractional protein degradation rate, as well as increased lysosomal cathepsin

D activity compared to muscle tissue from noncancerous control patients.[11,12] In addition, it appears that tumors may "trap" nitrogen. Isotope enrichment of free amino acid pools is greater in malignant intestinal tissue than in nonmalignant tissue after a constant infusion of [15]N-glycine is given to cancer patients.[13] In this experiment, the tumor exhibited a higher fractional synthetic rate than did the normal gut tissue, despite the fact that the ability of the tumor to reutilize amino acids was less efficient. Body composition studies also reflect progressive catabolism of skeletal muscle in the host. Arm muscle circumference measurements suggest moderate depletion of muscle mass.[14] Both body cell mass, measured by isotope dilution,[15] and lean body mass, measured by whole body [40]K counting,[16] are reduced in cancer patients as the disease progresses. Moreover, an abnormal metabolic state is apparent during attempts to rehabilitate the cachectic cancer patient by nutritional means. Repletion either by enteral or parenteral nutrition leads to body weight gain but not necessarily gain of lean body mass.[17]

Other issues aside, however, malnutrition is a poor prognostic sign in adults and children with cancer, which is the overriding reason why so much emphasis has been placed on nutritional support.[18] However, results to date seem to indicate that dietary manipulation is unlikely to be successful as primary therapy for most patients with the cancer cachexia syndrome. Certainly, however, it is rational to consider that since so little is known about the nutritional requirements in parenteral alimentation, an appropriate combination of nutrients might somehow be assembled to counteract the cancer cachexia syndrome.

### B. Localized Tumor Effects

In addition to systemic effects, various neoplasms may have a number of more localized effects leading to nutritional problems.[19] By far the most common cause of malnutrition in this category is interference with food intake as a result of partial or complete obstruction of some portion of the GI tract. Involvement of the pancreas or common bile duct may lead to impaired digestion and absorption of fat and fat-soluble vitamins. The decreased fat absorption may also lead to increased fecal losses of Ca and Mg. Lymphoma involving small bowel mesentery lymph nodes may present as malabsorption. Protein-losing enteropathy has also been described with cases of gastric carcinoma. The blind loop syndrome in the upper small bowel secondary to partial obstruction with bacterial overgrowth may also result in steatorrhea and vitamin deficiencies.

The major causes of fluid and electrolyte disturbances in cancer patients are vomiting and diarrhea secondary to partial and complete obstruction and gastric and small bowel losses through a fistula. Severe diarrhea may also occur with the gastrin-secreting islet cell adenoma (Zollinger-Ellison Syndrome).

### C. Nutritional Problems Arising from the Treatment of Cancer

Significant nutritional problems arise as the result of specific treatments given to control the neoplastic disease.

#### 1. Chemotherapy

Chemotherapy has come to play a very important role in the outcome of many tumors. However, many of the newer drugs are quite toxic, and higher concentrations of the agents either alone or in combination to obtain a more effective response increase the toxicity, which results in progressive weight loss and malnutrition.[19,20] Loss of as much as 15 to 20% of body weight might be expected among patients undergoing intensive chemotherapeutic regimes. Many of the nutritional consequences of chemotherapy can be explained as the result of interference in necessary metabolic reactions and pharmacological effects on the target tissues. Moderate to severe anemias, particularly megaloblastic types, are nutritionally important consequences of chemotherapy. The megaloblastic anemia produced by cytosine arabinoside can be readily explained by its biochemical effects on DNA synthesis of erythrocyte precursors. The profound host effects of methotrexate may be related to impaired folate metabolism.

Nausea and vomiting occur with almost every major class of chemotherapeutic compound, including the alkylating agents, nitrosoureas, folate analogues, purine analogues, pyrimidine analogues, and anthracycline antibiotics. In general, the effects of the chemotherapeutic agents on the GI tract include stomatitis, constipation, malabsorption, and diarrhea. Severe hepatocellular damage after some agents, such as asparaginase, leads to hypoalbuminemia along with the anorexia. Only a few chemotherapeutic agents are not associated with nausea and vomiting. These are certain alkylating agents such as L-phenylalanine mustard, chlorambucil, vincristine, and steroids.

Examples of nutritional manipulation used in the treatment of human cancer include intervention with agents affecting amino acid availability, vitamin availability, glucose utilization, nucleotrophic nutrient availability such as purines, pyrimidines, and their nucleosides, and trace element availability.[21] Not surprisingly, specific nutritional approaches to chemotherapy lead to severe nutritional consequences. Amino acid antagonists such as cycloleucine (which antagonizes the incorporation of valine into protein) produce severe nausea, vomiting, anorexia, diarrhea, and weight loss associated with a fall in serum protein levels. Selenocystine, thought to act by competing with L-cystine for incorporation into protein, also produces severe nausea and vomiting, thereby preventing the continued administration of this compound. Ethionine, a structural analogue of methionine, produces diarrhea, liver damage, as well as other toxic manifestations. In general, the deprivation of amino acids for protein synthesis has not proven successful as therapy for retarding tumor growth rate.

Galactoflavin, a potent riboflavin antagonist, produces severe riboflavin deficiency manifested by weight loss, cheilosis, glossitis, seborrheic dermatitis, and normochromic, normocytic anemia. A vitamin $B_6$ antagonist, 4-deoxypyridoxine, produces evidence of vitamin $B_6$ deficiency when used as therapy. Similarly, 6-aminonicotinamide, an antagonist of niacin, produces stomatitis, ulceration, cheilosis, and glossitis, which are typical of this vitamin deficiency. Methotrexate, a classic folate antagonist, can produce severe bone marrow depression and other manifestations of folic acid deficiency. With refinements in chemotherapeutic management most drastic reactions described rarely occur in day-to-day management. Many nutritionally important untoward effects can be prevented or treated nonspecifically with antiemetics, analgesics, antidiarrheal agents, antibiotics and transfusion during the initial critical phase of therapy to maintain appropriate nutritional status.[19,20]

### 2. Radiation Therapy

Approximately two thirds of all cancer patients undergo radiation therapy at some point during their illness.[22] Nutritional debilitation from radiotherapy is related to multiple factors including type of radiation therapy used, area being irradiated, size of field, dose delivered, and status of the patient. As would be expected, the patient's chance of experiencing toxicity is increased, the larger the dose and extended field being irradiated. Next to the bone marrow, the cells that line the alimentary tract have one of the highest turnover rates in the body; therefore, toxic effects resulting from radiation to the abdomen and pelvis area are most intense, and small bowel dysfunction may result. Under such circumstances, radiation therapy has a much more severe effect in the patient with GI cancer, causing a drop in body weight, arm muscle circumference, and peripheral lymphocytes.[23] Of more severe nutritional consequence is the development of a fistula as an aftermath to bowel irradiation. The current recommendation for malnourished patients with radiation-related fistulas of the GI tract is to prepare them for surgical procedure by utilizing intravenous hyperalimentation preoperatively for 2 to 3 weeks to stimulate weight gain and induce anabolism.[24]

Radiation of the head and neck can result in loss of taste sensation and dryness of the mouth, adding to the already serious situation of anorexia. Radiation to the lower neck and mediastinum may induce esophagitis, also adversely affecting the patient's desire for and ability to attain oral nutrition.

### 3. Surgery

In addition to the hypercatabolic state imposed by the stress of surgery per se, the malnutrition attending surgery in cancer patients is often due to extensive resection in bowel cancers. Each area of the GI tract presents its own specific problem when removed, and malnutrition can result from numerous factors varying from mechanical problems to malabsorption secondary to inadequate gastric, biliary, or pancreatic secretions, as well as loss of the intestine itself.

## III. NUTRITIONAL SUPPORT OF THE CANCER PATIENT

The primary objective of nutritional support of the cancer patient is to rehabilitate or maintain nutritional status so that appropriate antineoplastic therapy can be attempted or completed with maximal safety. Additionally, as antineoplastic treatments become more effective, cancer becomes a more chronic, protracted illness, and provision of adequate nutritional support may assist in improving the patient's quality of life both physiologically and psychologically. As summarized previously, both the tumor itself and the treatment modalities impart specific untoward effects on metabolism and nutritional needs. Therefore, a rational approach to devising a nutritional support program must begin with a systematic analysis of factors that contribute to nutritional depletion. A list of factors that should be evaluated in the initial patient work-up includes (1) nutritional status, (2) adequacy of GI function, (3) implications of staging of extent of disease, (4) magnitude and duration of treatment modality planned, (5) common nutritional side effects of proposed treatment, and (6) goal of planned anticancer therapy to determine the goal of nutritional support.[25]

### A. Nutritional Assessment

Objective criteria are needed to recognize malnutrition, particularly in the early stages, and then to evaluate the effectiveness of nutritional support. A combination of anthropometric, biochemical, and immunocompetence tests can provide an estimate of nutritional status in these patients.

### 1. Anthropometric Measurements

Weight for height, recent weight loss, triceps skinfold thickness, and arm muscle circumference are readily obtainable measurements that have prognostic value for the cancer patient. Reference standards for these types of measurements in the elderly are inadequate at present, but some values are available for the healthy aged 65 and older.[26,27] Weight for height and arm muscle circumference, estimators of skeletal muscle mass, have been significantly lower at the time of initiation of nutritional support in those cancer patients who subsequently died than in those who survived.[28] The degree of recent weight loss has correlated inversely with both tumor response therapy[29,30] and survival.[28,29] Also, the triceps skinfold thickness, an indicator of body fat stores, has been reduced in nonsurviving cancer patients compared to surviving cancer patients.[31]

### 2. Biochemical Tests

Protein-calorie malnutrition can be characterized further by determination of the creatinine-height index[32] upon collection of a 24-hr urine specimen. The creatinine-height index is a sensitive indicator of protein-calorie malnutrition, as it is reflective of skeletal protein mass, and it is also a predictor of survival in cancer patients.[28,31] Visceral protein status can be estimated by the serum levels of two transport proteins, albumin and transferrin. Again, levels of these proteins are predictors of survival.[28,31,33] Hypoalbuminemia occurs frequently in elderly cancer patients.[33] Although protein-calorie malnutrition may contribute substantially to this reduction, tumor effects, protein-losing enteropathy, or a dilutional effect of an expanded extracellular compartment may also contribute.[19] The fact that the albumin level does not always respond to aggressive nutritional support[28,33] further substantiates that other factors in the disease process or mode of tumor therapy may be affecting the metabolism of this protein.

### 3. Tests of Immunocompetence

Total lymphocyte counts have been used to assess immunocompetence, but the delayed hypersensitivity reaction to a battery of common recall antigens is preferable since positive reactions correlate with survival in cancer patients.[28] Anergy is seen more frequently in nonsurvivors, and survival rate improves when patients are immunocompetent or regain immunocompetence through hospitalization.[28,34,35] Anergic cancer patients are not always able to regain immunocompetence with seemingly adequate nutritional support.[14] However, in addition to the malnutrition component of depressed cell-mediated immunity, other factors may act to depress the delayed hypersensitivity reaction in cancer patients. These include advancing age, presence of tumor, previous or current immunosuppressive drug therapy, or underlying primary or acquired immunodeficiency.[36] All of these factors can complicate the interpretation of this test in elderly cancer patients.

## B. Planning Nutritional Care

The optimal nutritional needs of elderly cancer patients are unknown at present. The metabolic inefficiency imparted by the tumor complicates prediction of energy and protein requirements, and undoubtedly influences the need for other nutrients as well. Calorie needs can be estimated, however, from the following guidelines: provide 35 kcal/kg of body weight per day to maintain body weight and 45 kcal/kg of body weight per day to achieve anabolism.[24] A more individualized method makes use of the Harris-Benedict equations to determine basal energy expenditure.[37] This basal energy expenditure is amplified by a factor of 1.2 to 2.0, depending on whether the patient is afebrile or febrile, scheduled for nutritional maintenance or repletion, and to be nourished enterally or parenterally.[25] Protein needs have been recommended at 1.5 to 2.5 times the recommended dietary allowance of 0.8 g of protein per kilogram of weight per day.[14] Nitrogen balance studies are useful for adjusting protein intake to an adequate level.

Preference must be given to administering nutritional support via the GI tract if normal digestion or absorption is unimpaired and anorexia is not severe. In the cancer patient several obstacles may deter the consumption of an adequate oral diet, including chronic pain, difficulty in swallowing, stomatitis, esophagitis, loss or thickening of saliva, organoleptic aversions, early satiety, nausea, and vomiting.[25] Consequently, several precautions should be taken to assure that oral intake is adequate. Daily food intake should be carefully monitored. A diet plan should be constructed, considering the patient's personal food tolerances and preferences. Finally, the patient should be educated on the general principles of a balanced diet and given specific suggestions on how to implement his/her individual diet plan. Emphasis should be placed on choosing high-protein, energy-dense foods and consuming frequent, small meals and nutritious snacks throughout the day. Several commercial dietary supplements, milk-, casein-, or soy-based, have been used successfully to boost the oral intake of cancer patients.[38]

### 1. Enteral Nutritional Support

When the patient is unable to consume an adequate oral diet and moderate to severe caloric deficits are likely, enteral nutritional support via a feeding tube is indicated. Again, gut function must be adequate and nausea and vomiting induced by tumor and treatment must be absent or controllable. If digestive capacity is decreased, elemental tube feedings are advisable. The advantages of using enteral nutrition, when possible, over parenteral nutrition are numerous. Tube feedings and elemental diets are relatively economical and can be prepared simply without the need for sterilization. The risk of complications associated with enteral alimentation is minimal; the chief potential complication is pulmonary aspiration. Other possible problems include nasopharyngeal and mucosal irritation, fluid overload, hypertonic dehydration, and abdominal cramping and diarrhea.[39] Therefore, care should be taken to monitor fluid and electrolyte balance, especially in elderly patients whose declining renal and cardiac function renders them less able to handle large fluid loads. Also, it is often advisable to introduce enteral feedings at a diluted concentration and reduced delivery rate, then slowly escalate to the final desired dose schedule. These management practices will help to reduce the incidence of minor complications associated with enteral feedings.

It must be emphasized again that the primary goal of enteral hyperalimentation is to rehabilitate nutritional status when aggressive anticancer therapy is planned. There are a limited number of clinical studies that have tested the efficacy of enteral nutritional support to improve nutritional status. Preservation of body protein has been associated with high caloric, elemental jejunostomy feedings in a small group of undernourished patients with esophageal cancer.[40] The additional stress imposed by aggressive antitumor therapy may limit the effectiveness of enteral feeding in ensuring optimal nutritional rehabilitation. Hypercaloric, high-protein tube feeding administered postoperatively to surgical patients with head and neck cancer led to overall nitrogen balance without attendant weight gain.[41] Similarly, nitrogen retention was observed with administration of a chemically defined diet for patients with various malignancies undergoing chemotherapy, although retention of other elements suggested that the nitrogen was not utilized for formation of lean body mass.[42] During radiation therapy, weight gain was affected positively when an elemental diet was given,[43] but weight gain should not be equated with gain in lean body mass.[39] There is a paucity of information regarding the ability of enteral nutritional support to affect a change in response to antineoplastic therapy. A reduction in incidence of diarrhea has been reported when an elemental diet was combined with conventional radiation therapy.[43] Other studies do not support the contention that enteral feedings improve the outcome of treatment as reflected by disease status[38] or survival statistics.[44] Clearly, adequate studies of enteral nutritional support have not been conducted to test the general usefulness of this form of adjunctive therapy. More specific studies are needed to devise optimal levels of nutrients that might prove beneficial.

### 2. Parenteral Nutritional Support

When enteral feeding is ruled out due to GI obstruction, fistula formation, or severe dysfunction, the parenteral route of hyperalimentation is indicated. Also, antitumor therapy may precipitate unmanageable nausea, vomiting, diarrhea, and/or GI bleeding, conditions in which bowel rest may be desirable. Parenteral hyperalimentation is commonly delivered via a large-bore, central vein; hypertonic solutions containing 20 to 35% dextrose and 3 to 5% amino acids are well tolerated with less risk of thrombophlebitis at this site. Infusion of nutrients into peripheral veins is feasible, but requires a solution with lower osmolarity. Usually 5% dextrose, 2 to 3% amino acid solution is tolerated peripherally with lipid emulsion included to provide about 50 to 60% of calories.

The potential for complications arising as a result of aggressive intravenous feeding programs could limit the feasibility of this form of nutritional support. However, 1500 patients undergoing parenteral nutritional support during antineoplastic therapy were observed to have a low incidence of complications despite the often severe concomitant illnesses and considerable immunodeficiency.[45] Other groups have also shown long-term venous access for nutritional support is possible with minimal complications in severely compromised patients with malignancy.[41,46] An equal number of septic episodes developed in two randomized groups of patients with Ewing's sarcoma, rhabdomyosarcoma, or osteosarcoma selected to receive either conventional oral nutrition or total parenteral nutrition concurrently with intensive chemotherapy. Although three patients with positive blood cultures were in the group receiving total parenteral nutrition, the central lines were not believed to be the source of the bacteremia. The authors, however, recognize that the small number of patients in this study makes definitive statements regarding the effects of infectious complications of parenteral nutrition in this setting difficult.[47] In a number of studies using strict aseptic protocols, infections from central venous lines were less than 2%.[48,49] The incidence of catheter-related sepsis increases in patients with cancer of the head and neck region because they generally have open wounds or tracheostomy stomas nearby that constantly contaminate the catheter dressings.[50] As a rule the patient who develops fever during intravenous hyperalimentation should be presumed to have catheter-related sepsis unless another primary focus of infection becomes apparent. Diagnosis of catheter-related sepsis is confirmed by a blood culture and the catheter culture positive for the same organism. If catheter-related sepsis is suspected, the catheter should be removed immediately. However, if the primary focus of infection other than the catheter is identified and blood cultures are negative, the primary focus should be treated appropriately and the catheter left in place. A practical approach to reinsertion of the feeding catheter after an episode of catheter-related sepsis is to allow 48 hr to elapse after the temperature has returned to normal and blood cultures have become negative before reinsertion of the feeding catheter into the superior vena cava through the opposite subclavian vein.[51] Seemingly high incidences of subclavian vein and superior vena cava thrombosis have been observed in parenterally supported cancer patients and may be the result of alterations in the intrinsic clotting mechanism.[51] This risk may be reduced by using Silastic® catheters instead of the more thrombogenic polyvinylchloride catheter or by intravenously administering small doses of heparin.[52] An increased frequency of clinically significant anemia in the randomized parenteral nutrition cancer group has been noted and has been attributed to increased phlebotomies, unrecognized nutrient deficiencies (folic acid, vitamin $B_{12}$, Fe, Cu), and a dilutional effect secondary to expansion of plasma volume associated with total parenteral nutrition.

Most total parenteral nutrition-related complications are iatrogenic, avoidable, and in reality, an indictment of the quality of care rather than the technique itself. In a multi-institutional study the data suggest that total parenteral nutrition is a safe therapeutic modality in a patient mix undergoing aggressive antineoplastic therapy and can be accomplished on a multi-institutional basis under differing team formats and varying levels of expertise, education, experience, and training. The institution of parenteral nutrition resulted in little serious morbidity and mortality in the cancer subject group.[51]

An added concern associated with the use of hyperalimentation in cancer patients is the possibility that tumor growth might be enhanced in a nutrient-rich milieu. Data from several studies indicate that in cancer-bearing man, tumor growth is not preferentially stimulated by parenteral nutrition support.[30,53,54]

The evaluation of parenteral nutrition as an adjunct to current antineoplastic regimens should include testing the advantage of this nutritional support modality in repleting the malnourished patient. Also, it would be desirable to demonstrate that parenteral nutrition can reduce the toxicities or complications associated with therapy, or that it can improve the response to therapy with the possibility that survival may be positively affected as well.

### a. Parenteral Nutrition as Adjunct to Chemotherapy

From early retrospective studies, positive correlations were drawn between nutritional status, usually defined in terms of body weight changes, and chemotherapeutic tumor response.[55,56] However, the most valuable information gained from these studies was that parenteral nutrition could be conducted safely in this group of patients whose disease state and tumor therapy placed them at increased risk for septic complications associated with central venous alimentation.

Good patient tolerance to parenteral nutrition has been reaffirmed in several reports of randomized, prospective trials testing the efficacy of total parenteral nutrition as an adjunct to chemotherapy.[57-61] In general nutritional status improves during chemotherapy cycles with an intravenous hyperalimentation adjunct, and this form of nutritional support may have a slight, though statistically insignificant, advantage compared to enteral nutritional support.[42] A favorable effect of parenteral nutrition on body weight has been observed, although it has not been established whether body weight gain represents lean body mass[57,59] or merely gain of fat and water compartments.[17,42,62] Immunocompetence has been regained successfully when parenteral nutrition was added to the chemotherapeutic treatment for small cell bronchogenic carcinoma.[59]

The ability of parenteral nutrition to improve tumor response to chemotherapy is not well founded in the few randomized prospective studies conducted. The majority of trials report no significant beneficial effect on tumor response.[57-61,63,64] Similarly, parenteral nutrition provides no clear advantage in reducing chemotherapy-related toxicities — either in decreasing the duration or frequency of gut toxicities such as nausea and vomiting[58-60] or shortening periods of myelosuppression.[47,58-60,63,64]

### b. Parenteral Nutrition as Adjunct to Radiation Therapy

Radiation therapy can have severe debilitating effects on the GI tract; if prolonged, severe weight loss and malnutrition may result. In this situation, aggressive parenteral nutrition seems warranted as an adjunct to radiation treatment. In fact, the bowel rest afforded by total parenteral nutrition has reduced or eliminated symptoms of radiation stomatitis and enteritis.[65] The nutritional status of patients can be improved over the course of radiation therapy if total parenteral nutrition is provided.[65,66] Further, in both retrospective[65] and prospective[66] studies, a positive tumor response was associated with improvement in nutritional status.

Although it has not been shown that parenteral nutrition improves tolerance to therapy or leads to escalation of therapy, it has allowed for completion of a planned course of therapy for cancer patients who otherwise were considered poor risks due to debilitated nutritional status.[65]

### c. Parenteral Nutrition and the Surgical Treatment of Cancer

In addition to improving the nutritional status of the malnourished patient with a surgically resectable tumor, administration of parenteral nutrition might minimize post-surgical complications. A review of patients treated perioperatively with parenteral nutrition suggests that nutritional status can be improved,[67-69] and wound healing and/or wound infection rates have been affected positively.[67,70] Additional experience with esophageal and GI cancer patients indicates that the use of total parenteral nutrition preoperatively leads to reductions in major wound and infectious complications compared to either an orally fed control group or a group nourished parenterally but only postoperatively.[69] The authors recommend a 5-day course of preoperative total parenteral nutrition as a prophylactic measure. Short-term survival rates have not been influenced by parenteral nutrition in the small number of patient cases followed.[41] The most successful application of perioperative total parenteral nutrition in malnourished patients is when the lesion is completely resectable and nutritional support is provided preoperatively and continued postoperatively until adequate oral intake is possible.[48]

In conclusion, total parenteral nutrition should not be used indiscriminately as an adjunct to anticancer therapy. The use of parenteral nutrition is justified in malnourished patients who are unable to be repleted by the enteral route and have potentially responsive tumors. It is reasonable to expect that parenteral nutrition will improve nutritional status in such patients, and appropriate nutritional assessment should be performed periodically to monitor progress.

## IV. CONCLUDING REMARKS

At the present time it is not possible to make definitive recommendations for nutritional support of cancer patients. Ideal nutrient support systems for the elderly patient are even more nebulous, as rigorous studies aimed specifically at the metabolic changes and nutritional needs in this cancer-bearing subgroup are yet to be conducted. Most of the studies reviewed herein describe nutritional status and support of adult populations with mean age in the 5th or 6th decade; often the patients range as widely as 30 to 90 years of age. Certainly, these are heterogeneous populations with respect to metabolic rate and physiological organ function.[71] Some metabolic changes that occur during the neoplastic process also are natural consequences of aging, namely loss of lean body mass, decreased rate of protein synthesis, and depressed immune response.[72,73] Impairment of bone marrow activity, a common sequela of chemotherapy, also occurs with advancing age.[72] Consequently, sorting out the contributions of the aging process, disease process, and nutritional depletion process is a tedious task. Due to the small numbers of patients studied to date, it is impossible to ascertain if the age factor influenced the success or failure of nutritional support regimes to rehabilitate nutritional status and to improve outcome. One descriptive study of nutritional support of elderly cancer patients concludes that the nutritional deficits of the elderly are the same as other cancer patients but are more frequently encountered. The authors recommend intensive nutritional support, encourage the use of a regular oral diet whenever possible, and suggest careful monitoring and special encouragement.[33] As more information accrues on the metabolic anomalies of the host, a better approach to replenishing nutritional deficiencies of cancer patients will emerge. The nutrient composition of both enteral and parenteral formulations must be optimized with consideration of the special needs of the elderly patient.

## ACKNOWLEDGMENT

The authors would like to acknowledge the University of Wisconsin Medical School, the College of Agricultural and Life Sciences, and NIH grants 5 P30 AM26659 and AM 30843 for supporting, in part, the preparation of this manuscript.

## REFERENCES

1. **DeWys, W. D.,** Pathophysiology of cancer cachexia: current understanding and areas for future research, *Cancer Res.,* 42, 721S, 1982.
2. **Theologides, A.,** Pathogenesis of cachexia in cancer: a review and a hypothesis, *Cancer,* 29, 484, 1972.
3. **Berstein, I. L.,** Physiological and psychological mechanisms of cancer anorexia, *Cancer Res.,* 42, 715S, 1982
4 **Brennan, M. F.,** Uncomplicated starvation versus cancer cachexia, *Cancer Res.,* 37, 2359, 1977.
5. **Lundholm, K., Holm, G., and Scherstén, T.,** Insulin resistance in patients with cancer, *Cancer Res.,* 38, 4665, 1978.
6. **Lundholm, K., Edstrom, S., Ekman, L., Karlberg, I., and Scherstén, T.,** Metabolism in peripheral tissues in cancer patients, *Cancer Treat. Rep.,* 65(Suppl. 5), 79, 1981.
7. **Schein, P. S., Kisner, D., Haller, D., Blecher, M., and Hamosh, M.,** Cachexia of malignancy. Potential role of insulin in nutritional management, *Cancer,* 43, 2070, 1979.

8   Waterhouse, C., How tumors affect host metabolism, *Ann. N Y Acad. Sci.*, 230, 86, 1974.

9   Shapot, V. A., On the multiform relationships between the tumor and the host, *Adv. Cancer Res*, 30, 89, 1979.

10  Monakhov, N. K., Neistadt, E. L., Shavolovskil, M. M., Shvartsman, A. L., and Neifakh, S. A., Physicochemical properties and isoenzyme composition of hexokinase from normal and malignant human tissues, *J. Natl. Cancer Inst.*, 67, 27, 1978.

11. Lundholm, K., Bylund, A. C., Holm, J., and Scherstén, T., Skeletal muscle metabolism in patients with malignant tumor, *Eur. J. Cancer*, 12, 465, 1976.

12. Lundholm, K., Edstrom, S., Ekman, L., Karlberg, I., Bylund, A. C., and Scherstén, T., A comparative study of the influence of malignant tumor on host metabolism in mice and man, *Cancer*, 42, 453, 1978.

13  Stein, T. P., Mullen, J. L., Oran-Smith, J. C., Rosato, E. F., Wallace, H. W., and Hargrove, W. C., III, Relative rates of tumor, normal gut, liver and fibrinogen protein synthesis in man, *Am J. Physiol.*, 234, E648, 1978.

14. Blackburn, G. L., Maini, B. S., Bistrian, B. R., and McDermott, W. V., Jr., The effect of cancer on nitrogen, electrolyte, and mineral metabolism, *Cancer Res.*, 37, 2348, 1977.

15. Warnold, I., Lundholm, K., and Scherstén, T., Energy balance and body composition in cancer patients, *Cancer Res*, 38, 1801, 1978.

16. Cohn, S. H., Gartenhaus, W., Vartsky, D., Sawitsky, A., Zanzi, I., Vaswani, A., Yaumura, S., Rai, K., Cortes, E., and Ellis, K. J., Body composition and dietary intake in neoplastic disease, *Am J. Clin. Nutr.*, 34, 1997, 1981

17. Nixon, D. W., Hyperalimentation in the undernourished cancer patient, *Cancer Res.*, 42, 727S, 1982.

18  DeWys, W. D., Nutritional care of the cancer patient, *JAMA*, 244, 374, 1980

19. Shils, M. E., Nutritional problems induced by cancer, *Med Clin. N Am*, 63, 1009, 1979

20  Ohnuma, T. and Holland, J. F., Nutritional consequences of cancer chemotherapy and immunotherapy, *Cancer Res*, 37, 2395, 1977.

21. Demetrakopoulos, G. E. and Brennan, M. F., Tumoricidal potential of nutritional manipulations, *Cancer Res.*, 42, 756S, 1982.

22  Welch, D., Nutritional consequences of carcinogenesis in radiation therapy, *J. Am. Dietet Assoc.*, 78, 467, 1981.

23. Bozzetti, F., Migliavacca, S., Scotti, A., Bonalumi, M. G., Scarpa, D., Baticci, F., Ammatuna, M., Pupa, A., Terno, G., Sequeira, C., Masserini, C., and Emanuelli, H., Impact of cancer, type, site, stage and treatment on the nutritional status of patients, *Ann Surg.*, 196, 170, 1982.

24  Copeland, E. M., III, Nutritional concepts in the treatment of cancer, *J. Fla Med. Assoc.*, 66, 373, 1979.

25. Gildea, J. L., Motz, K., Costlow, N., Markley, E., and Pratt, G., A systematic approach to providing nutritional care to cancer patients, *Nutr. Suppl. Serv.*, 2, 24, 1982.

26. Abraham, S., Weight by height and age for adults 18—74 years, United States, 1971—1974, *Vital and Health Statistics Data from National Health Survey*, DHEW Publ. No (PHS)79-1656, Department of Health, Education and Welfare, Washington, D C., 1979, 17.

27  Bishop, C. W., Bowen, P. E., and Ritchey, S. J., Norms for nutritional assessment of American adults by upper arm anthropometry, *Am. J. Clin. Nutr.*, 34, 2530, 1981

28. Harvey, K. B., Bothe, A., Jr., and Blackburn, G. L., Nutritional assessment and patient outcome during oncologic therapy, *Cancer*, 43, 2065, 1979.

29. DeWys, W. D., Begg, C., Lavin, P. T., Band, P. R., Bennett, J. M., Bertino, J. R., Cohen, M. H., Douglas, H. O., Jr., Engstrom, P. F., Ezdinli, E. Z., Horton, J., Johnson, G. J., Moertel, C. G., Oken, M. M., Perlia, C., Rosenbaum, C., Silverstein, M. N., Skeel, R. T., Sponzo, R. W., and Tormey, D. C., Prognostic effect of weight loss prior to chemotherapy in cancer patients, *Am. J. Med.*, 69, 491, 1980.

30. Copeland, E. M., Daly, J. M., and Dudrick, S. J., Nutrition and cancer, *Int. Adv. Surg. Oncol*, 4, 1, 1981.

31. Nixon, D. W., Heymsfield, S. B., Cohen, A. E., Kutner, M. H., Ansley, J., Lawson, D. H., and Rudman, D., Protein-calorie undernutrition in hospitalized cancer patients, *Am. J. Med.*, 68, 683, 1980.

32. Blackburn, G. L., Bistrian, B. R., Maini, B. S., Schlamm, H. T., and Smith, M. F., Nutritional and metabolic assessment of the hospitalized patients, *J. Parenteral Enteral Nutr.*, 1, 11, 1977.

33. Ching, N., Grossi, C., Zurawinsky, H., Jham, G., Angers, J., Mills, C., and Nealon, T. F., Jr., Nutritional deficiencies and nutritional support therapy in geriatric cancer patients, *J. Am. Geriatr. Soc*, 27, 491, 1979.

34. Ota, D. M., Copeland, E. M., Corriere, J. N., Jr., and Dudrick, S. J., The effects of nutrition and treatment of cancer on host immunocompetence, *Surg. Gynecol Obstet.*, 148, 104, 1979.

35. Kaminski, M. V., Jr., Nasr, N. J., Mos, A. J., Berger, R. L., and Sriram, K., Nutritional status, immunity and survival in neoplastic disease, *Nutr. Suppl. Serv.*, 2, 10, 1982.

36. **Dominioni, L., Dionigi, R., Dionigi, P., Nazari, S., Fossati, G. S., Prati, U., Tibaldeschi, C., and Pavesi, F.,** Evaluation of possible causes of delayed hypersensitivity impairment in cancer patients, *J. Parenteral Enteral Nutr.,* 5, 300, 1981.

37 **Harris, J. A. and Benedict, F. G.,** Biometric Studies of Basal Metabolism in Man, Publ. No 279, Carnegie Institute of Washington, Washington, D C , 1919

38. **Elkort, R. J., Baker, F. L., Vitale, J. J., and Cordano, A.,** Long-term nutritional support as an adjunct to chemotherapy for breast cancer, *J. Parenteral Ent. Nutr ,* 5, 395, 1981.

39. **DeWhy, W. D. and Kubota, T. T.,** Enteral and parenteral nutrition in the care of the cancer patient, *JAMA,* 246, 1725, 1981.

40. **Burt, M. E., Stein, T. P., and Brennan, M. F.,** A controlled randomized trial evaluating the effects of enteral and parenteral nutrition on protein metabolism in cancer-bearing man, *J. Surg. Res.,* 34, 303, 1983.

41. **Sako, K., Loré, J. M., Kaufman, S., Razack, M. S., Bakamjian, V., and Reese, P.,** Parenteral hyperalimentation in surgical patients with head and neck cancer: a randomized study, *J. Surg. Oncol ,* 16, 391, 1981.

42. **Nixon, D. W., Lawson, D. H., Kutner, M., Ansley, J., Schwarz, M., Heymsfield, S., Chawla, R., Cartwright, T. H., and Rudman, D.,** Hyperalimentation of the cancer patient with protein-calorie undernutrition, *Cancer Res.,* 41, 2038, 1981

43. **Bounous, G., LeBel, E., Shuster, J., Gold, P., Tahan, W. T., and Bastin, E.,** Dietary protection during radiation therapy, *Strahlentherapie,* 149, 476, 1975.

44. **Douglass, H. O., Milliron, S., and Nava, H.,** Elemental diet as adjuvant for patients with locally advanced gastrointestinal cancer receiving radiation therapy: a prospectively randomized study, *J Parenteral Enteral Nutr.,* 2, 682, 1978.

45. **Copeland, E. M., Daly, J. M., and Dudrick, S. J.,** Nutrition as an adjunct to cancer treatment in the adult, *Cancer Res.,* 37, 23, 1977.

46. **Abraham, J., Mullen, J. L., Jacobson, N., and Polomano, R.,** Chronic central venous access in patients with acute leukemia, *Cancer Treat. Rep.,* 63, 11, 1980.

47. **Shamberger, R. C., Pizzo, P. A., Goodgame, J. T., Lowry, S., Maher, M. M., Wesley, R. A., and Brennan, M. F.,** The effect of total parenteral nutrition on chemotherapy-induced myelosuppression. A randomized study, *Am. J Med.,* 74, 40, 1983

48. **Deitel, M., Vasic, V., and Alexander, M. A.,** Specialized nutritional support in the cancer patient. is it worthwhile? *Cancer,* 41, 2359, 1978.

49. **Copeland, E. M., MacFadyen, B. V., and Dudrick, S. J.,** The use of hyperalimentation in patients with potential sepsis, *Surg Gynecol. Obstet.,* 138, 377, 1974

50. **Copeland, E. M., III, Daly, J. M., and Dudrick, S. J.,** Intravenous hyperalimentation, bowel rest and cancer, *Crit. Care Med.,* 8, 21, 1980.

51. **Mullen, J. L.,** Complications of total parenteral nutrition in the cancer patients, *Cancer Treat. Rep ,* 65(Suppl. 5), 107, 1981

52. **Daly, J. M.,** Nutritional support of the cancer patient, *Clin Consult. Nutr. Supp.,* 3, 1, 1983

53. **Mullen, J. L., Buzby, G. P., Gertner, M. N., Stein, T. P., Hargrove, W. C., Oram-Smith, J., and Rosato, E. F.,** Protein synthesis dynamics in human gastrointestinal malignancies, *Surgery,* 87, 331, 1980.

54. **Schwartz, G. F., Gren, H. L., Bendon, M. L., Graham, W. P., III, and Blakemore, W. S.,** Combined parenteral hyperalimentation and chemotherapy in the treatment of disseminated solid tumors, *Am J. Surg.,* 121, 169, 1971.

55. **Copeland, E. M., MacFadyen, B. V., Lanzotti, V. J., and Dudrick, S. J.,** Intravenous hyperalimentation as an adjunct to cancer chemotherapy, *Am J Surg.,* 129, 167, 1975

56. **Lanzotti, V. J., Copeland, E. M., George, S. L., Dudrick, S. J., and Samuels, M. L.,** Cancer chemotherapeutic response and intravenous hyperalimentation, *Cancer Chemother. Rep.,* 59, 437, 1975

57. **Issell, B. F., Valdivieso, M., Zaren, H. A., Dudrick, S. J., Freireich, E. J., Copeland, E. M., and Bodey, G. P.,** Protection against chemotherapy toxicity by intravenous hyperalimentation, *Cancer Treat. Rep.,* 62, 1139, 1978.

58. **Jordan, W. M., Valdivieso, M., Frankmann, C., Gillespie, M., Issell, B. F., Bodey, G. P., and Freireich, E. J.,** Treatment of advanced adenocarcinoma of the lung with ftorfur, doxorubicin, cyclophosphamide and cisplatin (FACP) and intensive IV hyperalimentation, *Cancer Treat Rep ,* 65, 197, 1981

59. **Valdivieso, M., Bodey, G. P., Benjamin, R. S., Barkley, H. T., Freeman, M. B., Ertel, M., Smith, T. L., and Mountain, C. F.,** Role of intravenous hyperalimentation as an adjunct to intensive chemotherapy for small cell bronchogenic carcinoma, *Cancer Treat Rep.,* 65(Suppl. 5), 145, 1981.

60. **Serrou, B., Cupissol, D., Plagne, R., Boutin, P., Carcassone, Y., and Michel, F. B.,** Parenteral intravenous nutrition as an adjunct to chemotherapy in small cell anaplastic lung carcinoma, *Cancer Treat. Rep.,* 65(Suppl. 5), 151, 1981.

61. **Nixon, D. W., Moffitt, S., Lawson, D. H., Ansley, J., Lynn, M. J., Kutner, M. H., Heymsfield, S. B., Wesley, M., Chawla, R., and Rudman, D.,** Total parenteral nutrition as an adjunct to chemotherapy of metastatic colorectal cancer, *Cancer Treat Rep.,* 65(Suppl 5), 121, 1981.

62. **Popp, M. B., Fisher, R. I., Wesley, R., Aamodt, R., and Brennan, M. F.,** A prospective randomized study of adjuvant parenteral nutrition in the treatment of diffuse lymphoma, influence on survival, *Surgery,* 90, 195, 1981.

63. **Popp, M. B., Fisher, R. I., Simon, R. M., and Brennan, M. F.,** A prospective randomized study of adjuvant parenteral nutrition in the treatment of diffuse lymphoma: effect on drug tolerance, *Cancer Treat. Rep.,* 65(Suppl. 5), 129, 1981.

64. **Samuels, M. L., Selig, D. E., Ogden, S., Grant, C., and Brown, B., IV,** Hyperalimentation and chemotherapy for stage III testicular cancer: a randomized study, *Cancer Treat. Rep ,* 65, 615, 1981.

65. **Copeland, E. M., Souchon, E. A., MacFadyen, B. V., Rapp, M. A., and Dudrick, S. J.,** Intravenous hyperalimentation as an adjunct to radiation therapy, *Cancer,* 39, 609, 1977.

66. **Valerio, D., Overett, L., Malcolm, A., and Blackburn, G. L.,** Nutritional support for cancer patients receiving abdominal and pelvic radiotherapy: a randomized prospective clinical experiment of intravenous versus oral feeding, *Surg. Forum,* 29, 145, 1978

67. **Copeland, E. M., MacFadyen, B. V., MacComb, W. S., Guillamondegui, O., Jesse, R. H., and Dudrick, S. J.,** Intravenous hyperalimentation in patients with head and neck cancer, *Cancer,* 35, 606, 1975.

68. **Holter, A. R. and Fischer, J. E.,** The effects of perioperative hyperalimentation on complications in patients with carcinoma and weight loss, *J. Surg. Res.,* 23, 31, 1977.

69. **Daly, J. M., Masar, E., Giacco, G., Frazier, O. H., Mountain, C. F., Dudrick, S. J., and Copeland, E. M.,** Parenteral nutrition in esophageal cancer patients, *Ann. Surg.,* 196, 203, 1982

70. **Heatley, R. V., Williams, R. H. P., and Lewis, M. H.,** Pre-operative intravenous feeding — a controlled trial, *Postgrad. Med. J.,* 55, 541, 1979.

71. **Shock, N. W.,** Physiological aspects of aging, *J Am. Dietet. Assoc ,* 56, 491, 1970.

72. **Steffee, W. P.,** Nutrition intervention in hospitalized geriatric patients, *Bull. N.Y. Acad. Med.,* 56, 564, 1980.

73. **Munro, H. N.,** Major gaps in nutrient allowances, *J. Am. Dietet Assoc ,* 76, 137, 1980.

Chapter 5

# DIABETES MELLITUS

**Gerald M. Reaven and Eve Reaven**

## TABLE OF CONTENTS

## I. INTRODUCTION

The aim of this chapter is to focus attention on aspects of nutrition which are of particular importance in the regulation of carbohydrate metabolism in older individuals. As such, no attempt will be made to address in a comprehensive fashion all the relationships that exist between nutrition and the various metabolic disorders that are included under the general heading of diabetes mellitus. At the least, diabetes mellitus must be subdivided into two broad syndromes:[1] insulin-dependent diabetes mellitus (IDDM) and noninsulin-dependent diabetes mellitus (NIDDM). IDDM is characterized by an absence of endogenous insulin secretion, and these patients are *absolutely* dependent upon exogenous insulin for short-term survival. This syndrome is more likely to occur in individuals with certain HLA types,[2] and the nearly universal presence of circulating antibodies to pancreatic islet cells at the outset of the disease has led to the view that the insulin deficiency is secondary to an autoimmune process involving pancreatic β cells.[3] Patients with this form of diabetes were previously said to have "juvenile-onset" diabetes, but it is quite clear that older individuals can have what appears to be the identical syndrome. Hence, the designation of IDDM, which emphasizes that the genetic, immunological, and metabolic features, not the age of the patient, define the syndrome.

The impact of uncontrolled IDDM on the nutritional status of an individual is enormous, and maintenance of normal growth and development in young individuals with IDDM remains an important clinical problem. On the other hand, these effects are well understood, as they represent the inevitable consequences of an absolute deficiency of insulin. Furthermore, the nutritional issues important in IDDM have been considered previously on many occasions; they have no unique relevance to aging, do not appear to be involved in the development of IDDM, and to a large extent are repaired by the judicious use of exogenous insulin. For all of these reasons we will not discuss this topic any further.

The characteristics of patients with NIDDM are very different from those with IDDM. There is no particular HLA type associated with NIDDM, nor are islet cell antibodies present in the plasma of patients with this syndrome. Thus, there is no evidence that an autoimmune destruction of pancreatic β cells is responsible for the hyperglycemia in NIDDM. Indeed, some evidence of β-cell function, estimated by measurement of plasma insulin levels, is almost always seen in patients with NIDDM.[4] Although pancreatic β-cell function may not be normal in NIDDM, absolute insulin deficiency, the hallmark of IDDM, is not characteristic of NIDDM.[5] Consequently, patients with NIDDM are not dependent upon exogenous insulin for short-term survival. Patients with NIDDM were previously designated as having "maturity-onset" diabetes, but it is now evident that a similar form of diabetes can also occur in young individuals.[6] On the other hand, the fact that not all patients with NIDDM are old should not obscure the fact that the prevalence of NIDDM, the syndrome accounting for approximately 90% of all diabetics, increases with age. Finally, the ability of weight gain to accentuate,[7] and weight loss to ameliorate,[8] this form of diabetes has been known for a long time.

Given the above considerations, it seems apparent that the relationship between age and nutrition in patients with diabetes is of primary importance in NIDDM. Consequently, attention during the remainder of this presentation will be focused on this clinical syndrome. Initially, the effects of age and nutrition on insulin secretion and insulin action in animals and humans will be reviewed, followed by an effort to apply this information to understanding the pathogenesis of NIDDM.

## II. EFFECTS OF AGE AND NUTRITION ON INSULIN SECRETION

### A. Rats
Attempts to assess insulin secretion in the intact organism usually rely on measurement

of plasma insulin concentrations after the administration of an insulin secretagogue. Glucose is the stimulus usually used, and evidence has been published which suggests that insulin secretory function determined in this fashion tends to increase with age.[9] On the other hand, there are limitations to this method of analysis. In the first place, plasma insulin concentrations are a function of both insulin secretion and insulin clearance, and an increase in concentration could occur as the result of an increase in secretion and/or a decrease in clearance. Since it appears that insulin removal rate from plasma is prolonged in older rats,[10] elevated plasma insulin levels in response to a glucose challenge need not indicate enhanced β-cell function with age.

A second major problem in relating plasma insulin levels to β-cell secretory response stems from confusion as to the "correct" stimulus. The plasma insulin response to a glucose challenge in normal subjects is determined to a significant extent by the size of the glucose load, and insulin levels will increase proportionately to the administered glucose, despite the fact that plasma glucose concentrations may not increase as a function of dose.[11] Since older rats are usually fatter, basing the glucose challenge on a weight basis, which is usually done, will lead to the self-fulfilling conclusion that older rats are hyperinsulinemic, and that age leads to an increase in β-cell secretion.

In order to avoid both of these dilemmas we have conducted a series of studies in which we have used isolated islets of Langerhans to assess the effect of age on β-cell function in rats. In our initial studies we could demonstrate that islet size increased progressively as rats grew from 2 to 12 months of age, associated with an increase in β granule and insulin content.[12] However, the secretory function of the large islets was impaired, and both glucose-[12] and leucine-stimulated insulin secretion[13] (estimated per β cell) declined with age.

Comparable changes were also seen when we used the isolated perfused pancreas to evaluate the effect of age on β-cell function.[14] Thus, both pancreatic weight and pancreatic islet content increased with age, resulting in an approximate fourfold rise in β-cell pool size. However, as in the case of isolated islets, maximal glucose-stimulated insulin secretion per β cell was reduced with age. On the other hand, the β-cell hyperplasia seen with age was great enough to compensate for the loss of cellular function, and maximal glucose-stimulated insulin secretion by the entire perfused pancreas was the same in young and old rats.

As mentioned earlier, rats usually get fatter as they grow older, and this was particularly true of the strain used in these studies. Therefore, an attempt was made to differentiate between the effects of age, as distinguished from those due to obesity, on the structural and functional changes seen in the endocrine pancreas of older (and fatter) rats. Specifically, the weight gain of rats growing from 2 to 12 months of age was reduced: this was accomplished either by decreasing caloric intake or increasing the level of physical activity of the animals as they aged. Both manipulations prevented the increase in endocrine cell mass seen in 12-month-old control rats,[14-16] supporting the view that β-cell hyperplasia in older rats is secondary to age-related environmental factors, and not the result of age per se. It seems most likely that the increase in β-cell pool size represents an effort to compensate for the decline in insulin secretion per β cell associated with aging, and this formulation will be discussed in detail subsequently.

In contrast to the ability of calorie control and exercise training to prevent the changes in islet structure associated with aging, islet function remained grossly abnormal. Thus, maximal glucose-stimulated insulin secretion was similar when studied in either isolated islets or in perfused pancreas of older obese and nonobese rats.[14-16] The situation was less clear in the case of the effect of exercise training, and deterioration in β-cell secretory function with age may have been attenuated in rats treated in this manner.[15] However, the experimental results were not conclusive, and this matter requires further study.

In summary, it seems clear that there is a progressive loss of maximal insulin secretion per β cell as rats age, and the decline seems to be an inevitable consequence of the aging process. When rats are allowed to become obese and/or sedentary as they age, this change in β-cell function is associated with β-cell hyperplasia. However, either calorie restriction or exercise training prevents the age-related increase in β-cell pool size. The β-cell hyperplasia seen in older, fatter, and sedentary rats permits these rats to maintain insulin secretory capacity equal to that of younger rats, despite the decline in function of the individual β cell associated with aging. This phenomenon, coupled with the decline in insulin removal from plasma in older rats,[10] provides an explanation for the observation that hyperinsulinemia occurs in older rats.[9]

### B. Humans

It is obviously impossible to carry out in humans the kind of in vitro studies that have so clearly defined the effect of age on β-cell function in rats. Therefore, estimates of changes in insulin secretion with age in humans rely almost entirely on measurement of plasma insulin responses to various secretagogues. The vast literature relevant to this issue has been recently reviewed,[17] and the majority of published results indicate that plasma insulin levels remain the same, or may actually increase, as humans grow older. In this regard, age-related defects in insulin catabolism have also been reported in humans,[10] which may contribute to the ability of older humans to maintain levels of circulating insulin. Whether or not there is a decline in function of the individual β cell with age in humans, similar to that observed in rats, cannot be determined. Even if such a defect occurs, it appears that normal humans, as normal rats, can maintain total insulin secretion as they age.

## III. EFFECTS OF AGE AND NUTRITION ON INSULIN ACTION

### A. Rats

Evidence[9] that older rats have normal plasma glucose concentrations, associated with elevated plasma insulin levels, is consistent with the view that insulin sensitivity declines with age. This notion is supported by observations that insulin-stimulated glucose transport in adipocytes is reduced as rats get older.[18] However, an increase in plasma insulin concentration does not necessarily mean that insulin resistance is present. Furthermore, it must be appreciated that adipose tissue, although convenient to use for studies of insulin action, plays a minor role in total glucose utilization by the intact organism.[19] Thus, normal in vivo insulin-stimulated glucose utilization could occur despite a decline in the action of insulin on adipocytes. In order to clarify this issue we have assessed in vivo insulin action in 6-week-old and 12-month-old rats, and were able to document a loss in the ability of older rats to respond to insulin.[16]

Although there is now good evidence that insulin resistance occurs as rats get older, it must be realized that rats allowed to age under laboratory conditions also become sedentary and obese. Since both levels of physical activity[20] and obesity[21] can modulate insulin action, it seemed important to see how much of the age-related loss of insulin-stimulated glucose uptake was due to each of these factors. Although this is an area of continuing investigative interest, enough information exists at this time to permit certain generalizations. For example, calorie restriction, leading to a decrease in rate of weight gain, inhibits the hyperinsulinemia that develops when rats consume diets *ad libitum* as they grow from 2 to 12 months of age.[15,16] Modification of physical activity is equally effective, and 12-month-old rats, allowed to eat both *ad libitum* and run spontaneously, have plasma insulin levels equal to those of 2-month-old rats.[15] Furthermore, the deterioration in in vivo insulin-stimulated glucose utilization seen in 12-month-old rats can be prevented by modest caloric restriction.[16] These observations are all consistent with the view that age-related changes in body weight and physical activity, rather than age itself, accounts for the insulin resistance seen in older rats.

The conclusion that environmental factors, not age, are responsible for the insulin resistance of older rats is supported by observations that insulin resistance occurs relatively early in the life span of rats. This point was first established by Goodman and associates,[22] who demonstrated resistance to insulin-stimulated glucose uptake by perfused rat hind limbs as early as 4 months of age, with no further progression over the next 20 months. We have examined the same question in vivo,[23] and found that marked insulin resistance was present in 4-month-old rats, and that in vivo insulin action did not decline appreciably as rats grew from 4 to 12 months of age.

Based upon the above information it appears that the insulin resistance that characterizes the older rat develops early in the process of maturation and, as contrasted to the decline in insulin secretory capacity of the older rat, is not a true aging phenomenon. Of great pragmatic importance is the clinical implication of this conclusion, which supports the view that insulin action can be maintained as individuals age by manipulation of the environment.

## B. Humans

Several investigators have studied the effect of age on in vivo insulin-stimulated glucose utilization,[24-28] and there appears to be general agreement that there is increased insulin resistance in older subjects. However, the magnitude of the change has varied from study to study. For example, in a study conducted several years ago,[24] we were unable to document any difference in insulin resistance when we compared healthy and nonobese young and old individuals. We have more recently finished a second study,[26] using a somewhat different method to assess in vivo insulin action, in which we could discern a modest loss of insulin-stimulated glucose disposal in older normal subjects, independent of differences in obesity. However, this change was confined to a group of individuals who were more than 54 years old, and there was no loss of insulin sensitivity when subjects 21 to 34 years of age were compared to individuals between 35 and 54 years of age. Consequently, we could not define a progressive decline in insulin sensitivity with increasing age, and similar results were noted by Fink and associates.[27] Of great interest to us was the fact that the variance in insulin action of individuals in the oldest age group (>54 years) was approximately twice that seen in the other two groups (21 to 34 and 35 to 54 years of age).

Although the data in humans as to the effect of age on insulin action seems to be somewhat at odds with the rat studies, the differences may be more apparent than real. In the first place, the changes in insulin action that have been described are, at best, moderate in magnitude. Furthermore, as anticipated by the increased variance in the insulin sensitivity of older subjects, there are many older individuals who are quite insulin sensitive. Finally, the changes in insulin action that have been seen in association with aging may be related to the experimental methods used. For example, muscle is the primary site of insulin-stimulated glucose utilization during the kinds of intravenous infusion used to assess in vivo insulin action. Thus, an age-related decline in the ability of insulin to induce glucose uptake in older subjects could simply be due to a decreased muscle mass in these individuals. Given prior evidence of loss of muscle mass with age,[29] the above explanation for the decline in insulin action cannot be dismissed out of hand.

In a somewhat related fashion it is also important to take into account the possible confounding effect that variations in physical conditioning have on the age-related loss of insulin sensitivity. Studies in rats have documented the fact that insulin sensitivity is enhanced by exercise training, and that this is due to a change in the response of muscle to the cellular effect of insulin.[30] More recently,[20] we have described a very significant direct relationship between level of physical conditioning (as estimated by maximal oxygen consumption during bicycle ergometry) and efficiency of in vivo insulin-stimulated glucose uptake. In light of these observations it is possible that age per se has relatively little effect on insulin-stimulated glucose utilization. Obviously, a good deal more needs to be learned about the effects of age on insulin action in humans. However, it is already clear that the in vivo insulin action is relatively well maintained in healthy, nonobese older subjects. Indeed, when the effects of variations in habitual level of physical activity and body composition are taken into consideration, it may turn out that the impact of age, per se on insulin action is quite moderate in degree.

## IV. RELATIONSHIP BETWEEN AGE, NUTRITION, AND THE DEVELOPMENT OF NIDDM

Hyperglycemia can result from decreased insulin secretion and/or insulin action, and it is apparent that both of these defects occur in patients with NIDDM.[11] Furthermore, there is evidence that hypoinsulinemia can lead to a loss of normal insulin sensitivity,[32] and that a decrease in insulin secretion can occur secondary to insulin resistance.[33] As a result of these considerations, it is difficult in any given patient with NIDDM to define the sequence of events that culminates in the development of fasting hyperglycemia. On the other hand, given our current understanding of the changes in insulin secretion and action associated with aging, certain generalizations can be made as to how age and nutrition might interact to decrease glucose tolerance.

As discussed earlier, age is often associated with the development of obesity and/or physical inactivity. Both of these changes will reduce the ability of endogenous insulin to stimulate glucose disposal, and aggravate any loss of insulin sensitivity that is due to age per se. As a consequence, endogenous insulin secretion must increase in order to prevent the plasma glucose level from rising, and the greater the age-related loss of insulin action, the more difficult it becomes for the β cells to maintain glucose tolerance.

The ability of the endocrine pancreas to compensate for the insulin resistance seen as obese and physically inactive rats age is hampered by the inexorable decline in maximal glucose-stimulated insulin secretion per individual β cell. The rat responds to this stress by producing more β cells, thereby managing to secrete the amount of insulin needed to overcome the age-related development of insulin resistance. Whether or not similar changes take place in humans remains to be determined, but for the sake of this presentation let us assume that it does.

Based upon our experiences with aging rats, and the fact that only a minority of older humans are frankly diabetic, it appears that the endocrine pancreas in the majority of older subjects is capable of responding to the need for increased insulin that is likely to take place as individuals age. However, this compensation does not always occur, and it is obvious that NIDDM does develop in a significant number of older humans. This could happen for a variety of reasons. For example, genetic differences in insulin sensitivity and/or insulin secretory capacity could exist and, in conjunction with the age-related changes in both functions, lead to further decompensation of glucose tolerance. Alternatively, other environmental insults could magnify the age-related defects in insulin secretion and action. One example of this might be the loss of a significant portion of the endocrine pancreas as the result of a viral infection in childhood, which only produces clinical changes when age leads to the need to secrete increased amounts of insulin. Indeed, there are an almost infinite number of combinations of events that could impair the ability of humans to maintain normal glucose tolerance as they get older.

It is obvious that there is insufficient data available at the present time to decide which of these factors is most responsible for the increased incidence of NIDDM in older humans. On the other hand, it is obvious that environmentally induced changes in insulin action play an important role. As a corollary, it seems reasonable to suggest that every effort be made to encourage individuals to remain thin and physically active as they get older. Although achieving this goal does not guarantee that NIDDM will not develop, there is every reason to believe that it will significantly reduce the likelihood of this event.

# REFERENCES

1. National Diabetes Data Group, Classification and diagnosis of diabetes mellitus and other categories of glucose intolerance, *Diabetes,* 28, 1039, 1979.
2. **Platz, P., Jakobsen, B. K., Morling, N., Ryder, L. P., Svejgaard, A., Thomsen, M., Christy, M., Kromann, H., Benn, J., Nerup, J., Green, A., and Hauge, M.,** HLA-D and -DR antigens in genetic analysis of insulin dependent diabetes mellitus, *Diabetologia,* 21, 108, 1981.
3. **Lernmark, A. and Baekkeskov, S.,** Islet cell antibodies — theoretical and practical implications, *Diabetologia,* 21, 431, 1981
4. **Yalow, R. S. and Berson, A. S.,** Immunoassay of endogenous plasma insulin in man, *J. Clin. Invest.,* 39, 1157, 1960.
5. **Liu, G., Coulston, A., Chen, Y.-D. I., and Reaven, G. M.,** Does day-long absolute hypoinsulinemia characterize the patient with non-insulin-dependent diabetes mellitus? *Metabolism,* 32, 754, 1983
6. **Tattersall, R. B. and Fajans, S. S.,** A difference between the inheritance of classical juvenile-onset and maturity-onset type diabetes of young people, *Diabetes,* 24, 44, 1975
7. **Sims, E. A., Horton, E. S., and Salans, L. B.,** Inducible metabolic abnormalities during development of obesity, *Ann. Rev. Med.,* 22, 235, 1971
8. **Newburgh, L. H., Conn, J. W., Johnston, M. W., and Conn, E. S.,** A new interpretation of diabetes mellitus in obese, middle-aged persons: recovery through reduction of weight, *Transactions of the Association of American Physicians,* First Report, Vol. 53, Dornan, Philadelphia, 1938, 245.
9. **Bracho-Romero, E. and Reaven, G. M.,** Effect of age and weight on plasma glucose and insulin responses in the rat, *J. Am Geriatr. Soc.,* 25, 299, 1977.
10. **Reaven, G. M., Greenfield, M. S., Mondon, C. E., Rosenthal, M., Wright, D., and Reaven, E. P.,** Does insulin removal rate from plasma decline with age? *Diabetes,* 31, 670, 1982.
11. **Castro, A., Scott, J. P., Grettie, D. P., McFarlane, D., and Bailey, R. E.,** Plasma insulin and glucose responses of healthy subjects to varying glucose loads during three-hour oral glucose tolerance tests, *Diabetes,* 19, 842, 1970
12. **Reaven, E. P., Gold, G., and Reaven, G. M.,** Effect of age on glucose-stimulated insulin release by the β-cell of the rat, *J. Clin Invest.,* 64, 591, 1979
13. **Reaven, E., Gold, G., and Reaven, G. M.,** Effects of age on leucine-induced insulin secretion by the β-cell, *J. Gerontol.,* 34, 324, 1980
14. **Reaven, E., Curry, D., Moore, J., and Reaven, G.,** Effect of age and environmental factors on insulin release from the perfused pancreas of the rat, *J Clin. Invest ,* 71, 345, 1983.
15. **Reaven, E. P. and Reaven, G. M.,** Structure and function changes in the endocrine pancreas of aging rats with reference to the modulating effects of exercise and caloric restriction, *J. Clin Invest.,* 68, 75, 1981
16. **Reaven, E., Wright, D., Mondon, C. E., Solomon, R., Ho, H., and Reaven, G. M.,** Effect of age and diet on insulin secretion and insulin action in the rat, *Diabetes,* 32, 175, 1983.
17. **Reaven, G. M. and Reaven, E. P.,** Effects of age on various aspects of glucose and insulin metabolism, *Mol. Cell. Biochem.,* 31, 37, 1980.
18. **Hissin, P. J., Foley, J. E., Wardzala, L. J., Karnieli, E., Simpson, I. A., Salans, L. B., and Cushman, S. W.,** Mechanism of insulin-resistant glucose transport activity in the enlarged adipose cell of the aged, obese rat, *J Clin. Invest.,* 70, 780, 1982.
19. **Bjorntorp, P. and Sjostrom, L.,** Carbohydrate storage in man. speculations and some quantitative considerations, *Metabolism,* 27, 1853, 1978.
20. **Rosenthal, M., Haskell, W. L., Solomon, R., Widstrom, A., and Reaven, G. M.,** Demonstration of a relationship between level of physical training and insulin-stimulated glucose utilization in normal humans, *Diabetes,* 32, 408, 1983.
21. **DeFronzo, R. A., Soman, V., Sherwin, R. S., Hendler, R., and Felig, P.,** Insulin binding to monocytes and insulin action in human obesity, starvation, and refeeding, *J. Clin. Invest.,* 62, 204, 1978.
22. **Goodman, M. N., Dluz, S. M., McElaney, M. A., Belur, E., and Ruderman, N. B.,** Glucose uptake and insulin sensitivity in rat muscle· changes during 3-96 weeks of age, *Am J Physiol.,* 244, E93, 1983.
23. **Narimiya, M., Azhar, S., Dolkas, C. B., Mondon, C. E., Sims, C., Wright, D. W., and Reaven, G. M.,** Insulin resistance in older rats, *Am. J. Physiol ,* 246(9), E397, 1984
24. **Kimmerling, G., Javorski, W. C., and Reaven, G. M.,** Aging and insulin resistance in a group of nonobese male volunteers, *J Am. Geriatr. Soc.,* 25, 349, 1977.
25. **DeFronzo, R. A.,** Glucose intolerance and aging. Evidence of tissue insensitivity to insulin, *Diabetes,* 28, 1095, 1979
26. **Rosenthal, M., Doberne, L., Greenfield, M., Widstrom, A., and Reaven, G. M.,** Effect of age on glucose tolerance, insulin secretion, and in vivo insulin action, *J Am Geriatr. Soc ,* 30, 562, 1982.
27. **Fink, R. I., Kolterman, O. G., Griffin, J., and Olefsky, J. M.,** Mechanisms of insulin resistance in aging, *J Clin. Invest.,* 71, 1523, 1983.

28  **Rowe, J. W., Minaker, K. L., Pallotta, J. A., and Flier, J. S.,** Characterization of the insulin resistance of aging, *J. Clin. Invest.,* 71, 1581, 1983.

29. **Rossman, I.,** Anatomic and body composition changes with aging, in *Handbook of the Biology of Aging,* Finch, C. E and Hayflick, L., Eds., Van Nostrand Reinhold, New York, 1977, chap. 8.

30. **Mondon, C. E., Dolkas, C. B., and Reaven, G. M.,** Site of enhanced insulin sensitivity in exercise-trained rats at rest, *Am J. Physiol.,* 239, E169, 1980.

31. **Reaven, G. M.,** Insulin-independent diabetes mellitus: metabolic characteristics, *Metabolism,* 29, 445, 1980

32  **Reaven, G. M., Sageman, W. S., and Swenson, R. S.,** Development of insulin resistance in normal dogs following alloxan-induced insulin deficiency, *Diabetologia,* 13, 459, 1977.

33. **Savage, P. J., Bennion, L. J., Flock, E. J., Nagulesparan, M., Motte, D., Roth, J., Unger, R. H., and Bennett, P. H.,** Diet-induced improvement of abnormalities in insulin and glucagon secretion and in insulin receptor binding in diabetes mellitus, *J. Clin. Endocrinol. Metab.,* 48, 999, 1979.

Chapter 6

# OSTEOPOROSIS

**Harold H. Draper**

## TABLE OF CONTENTS

## I. INTRODUCTION

Elucidating the role of nutrition in osteoporosis has proven to be as difficult as defining its role in cardiovascular disease, and for much of the same reasons. While there is substantial evidence for an involvement of nutrition in both diseases, it is difficult to isolate the effects of nutrition from those of other factors which clearly have a concurrent influence: heredity, gender, physical activity, and hormonal status.

In the present context, the term "osteoporosis" is applied to a condition of aging osteopenia which has progressed to the point of a bone fracture or to an increased risk of fracture arising from the trauma of normal daily activities. It is most prevalent in post-menopausal women, in whom it is clearly related to decreased estrogen production, and in individuals with a small bone mass. A greater bone mass among blacks than among whites, rather than a difference in the rate of aging bone loss, evidently is the explanation of the lesser susceptibility of blacks to osteoporotic bone fractures. There are also familial differences in vulnerability which are related to inherited skeletal mass. Weight-bearing stress has a stabilizing influence on bone, which is exemplified by the osteopenia caused by physical inactivity, bed rest, and the weightless state. Obesity is associated with a lower rate of bone loss, in part because of increased weight-bearing stress and in part because of the presence, in the adipose tissue, of enzymes which are capable of synthesizing estrogens from androgenic precursors. The pathogenesis and treatment of post-menopausal osteoporosis have been discussed elsewhere.[1]

Although several nutrients have been implicated as contributory factors over the past several decades, the relationship between Ca metabolism and osteoporosis has been the central theme of research on the role of nutrition in the etiology of this disease. Research on other nutrients (vitamin D, phosphorus, protein) has centered on their influence on Ca metabolism and thereby on the Ca requirement. The question of prime interest with respect to the role of nutrition in osteoporosis remains whether susceptibility to this disease is affected by chronic Ca intake.

## II. CALCIUM

Although there are marked differences in the amount of Ca furnished by national food supplies, cross-cultural studies have failed to reveal significant differences in aging osteopenia. Osteoporotic bone disease appears to be no more prevalent in cereal-based food economies where the diet provides 400 to 500 mg of Ca per day than in Western cultures where the consumption of dairy foods raises the per capita Ca supply to twice this amount. Whereas a daily Ca intake of 450 to 500 mg has been cited as contributory to osteoporosis in post-menopausal women in the U.S.,[2] it is the amount recommended for adult females by the WHO/FAO. There is a clear tendency for recommended Ca intakes to reflect the supply available.

Measurements of metacarpal bone loss in adults of three countries differing markedly in Ca intake (Guatemala, El Salvador, and the U.S.) have failed to reveal any differences in aging osteopenia.[3] It was therefore concluded that aging bone loss is universal and independent of diet. However, reliable information on age-related bone loss in many populations, based on the use of modern techniques of assessing bone mineral such as dual photon absorptiometry or neutron activation analysis, is still not available.[4]

The validity of cross-cultural comparisons of calcium intake alone has also been brought into question by the demonstration that other dietary factors, as well as differences in lifestyle and sunlight exposure, can influence the Ca requirement. An intake of Ca which may be adequate to maintain bone homeostasis in adults consuming a neutral to alkaline cereal-based diet low in protein and available P may not be adequate for adults consuming a high-P, high protein, acid-producing mixed diet.

An unusually rapid rate of bone loss has been recorded in Alaskan and Canadian Eskimo adults.[5,6] In the national nutrition survey conducted in Canada in the 1960s, the Ca intake of Eskimos was found to be markedly lower than that of the national population.[7] There was also evidence of a higher serum P and lower serum Ca level, suggestive of parathyroid stimulation and increased bone resorption. Acculturation has led to an Arctic Eskimo diet which is unusually low in Ca and high in protein and P, a combination of the main nutritional factors which have been implicated in bone loss. Historically, the Ca intake of Eskimos was maintained, probably at a high level, by consumption of the trabecular bone of land and sea mammals and the soft bones of fish. The abandonment of bone chewing among modern Eskimos and a limited consumption of imported dairy foods has resulted in a low intake of Ca. However, the intakes of protein and P frequently remain high because of the persistence of the hunting culture among many older adults. Whether there is an unusual rate of osteoporotic bone fractures among Eskimos cannot be determined from the available health statistics.

Evidence for an association between chronic Ca intake and osteoporosis has been strengthened by two recent epidemiological studies. In one, metacarpal cortical area was assessed by radiography in adults living in two communities of Yugoslavia differing markedly in Ca intake (viz., approximately 400 to 500 mg vs. 800 to 900 mg/day).[8] Significant differences in metacarpal bone area and in incidence of bone fractures, especially among women, were found between the two regions. It is noteworthy that evidence for a difference in metacarpal area already was in evidence at 30 years, the earliest age at which measurements were recorded, indicating that Ca intake during growth and/or early adulthood had influenced mature skeletal mass, a critical factor in predisposition to osteoporotic bone fractures in later life.

The second study involved a longitudinal assessment of bone mass and diet in a free-living population of women living in the U.S.[9] A significant relationship was found between Ca intake and retention, from which it was estimated that 1500 mg of Ca per day was required to prevent post-menopausal bone loss in this cohort. If this is a true indication of the amount of dietary Ca required to prevent aging osteopenia, it appears likely that differences in Ca intake within the normal range of 500 to 1000 mg/day have a significant influence on bone loss and risk of bone fractures in the general population of post-menopausal women.

Full skeletal development occurs at an older age (30 to 35 years) than previously believed. Whether the vitamin D- and parathyroid hormone-dependent adaptive mechanism of Ca absorption is capable of fully compensating, in terms of maximizing skeletal development, for low Ca intakes during growth and early adulthood is unknown. The proposition that the recommended adult intake of vitamin D should be increased as a means of enhancing the amount of Ca absorbed[10] implies that the same result could be obtained by increasing dietary Ca, which at high intakes is less dependent on vitamin D for its absorption.

Ca supplements taken prophylactically significantly reduce the bone loss which follows menopause or oophorectomy. A supplement of 1000 mg of Ca per day is nearly as effective as estrogen in redressing the imbalance between bone formation and bone resorption which follows menopause.[11] The U.S. and Canadian recommended Ca intakes of 800 mg/day are clearly inadequate to prevent bone loss in the estrogen-deficient state, and most post-menopausal women fail to achieve even this level of intake. A total intake of 1000 to 1500 mg appears to be necessary for maximal suppression of bone loss in the post-menopausal period.[10-14]

The effectiveness of Ca supplementation presumably is attributable to maintenance of serum Ca at a level which suppresses the synthesis of parathyroid hormone. The consequent reduction in parathyroid hormone-induced bone resorption compensates for the decreased bone formation associated with the post-menopausal decline in estrogen production. Like most other therapeutic measures, Ca administration has not been found to be effective in the reversal of bone loss, although it is beneficial in the prevention of further fractures in established osteoporosis.[15]

The fact that 1000 to 1500 mg of Ca per day is not readily obtainable from the diet raises a question concerning the prophylactic use of Ca supplements for the prevention of post-menopausal osteoporosis. Although this affliction represents the most prevalent bone disease of older women, it nevertheless occurs in a minority of females, and therefore, prescription of Ca supplements for the entire post-menopausal population is unjustified. While an increase in the current U.S. RDA of 800 mg Ca has been proposed, it is unlikely that such a recommendation would lead to much increase in Ca consumption, which is already high by international standards. It may be feasible to screen women presenting at the onset of menopause for factors predisposing to osteoporosis — small bone structure, lack of body fat, sedentary lifestyle, low Ca intake — as a basis for prescribing Ca prophylaxis.

## III. VITAMIN D

There have been several recent reports of low vitamin D status, assessed on the basis of serum 25-hydroxy vitamin D levels, among middle-aged and elderly women. Hypovitaminosis D appears to be due mainly to insufficient exposure to solar radiation for adequate vitamin D generation in the skin, which is normally the main source of the vitamin in the body.[16] Confinement indoors, envelopment in clothing for cultural or climatic reasons, and a low intake of vitamin D-fortified foods are predisposing factors. For example, the serum level of 25-hydroxy vitamin D in Britain is lower than it is in the U.S. and Canada, where solar radiation is greater and more foods are fortified.[17]

It has been suggested that the decrease in absorption of dietary Ca which occurs with advancing age may be due to a declining vitamin D status. There is some evidence of impaired responsiveness to vitamin D in the elderly in terms of reduced efficiency of conversion of the vitamin to its hydroxylated metabolites. In addition to a tendency among older adults to have low serum levels of 25-hydroxy vitamin D (an indicator of vitamin D status), some elderly persons may have an impaired capacity to synthesize the active metabolite 1,25-dihydroxy vitamin D (calcitriol).[18] To enhance the efficiency of Ca absorption, it has been proposed that the U.S. RDA for vitamin D be raised from 400 to 600 or 800 IU.[10]

If vitamin D is a significant factor in aging osteopenia, a difference in bone loss should be discernible between women living in climates differing markedly in solar radiation. The possible risk of hypervitaminosis D in individuals with a high intake of vitamin D-fortified foods combined with a high exposure to sunlight is a consideration in further fortification. If the desired effect of vitamin D is limited to increased Ca absorption, Ca supplementation may be a more selective and safer method of increasing the amount of Ca absorbed. Vitamin $D_3$ and 1-α-hydroxy vitamin $D_3$ (which is converted to calcitriol in the body), as well as calcitriol itself, have been reported to have no prophylactic value as alternatives to estrogen replacement in the prevention of bone loss in post-menopausal women.[19,20] Other investigators have observed an increase in Ca absorption and Ca balance in osteoporotic women given calcitriol supplements, but patients must be closely monitored for hypercalcemia.[1]

## IV. PHOSPHORUS

The food supply of some industrialized countries has a high P content because of the association of this element with animal protein and the widespread use of phosphate food additives. Additives contribute an estimated 300 mg of P per day to the average per capita diet in the U.S.[21] For significant numbers of adults, additives furnish 500 mg/day, or about one third of the total P intake.[22] Diets high in P, unless they are also high in Ca, have been found to increase bone loss in several species of adult animals. Although this effect frequently has been attributed to a decrease in Ca absorption, it is now apparent that (except in the case of some plant phosphates such as phytin and cellulose phosphates) the interaction between phosphate and Ca occurs in the blood and not in the intestine. At high serum phosphate levels, there is an increased secretion of Ca into the gut, an effect which is evident from the results of radiocalcium studies on adult animals and recently has been confirmed in adult humans. The resulting depression in serum Ca stimulates parathryoid hormone synthesis, which leads to increased bone resorption. At a moderate excess of dietary P there is also an increase in bone formation and bone homeostasis is maintained, but a large excess, at least in animals, results in bone loss.

Experiments on human adults indicate that they may be more tolerant of phosphate loads than adult animals. Even at a modest daily intake of Ca (500 mg), Ca balance has been found to be maintained at P intakes as high as 2525 mg/day. Increases in urinary hydroxy-proline and cyclic AMP (cAMP) indicate that, as in animals, high-phosphate diets stimulate parathyroid activity and bone resorption in man, but the observation that Ca balance is maintained over a broad range of P intakes indicates that the coupling of bone formation and bone resorption in humans is more resistant to the stress of excess dietary phosphate. The comparative response of animals and man to factors affecting Ca needs is discussed in a recent review.[23]

Much attention has been given to the dietary Ca:P ratio and its significance for bone homeostasis in man. This interest stems from the fact that bone growth and homeostasis in animals is well known to be adversely affected by either an inadequate Ca intake or an excess of P relative to Ca. Consequently, in the formulation of diets for both growing and adult animals, the Ca:P ratio is maintained within established limits.

It is difficult to define an optimal Ca:P ratio in the human diet because of wide variability in the intake of both elements. Any such ratio depends upon the absolute intake of P and, to a lesser extent, of Ca. This is due to the fact that the efficiency of Ca absorption decreases markedly at high intakes, whereas the efficiency of P absorption remains essentially constant (about 70%) even at very high intakes. Consequently, the ratio of absorbed Ca to absorbed P (the important relationship in any effect on bone) shifts in favor of P as the intake of both elements is increased proportionately. In the absence of information about the absolute intake of both elements, the Ca:P ratio in the human diet therefore is of limited value as an indicator of the adequacy of the diet with respect to these nutrients.[22]

## V. PROTEIN

The calciuretic action of high-protein diets has been the subject of extensive investigation as a possible factor in bone loss in Western societies. The increase in urinary Ca associated with a high-protein intake, once assumed to be due to an increase in Ca absorption, has been related to increased acid production arising from the oxidation of excess sulfur amino acids. Acid excretion is associated with a decrease in reabsorption of Ca from the renal tubules and an increase in glomerular filtration rate. Adult subjects fed high intakes of purified proteins exhibit a persistent hypercalciuria and a negative Ca balance which can be corrected only by increasing Ca intake.[24]

Recent studies indicate, however, that the effect of protein supplements on Ca excretion differs from that of high-protein diets as ordinarily consumed. A high-meat diet reportedly produces neither an increase in urinary Ca nor a negative Ca balance.[25] The explanation of the difference in response to purified proteins and to high-protein diets evidently lies in the increase in P intake which normally accompanies a high intake of protein. The parathyroid stimulation produced by excess dietary phosphate results in an increase in reabsorption of Ca from the renal tubules, thereby counteracting the decrease in Ca reabsorption caused by excretion of excess acid produced in protein metabolism. High-protein, high-P diets are associated with increases in urinary cAMP and hydroxyproline, indicative of increased bone resorption, but Ca balance data indicate that at least to a large degree, this increase in resorption is offset by an increase in bone formation (i.e., that such diets result in increased bone turnover with little or no net loss of bone). It is not clear why parathyroid stimulation caused by inadequate Ca or vitamin D should cause bone loss (i.e., uncoupling of bone formation and bone resorption) if parathyroid stimulation caused by excess phosphate ("secondary hyperparathyroidism") does not.

These studies have introduced a novel consideration into the estimation of the P requirement, namely, that at high protein intakes more P is required to stimulate parathyroid-dependent renal Ca reabsorption as a means of counteracting the hypercalciuria caused by increased acid excretion. Whether the stimulus to bone resorption caused by excess P and the stimulus to Ca excretion caused by excess protein are always commensurate with Ca balance on high-protein, high-P diets is unknown. The effect of dietary protein on Ca metabolism recently has been reviewed.[26]

## VI. FLUORIDE

Although fluoride has been used therapeutically in the treatment of osteoporosis, usually in combination with estrogen or Ca, there is little evidence that fluoridation of drinking water for the prevention of tooth decay or that the fluoride levels normally present in the diet have any prophylactic effect on aging bone loss. It is possible, however, that in localities where the drinking water contains an unusually high level of natural fluoride, this element may be beneficial in the prevention of osteoporosis.[27]

## VII. CONSPECTUS

From the foregoing review it is apparent that much of the research on nutrition and osteoporosis has been concerned with defining the Ca requirement of adults and the factors which affect it. This emphasis lends substance to the statement that "few issues in modern nutrition have been as genuinely controversial and vexing as the intake requirement of calcium in the mature adult".[28]

There is stronger evidence now than at any previous time that the Ca intake of adults in some populations has a significant influence on aging osteopenia and the risk of osteoporotic bone fractures. Yet the amount of Ca required to effectively inhibit aging bone loss apparently exceeds the Ca content of even Ca-rich food supplies, and far exceeds that of the diet of much of the adult population of the world. Why there is not clearer evidence of cross-cultural differences in bone loss, relatable to known differences in Ca intake, is unknown.

Undoubtedly, in concert with other degenerative processes such as senile muscle dystrophy, some degree of senile osteopenia is an inevitable accompaniment of aging. Although a majority of adults accumulate sufficient bone by the age of skeletal maturity to withstand this loss without adverse consequences, risk of fractures among women nevertheless increases markedly in the 7th and 8th decades. The proposition that this risk could be virtually eliminated by an adequate Ca intake (i.e., 1500 mg/day or more) leads to the further proposition that most of the global population of older adults is deficient in Ca.

Anthropological evidence demonstrates that many different food cultures have successfully sustained reproduction and survival for thousands of years. Does the fact that osteoporosis occurs at a post-reproductive age indicate that there has been no selection pressure favoring a Ca requirement that is commensurate with the Ca supply available? What will be the consequences for bone health of any further prolongation of the life span? Although Ca prophylaxis appears to be the most promising intervention strategy from the overall standpoint of efficacy, economy, and safety, there is an obvious need for additional epidemiological research to define more clearly the general relationship between Ca intake and osteoporosis and the reasons for the apparent differences in this relationship between contemporary food cultures.

# REFERENCES

1. **Milhaud, G., Christiansen, C., Gallagher, C., Reeve, J., Seeman, E., Chesnut, C., and Parfitt, A.,** Pathogenesis and treatment of postmenopausal osteoporosis, *Calcif. Tissue Int.,* 35, 708, 1983.
2. **Avioli, L. V.,** Postmenopausal osteoporosis: prevention versus cure, *Fed. Proc. Fed Am Soc Exp Biol.,* 40, 2418, 1981.
3. **Garn, S. M., Rohmann, C. G., and Wagner, B.,** Bone loss as a general phenomenon in man, *Fed. Proc. Fed. Am. Soc Exp. Biol.,* 26, 1729, 1967
4. **Mazess, R. B.,** On aging bone loss, *Clin. Orthopaed. Rel. Res.,* 165, 239, 1982.
5. **Mazess, R. B. and Mathur, W.,** Bone mineral content of northern Alaskan Eskimos, *Am. J. Clin. Nutr.,* 27, 916, 1974.
6. **Mazess, R. B. and Mathur, W.,** Bone mineral content of Canadian Eskimos, *Human Biol.,* 47, 45, 1975
7. Health and Welfare Canada, *Nutrition Canada, The Eskimo Survey Report,* Ottawa, 1975.
8. **Matkovic, V., Kostial, K., Simonovic, I., Buzina, R., Brodarec, A., and Nordin, B. E. C.,** Bone status and fracture rates in two regions of Yugoslavia, *Am. J. Clin Nutr.,* 32, 540, 1979.
9. **Heaney, R. P., Recker, R. R., and Saville, P. D.,** Menopausal changes in calcium balance performance, *J. Lab. Clin. Med.,* 92, 953, 1978.
10. **Parfitt, A. M., Gallagher, J. C., Heaney, R. P., Johnston, C. C., Neer, R., and Whedon, G. D.,** Vitamin D and bone health in the elderly, *Am. J. Clin. Nutr ,* 36, 1014, 1982.
11. **Recker, R. R., Saville, P. D., and Heaney, R. P.,** Effect of estrogens and calcium carbonate on bone loss in postmenopausal women, *Ann. Intern. Med.,* 87, 649, 1977.
12. **Heaney, R. P., Recker, R. R., and Saville, P. D.,** Calcium balance and calcium requirements in middle-aged women, *Am. J. Clin. Nutr.,* 30, 1603, 1977.
13. **Albanese, A. A., Lorenze, E. J., Edelson, A. H., Wein, E. H., and Carroll, L.,** Effects of calcium supplements and estrogen replacement therapy on bone loss of postmenopausal women, *Nutr. Rep. Int.,* 24, 403, 1981.
14. **Horsman, A., Gallagher, J. C., Simpson, M., and Nordin, B. E. C.,** Prospective trial of oestrogen and calcium in postmenopausal women, *Br Med. J.,* 2, 789, 1977.
15. **Riggs, B. C., Seeman, E., Hodgson, S. F., Taves, D. R., and O'Fallon, W. M.,** Effect of the fluoride/calcium regimen on vertebral fracture occurrence in postmenopausal women, *N. Engl. J. Med ,* 306, 446, 1982.
16. **Lawson, D. E. M., Paul, A. A., Black, A. E., Cole, T. J., Mandal, A. R., and Davie, M.,** Relative contributions of diet and sunlight to vitamin D state in the elderly, *Br. Med. J.,* 2, 303, 1979.
17 **Poskitt, E. M. E., Cole, T. J., and Lawson, D. E. M.,** Diet, sunlight, and 25-hydroxyvitamin D in healthy children and adults, *Br. Med J.,* 1, 221, 1979.
18. **Heaney, R. P., Gallagher, J. C., Johnston, C. C., Near, R., Parfitt, A. M., and Whedon, G. D.,** Calcium nutrition and bone health in the elderly, *Am. J Clin. Nutr.,* 36, 986, 1982.
19. **Christiansen, C., Mazess, R. B., Transbol, I., and Jensen, G. F.,** Factors in response to treatment of early postmenopausal bone loss, *Calcif. Tissue Int.,* 33, 575, 1981.
20. **Christiansen, C., Christensen, M. S., Rodboro, P., Hagen, C., and Transbol, I.,** Effect of 1,25-dihydroxy-vitamin $D_3$ in itself or combined with hormonal treatment in preventing postmenopausal osteoporosis, *Eur. J. Clin. Invest.,* 11, 305, 1981.
21. Federation of American Societies for Experimental Biology, Life Sciences Research Office, *Effects of Dietary Factors on Skeletal Integrity in Adults. Calcium, Phosphorus, Vitamin D and Protein,* Chinn, A. I., Ed., Bethesda, Md , 1981
22. **Draper, H. H. and Scythes, C. A.,** Calcium, phosphorus and osteoporosis, *Fed. Proc. Fed. Am. Soc. Exp. Biol.,* 40, 2434, 1981.
23. **Draper, H. H.,** Similarities and differences in the response of animals and man to factors affecting calcium needs, in *Calcium in Biological Systems,* Rubin, R. P., Weiss, G., and Putney, J. W., Jr., Eds., Plenum Press, New York, in press.
24. **Linkswiler, H. M., Zemel, M. B., Hegsted, M., and Schuette, S. A.,** Protein induced hypercalciuria, *Fed. Proc. Fed. Am. Soc. Exp. Biol.,* 40, 2429, 1981.

25. **Spencer, H., Kramer, L., Debartolo, M., Norris, C., and Osis, D.,** Further studies on the effect of high protein diet as meat on calcium metabolism, *Am. J. Clin. Nutr.,* 37, 924, 1983.
26. **Yuen, D. E., Draper, H. H., and Trilok, G.,** The effect of dietary protein on calcium metabolism in man, *Nutr Abstr Rev ,* 54, 447, 1984.
27 **Bernstein, D., Sadowski, N., Hegsted, D. M., Guir, D., and Stare, F. J.,** Prevalence of osteoporosis in high- and low-fluoride areas in North Dakota, *JAMA,* 198, 499, 1966.
28. **Heaney, R. P.,** Calcium intake requirement and bone mass in the elderly, *J. Lab Clin. Med.,* 100, 309, 1982.

Chapter 7

# PERIODONTAL DISEASES

**Olav F. Alvares**

## TABLE OF CONTENTS

## I. INTRODUCTION

Significant changes in the age structure of the U.S. population have occurred and will continue to take place in the next 2 decades. The median age of the American population increased from 28 years in 1970 to nearly 30 years in 1980, and it is projected to reach 35 years by the year 2000.[1][2] The number of individuals aged 45 to 64 years, which currently stands at approximately 45 million, is expected to climb to 59 million and constitute over 20% of the population by the turn of the century. Along with the increase in size of the somewhat older population, more individuals are maintaining their natural dentition for a longer period of time, primarily because of better control of dental caries. However, these teeth are vulnerable to a destruction of their supporting structures which can occur in a variety of ways, collectively referred to as "periodontal diseases". Periodontal diseases are endemic in the U.S. population. These diseases are a major cause of tooth loss after the age of 35 years, with gingivitis and periodontitis affecting 75% of the adults.[3] The pain and distress associated with this disease process, the expenditures (estimated in millions of dollars) incurred, and the time lost by the labor force in the course of its treatment and management have drawn attention to the tremendous socioeconomic impact of periodontal diseases. These diseases will continue to be a public health problem, particularly for the older individuals in our population.[4]

## II. PERIODONTAL DISEASES

The most common type of periodontal disease is referred to as "adult periodontitis". This condition usually begins as a gingivitis which can persist as a chronic disorder and is reversible if treated. If left untreated gingivitis can, but not necessarily, go on to involve the periodontium (cementum, periodontal ligament, and alveolar bone) of teeth. Until recently, periodontitis was regarded as a continually progressing destructive process involving the creation of periodontal pockets, loss of attachment of the tooth to supporting structures, and destruction of alveolar bone — processes eventually culminating in tooth loss. Recent experimental and clinical studies have altered this perception of the behavior of chronic adult periodontitis. It is now evident that despite significant accumulations of dental plaque and its attendant gingivitis, the destructive consequences may vary greatly between individuals and between sites in the same mouth.[5] Clinical studies have led to the proposal of models according to which the pattern of progress of human periodontal disease is characterized by "bursts", rather than a continuous process, of active disease of comparatively short duration followed by varying periods of quiescence.[6] Further, these "bursts" of destructive disease may occur asynchronously in different sites in the mouth. The circumstances that lead to the conversion of an active lesion to a quiescent stage and vice versa are not fully understood and continue to intrigue investigators in this field.

Perhaps the most significant advance in our understanding of periodontal diseases is with respect to its etiology. Human and animal studies, dating back to the 1950s, have clearly shown that periodontal disease is an infectious process.[5,7] The putative microbial agents reside in the deposits on the tooth surfaces referred to as "dental plaque". The microbial ecology of dental plaque situated above the gingival margin (supragingival) differs greatly from that below (subgingival) the gingival margin. Furthermore, in the same individual, there are differences in the microbial population of plaque from one tooth site to another. The significance of these differences are not fully understood. The majority of flora in dental plaque associated with clinically healthy gingival tissues are Gram positive aerobic cocci and rods, particularly *Streptococcus sanguis* and species of *Actinomyces*.[7] In contrast, the microorganisms in adult chronic periodontitis are predominantly asaccharolytic, Gram negative, anaerobic, and motile.[7] In particular, spirochetes and species of *Bacteroides, Acti-*

*nomyces,* and *Eikenella corrodens* are present in high numbers. Yet another form of periodontal disease, localized juvenile periodontitis, is associated with a predominantly different set of Gram negative microorganisms, i.e., *Actinobacillus actinomyctemcomitans* and *Capnocytophaga* species.[7] The implication of these bacteriological findings is that they open the way for the immunodiagnosis of the different forms of periodontal diseases.

The cellular events associated with chronic periodontitis that occur in response to the bacterial provocation have been classified into the "initial", "early", "established", and "advanced" lesions.[7] These cellular events appear to result from bacterial action, and indirectly, as a result of host responses to bacteria. Generally speaking, our immune system mounts an appropriate response to meet the challenge of invading microorganisms. The intriguing aspect of periodontal diseases is that the components of the immune system may be indirectly responsible for the tissue destruction. The complex interaction between periodontal pathogens and the cells and components of the immune system has held the interest of several laboratories. To date, the fruits of these research endeavors has drawn attention to the important role of the polymorphonuclear leukocyte as a significant host defense factor in the maintenance of periodontal health.[7] It is generally accepted that individuals with impaired neutrophil function (chemotaxis, phagocytosis, etc.) have a high prevalence of bacterial infections. In the present context, individuals with agranulocytosis, cyclic neutropenia, Chediak-Higashi syndrome, and insulin-dependent diabetes mellitus exhibit a defect in neutrophil function and an increased severity of periodontal disease. Further evidence in support of the protective role of neutrophils is derived from studies of individuals with localized juvenile periodontitis. The neutrophils in these individuals exhibit a defect in directed locomotion but not random migration.[8-10] This defect in movement is believed to be related to reduced cell surface chemotactic receptor binding sites.[10] Over the past 2 decades, much has been learned about the possible contributions of the other components of the inflammatory immune system in the tissue destruction that characterizes periodontal diseases. Polyclonal activation of B lymphocytes can lead to the production of a number of biologically potent molecules.[11] Certain lymphokines, such as lymphotoxin, can cause death of neighboring cells while osteoclast activating factor can lead to bone destruction. This bone destruction may also be facilitated by inflammatory mediators such as prostaglandin ($PGE_2$) synthesized by activated macrophages. Macrophage-derived enzymes (collagenase) and oxidizing agents (hydrogen peroxide) are also believed to participate in the degradation of collagen, basement membranes, and other connective tissue components of the periodontium. Nevertheless, owing to the present gaps in our knowledge, it is not possible to present a cohesive picture which would adequately resolve the question of the protective and/or destructive role of the immune system in the observed tissue destruction. In this regard, the reader may wish to refer to some recent texts[5,7] for a more in-depth discussion of possible pathways or mechanisms of tissue destruction that occur in periodontal diseases.

## III. NUTRITION AND PERIODONTAL DISEASES

In 1966, the World Workshop in Periodontics[12] addressed various issues pertaining to the etiology, pathogenesis, and treatment of plaque-associated periodontal disease. With respect to the role of nutrition in periodontal disease, the consensus was that while dental plaque is the major etiologic factor of inflammatory periodontal disease, inadequate nutrition may render the host susceptible to periodontal disease or modulate the progress of an existing condition.[12] Basically the same position was reiterated at the International Conference on Research in the Biology of Periodontal Disease[13] some 11 years later. Meanwhile, diametrically opposing points of view and conflicting evidence from poorly controlled studies enveloped the role of nutrition in periodontal disease with controversy.

That inadequate nutrition is invariably accompanied by an increase in the incidence or gravity of various illnesses can be traced back to antiquity, and one is led by the results of numerous studies to conclude that the ensuing undernourishment and malnutrition tend to increase the susceptibility of the organism to a number of different infections.[14,15] Invariably, this increased susceptibility to infection has been due to a compromise in host defense factors. Given the infectious nature of plaque-associated periodontal disease, it would seem reasonable to assume that inadequate nutrition could also influence periodontal health via a compromise in host defense. During this past decade efforts have been made to better understand the trinity of nutrition-host defense-periodontal disease. The discussion that now follows will deal with (1) comments on host defense factors that are of relevance to periodontal health followed by (2) animal studies, and (3) human investigations concerned with nutrition and periodontal health.

## A. Host Defense Factors Pertinent to Periodontal Health

The host defense factors that have an important bearing on periodontal health are (1) qualitative and quantitative characteristics of saliva and functional capacity of the salivary glands, (2) gingival fluid production, (3) repair, (4) the endocrine balance, (5) the integrity of the inflammatory-immune response, and (6) optimal function of oral epithelium. Adequate reviews[16-18] preclude the need for yet another in-depth discussion of all of these factors. The host defense factor that has been most extensively studied vis-a-vis nutrition and periodontal disease is the protective influence exerted by oral epithelia. A brief overview of this particular aspect of host defense as it pertains to the present topic is presented.

Oral epithelia are able to play a protective role, in part, by virtue of their multilayer topography and their capacity to replace cells rapidly. The generation of new cells in the progenitor compartment affords a means for compensating loss of cells at the surface due to various insults, e.g., friction and antigenic challenges (a situation that is ever-present in the gingival sulcus). This enables the oral epithelium to maintain a constant thickness in the steady state. The gingival sulcular and junctional epithelia have one of the fastest turnover rates. In nonhuman primates, the turnover time for junctional epithelium has been calculated to range from 4.6 to 10.9 days.[19] To maintain a rapid turnover time, this tissue probably exhibits a high rate of DNA, RNA, and protein synthesis. These processes undoubtedly require a continuous and steady supply of nutrients. To underscore the importance of an adequate supply of nutrients to the junctional and sulcular epithelia, it has been suggested that these tissues are in a "continuous critical period" analogous to the critical periods which occur during growth and development.[20,21] Further, the rapid turnover time coupled with the continuously ongoing repair of these epithelia may dictate a higher nutrient requirement. It is conceivable that if this potentially higher requirement is not met, the sulcular and junctional epithelia may exhibit an "end-organ deficiency" even in the face of adequate nutrients at the systemic level.

An event of considerable significance that occurs during the maturation or differentiation of oral epithelia cells is the development of the intraepithelial barrier. This barrier occupies the intercellular spaces in the superficial layers of keratinized and nonkeratinized oral epithelia and has been shown to impede the passage of tracer molecules, some of which are similar in size to dental plaque bacterial antigens.[22] The intraepithelial barrier is not the only barrier to molecules. In vivo and in vitro studies have shown that the interface between the epithelium and connective tissue acts as a rate-limiting barrier to several tracer molecules, including bacterial endotoxin.[23,24] In all likelihood, the barrier function of oral epithelium, and by implication its permeability, is an expression of the cumulative integrity of the intraepithelial barrier, basement membrane, and the plasma membranes of epithelial cells. Such a barrier serves the vital purpose of limiting the passage of antigenic material from the surface into the underlying connective tissue.

## B. Animal Studies

Protein-calorie malnutrition is the most commonly encountered deficiency state on a worldwide basis. In experimental animals, protein-calorie malnutrition has been linked to osteoporosis of alveolar bone, thinning of the periodontal ligament, degeneration of periodontal collagen fibers, retardation in deposition of cementum, and delayed gingival wound healing.[25-32] Cytological studies have shown decreased stainability, perinuclear vacuolization, and increased keratinization of buccal mucosa in protein-calorie-deficient pigs and dogs.[33] In addition, gingival and/or periodontal pathosis have been observed in animals fed diets deficient in one of the following nutrients: niacin, folic acid, vitamin A, ascorbic acid, P, or Ca.[34-41] Dreizen and co-workers[34-36] were probably the first to vaguely suggest that the periodontal and gingival changes may have been mediated via a compromise in the host defense function of oral epithelia. Their light microscopic observations, however, did not provide quantitative data. In the 1970s, Alfano and co-workers[42] undertook several studies designed to quantitate the effects of nutritional deficiencies on the barrier function of oral epithelium. These investigators employed an in vitro system to measure epithelial permeability. Employing this technique, Alfano and co-workers[43] showed that the lingual epithelium of guinea pigs exhibits significantly greater permeability when kept on an ascorbate-deficient diet for a period of 2 weeks or greater (Figure 1). Furthermore, tissue permeability was inversely related to the oral tissue level of ascorbate and this level dropped sooner and faster than the blood ascorbate level. However, the difference in permeability between the pair-fed and experimental group was not statistically significant. Consequently, much of the increase in permeability was attributed to the inanition that accompanies ascorbate deficiency. When the deficient animals were rehabilitated for 2 weeks after being on the ascorbate-deficient diet for 3 weeks, the penetration coefficients (i.e., permeability) declined slightly, thereby revealing that the effects of ascorbate deficiency on barrier function are not rapidly reversible (Figure 1). Somewhat similar findings are seen in folic acid-deficient developing guinea pigs.[44] After 3 weeks on a folate-deficient diet, the oral mucosal tissues of both pair-fed and experimental animals demonstrate significantly greater permeability as compared to ad libitum controls (Figure 2). Of even greater interest in this study was the finding of a significant decrease in oral epithelial permeability between baseline and week 1. This is consistent with the work of Sabin,[45] and more recently that of Sperber,[46] whose data suggest a "closing down" or "maturation" of the barrier of a variety of epithelial tissues with time. A similar situation is seen in the gut epithelium of the neonate which ceases to absorb large intact molecules such as maternal antibodies. More importantly, Sperber's[46] study showed that neonatal Fe deficiency prevented the "maturation" process in barrier function from taking place. The implication of these studies is that nutritional deficiencies during the developmental period may not only increase the duration of vulnerability of oral mucosa to antigen penetration, but may also influence the subsequent immunological response of the host.[44] Finally, Joseph et al.[47] have recently shown that the periodontal tissues of Zn-deficient rabbits exhibit an increase in uptake of $^{14}C$-phenytoin and $^{14}C$-bovine serum albumin.

The quantitative data generated by the aforementioned studies substantiate the qualitative findings seen in numerous studies which suggested that the barrier function of mucosal epithelium may be compromised in malnourished animals.[48-50] Further, it is apparent from this discussion that the growing body of evidence indicates that the barrier function of oral epithelium can be modulated by a variety of nutrients, nutrients that have an important role to play in several fundamental cellular processes in periodontal tissues. The next logical step was the demonstration of the existence of a link between a compromise in host defense and increased susceptibility to periodontal disease in the circumstance of inadequate nutrition.

Alvares and co-workers assessed the effects of clinical and marginal ascorbic acid deficiency on host defense factors relevant to periodontal health. In the first study,[51] young adult nonhuman primates *(Macaca fascicularis)* were initially fed an ascorbic acid-free diet and, subsequently, a diet with a suboptimal level of the vitamin. Following approximately 12 weeks of this dietary regime, 5 of the 7 experimental animals developed scorbutic gingivitis.

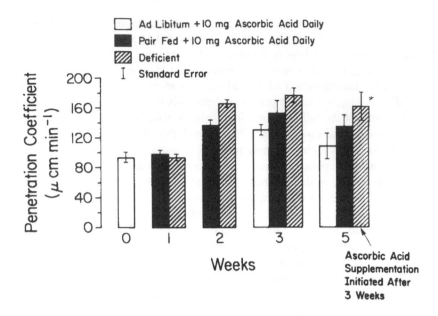

FIGURE 1    Penetration coefficients for oral mucosa obtained from guinea pigs subsisting on an ascorbic acid-deficient diet or the identical ascorbic acid-deficient diet supplemented with 10.0 mg L-ascorbic acid daily (pair-fed and ad libitum-fed control groups) At week 3, the deficient group was rehabilitated by supplementation with 10.0 mg L-ascorbic acid daily for 2 weeks. Tritiated bacteria endotoxin was employed as a marker. (Courtesy of Dr. M. Alfano.)

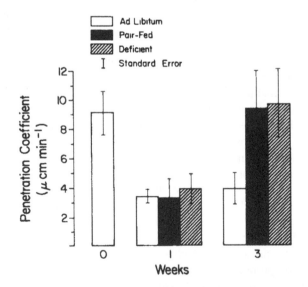

FIGURE 2.    Penetration coefficients for oral mucosa obtained from developing guinea pigs subsisting on a folic acid-deficient or adequate (ad libitum and pair-fed group) diet. $^{14}$C-Dextran (mol wt, 70,000) was employed as a marker. There is a significant decrease in permeability between weeks 0 and 1, a time period which corresponds to puberty in these animals. (Courtesy of Dr M Alfano.)

Between 2 and 3 weeks prior to this stage, the in vivo permeability of the gingival sulcular epithelium to $^{3}$H-inulin, but not to $^{14}$C-butyric acid or $^{14}$C-urea, had increased significantly (85%, $p < 0.01$) in the experimental group as compared to ad libitum or pair-fed controls

**Table 1**
**PERMEABILITY COEFFICIENTS OF**
**GINGIVAL SULCULAR**
**EPITHELIUM IN ASCORBIC ACID**
**DEFICIENCY**[a]

|  | Baseline | Week 9 |
|---|---|---|
| Ad libitum |  |  |
| Urea | — | — |
| Butyric acid[b] | 15 3 ± 1 10 | 13.8 ± 1 45 |
| Inulin[b] | 0.39 ± 0 04 | 0 35 ± 0.05 |
| Pair-fed |  |  |
| Urea[c] | 1 24 ± 0.12 | 1 09 ± 0.12 |
| Butyric acid[d] | 14 9 ± 1.66 | 14.6 ± 1.40 |
| Inulin[d] | 0 37 ± 0 06 | 0 34 ± 0.04 |
| Experimental |  |  |
| Urea[b] | 1.19 ± 0.11 | 1.42 ± 0 15 |
| Butyric acid[d] | 14 4 ± 1.36 | 17.3 ± 1 25 |
| Inulin[d] | 0.36 ± 0 05 | 0.67 ± 0 05 |

[a]  Values are expressed as cm $sec^{-1} \times 10^5$ (x ± SEM)
[b]  n = 6
[c]  n = 4.
[d]  n = 7

(Table 1). The finding with inulin was consistent with the previously reported increase in the in vitro permeability of oral epithelium in scorbutic guinea pigs.[43] Inulin has been traditionally used as a marker for the intercellular route of penetration of molecules. Thus, it appears that the degree of ascorbic acid deficiency achieved in this study chiefly affected this particular route of penetration in oral epithelia. The mechanism whereby ascorbic acid deficiency leads to an increase in permeability is not clear. The nature of the intraepithelial barrier could have been altered. Ascorbic acid is also known to influence the synthesis and structure of collagen.[52] A lack of continuity in the corneal epithelium basal lamina and impairment of oral epithelial basement membrane collagen synthesis have been observed in scorbutic guinea pigs.[43,53] These alterations could have resulted in the formation of basement membrane that permitted increased passage of large molecular weight compounds such as inulin. A subsequent study showed that monkeys maintained on an ascorbic acid-free diet for 9 weeks also exhibited a decrease in polymorphonuclear leukocyte (PMN) phagocytic activity.[54] These defects in host defense factors, namely oral epithelial barrier and PMN function, preceded the onset of clinical disease and could have played a role in the pathogenesis of scorbutic gingivitis.

In a second study, the effects of long-term subclinical ascorbic acid deficiency on periodontal health were determined.[54] Young adult *M. fascicularis* monkeys were fed suboptimal levels of ascorbic acid-free diet for approximately 6 months. Despite significantly reduced blood ascorbate levels, the experimental animals did not develop clinical signs of ascorbate deficiency. Further, there was little evidence of spontaneous gingivitis or periodontitis in either experimental or control groups between weeks 0 and 23. This was consistent with the clinical finding of very little plaque accumulation in the animals. It also confirms the fact that gingivitis or periodontitis will not develop in a nutritionally compromised animal in the absence of dental plaque. At week 23, periodontitis was experimentally induced to evaluate the response of the periodontal tissues to the underlying subclinical ascorbate deficiency. While the amount of calculus and debris scores were similar in the two groups of animals, periodontal pocket depth measurements and gingival inflammation were significantly greater in the ascorbate-deficient animals than in controls. Preliminary studies of leukocyte function suggested that this susceptibility to periodontitis might be related to impaired polymorphonuclear leukocyte chemotaxis and phagocytosis. An analysis of the gingival inflammatory infiltrate revealed significantly higher numbers of PMNs in the experimental animals as compared to controls.[55] These PMNs could have contributed to the greater tissue destruction noted in the experimental group via the extracellular release of lysosomal enzymes. In the experimental animals, the increased numbers of PMNs in the gingival inflammatory infiltrate must be contrasted with the reduced chemotactic response of peripheral PMNs in the same animals. These paradoxical findings raise the interesting possiblity that in subclinical vitamin C deficiency, a subpopulation of "intact" PMNs is capable of emigrating into the gingival connective tissue in response to chemoattractants, whereas a "defective" subpopulation with a reduced chemotactic response remains in circulation. There is speculation that their greater numbers in the inflammatory infiltrate as compared to controls may be related to the potency of microbially derived chemoattractants. In any event, the above study was of interest for two reasons. First, the nutritional deficiency induced was marginal and long term in nature — a situation which is more likely to be encountered in industrialized societies. Second, it seems to imply that the periodontium is indeed vulnerable to chronic inflammation as a consequence of subclinical ascorbic acid deficiency.

In summary, laboratory and animal studies have provided reasonable insights with respect to nutritional influences on periodontal health and reduced the ambiguity concerning the role of nutrition in plaque-associated periodontal disease.

## C. Human Investigations

There have been numerous clinical studies on the role of various nutrients in the maintenance of periodontal health. Unfortunately, a vast majority of these studies have been plagued with a failure to meet rigorous scientific standards. This failure has in turn contributed to the controversy that envelops the influence of nutrition and diet on periodontal health. For an in-depth discussion of the methodological problems encountered in human studies, the reader is referred to a review by Alfano.[18] The intent here is to deal only with those studies that have generated data in support of a conceptual basis for the modulating influence of nutrition on periodontal health.

### 1. Diet Consistency

It is noteworthy that the effects of diet consistency on oral tissues predates the current heightened interest in the physical character of the food we ingest. There was a time when it was accepted as dogma that the chewing of firm fibrous foods would keep the teeth free of dental plaque accumulation. Several studies have shown that the mastication of a variety of fibrous foods has no effect on dental plaque formation and gingival inflammation.[56,57] However, food consistency can have a profound effect on salivary gland physiology. This has been well documented in experimental animals.[58] In addition, Hall et al.[59] showed that young adult humans, when maintained on a liquid diet for 7 days, exhibited a significant decrease in the volume, total protein, and amylase activity of the parotid saliva. Individuals who undergo orthognathic surgery (wiring of jaws together) are also fed a liquid diet. Preliminary data indicates a drop in salivary pH in those individuals.[96] The decrease in salivary proteins which is seen when ingesting food requiring little or no mastication may be of importance in promoting oral disease. These proteins are of importance because some act as buffers while others influence the growth, colonization, and adherence of bacteria in the oral cavity. The effects of soft diets on salivary gland physiology do have some potential clinical applications. The elderly may opt for a diet requiring little or no mastication for various reasons — among them, ill-fitting dentures and failure to replace missing teeth. Such an option may lead to stomatological problems that could be mediated by altered salivary gland function.

## 2. Vitamins
### a. Folic Acid

The Health and Nutrition Examination Survey revealed that a high percentage of the population is at risk for folic acid deficiency.[60] The dentist is more likely to see a patient with secondarily induced ("conditioned") marginal or subclinical folic acid deficiency rather than one with overt manifestations of a primary deficiency state. A conditioned folic acid deficiency may be due to a variety of circumstances. Chronic use of drugs such as oral contraceptives and diphenylhydantoin or alcohol abuse represent situations that carry a risk for conditioned folic acid deficiency.[61] The statistics in terms of number of individuals potentially at risk due to these factors is rather impressive. It is estimated that between 10 to 50 million American women are using oral contraceptives for birth control. Epilepsy is the second most prevalent neurological disorder, affecting 1:200 of the world population. The anticonvulsant drug, diphenylhydantoin, has proven to be effective in the treatment of these individuals. It has been estimated that 7% of the adult U.S. population are alcohol abusers.[61] To this figure must be added the growing number of teenage alcoholics.

Several side effects are reportedly associated with the usage of oral contraceptives. These side effects include the possibility that folate metabolism may also be altered. The data from different studies are not always in agreement on this issue.[62] The problem one has to face is that conventional indexes of nutrient status may not always reveal a true picture of nutriture at the organ level. Studies have shown that oral contraceptive users may have normal serum folate levels and yet exhibit megaloblastosis and cytological changes in cervicovaginal cells, suggestive of local folate depletion.[63,64] These observations are consistent with the concept of "end-organ" deficiency. An increased tendency for gingival enlargement and gingivitis reportedly accompanies oral contraceptive usage. Vogel et al.[65] have reported a significant improvement in gingival health of oral contraceptive users following folic acid supplementation even though plasma folic acid levels in these individuals did not indicate folic acid deficiency. The findings from this study also indicated that oral contraceptive users exhibited megaloblastic changes in the gingival sulcular epithelium that were reversed after 60 days of folate supplementation. The reversal of cytologic abnormalities was associated with a decrease in the in vitro permeability of the gingival sulcular epithelium along with a significant reduction in gingival inflammation. Similar beneficial effects of folic acid supplementation on gingival health have been reported in nonoral contraceptive users and in pregnancy-associated gingivitis.[66,67] The implication of these studies must be viewed with some caution. These studies primarily dealt with gingivitis, not with chronic destructive periodontitis. Furthermore, as pointed out in Section II, the destructive phase of periodontal disease may occur in "bursts". The effect of folic acid supplementation during the active phase of periodontal destruction remains to be determined. In any event, the data from aforegoing studies serve to raise the question — is it possible that a deficiency of a given nutrient can exist at the organ level, i.e., gingiva and periodontium, despite adequate dietary intake of the nutrient? The findings from the folic acid supplementation studies together with clinical studies on vitamin C supplementation (see below) would tend to favor that possibility. The possible lack of a suboptimal reservoir of folic acid in the gingival and periodontal tissues is consistent with the "milieu" of the dentogingival area. The ever-present dental plaque (with its microbial flora) leads to micro-ulcerations of the gingival sulcular and junctional epithelium and provokes an inflammatory-immune response in the connective tissue. The repair of epithelial tissues by epithelial cell division and an adequate cell-mediated immune response require folic acid. If this nutrient is not available in adequate amounts to meet the increased demands imposed on the periodontal tissues by dental plaque, then these fundamental processes could conceivably be compromised. In such a circumstance, supplementation with folic acid could have a beneficial effect on pertinent host defense mechanisms.

Like many drugs, the usage of diphenylhydantoin (DPH) in the treatment of epileptics is not without adverse effects. The incidence of subnormal serum folate levels has been reported to be between 37 and 76% following DPH therapy.[68] In some of these individuals, megaloblastic changes are evident in the bone marrow cells, GI, vaginal, and buccal epithelia.[69] The mechanism whereby DPH can cause folic acid-altered metabolism remains in doubt. The evidence for an inhibitory effect of DPH on folic acid conjugase or an interference with the absorption of folic acid in the intestinal epithelium is equivocal.[68] It has also been proposed that serum folate levels may be reduced as a result of the displacement of the vitamin from the carrier protein.[70] One of the classical side effects associated with DPH therapy is gingival overgrowth. It is not clear why the gingiva is the site of predilection for the observed connective tissue changes. In phenytoin-treated ferrets, of the tissues examined, oral mucosa exhibits the highest level of the drug, followed by salivary glands.[71] The latter investigators have suggested that the influx of drug from both blood and saliva creates an inordinately high level of the drug in the gingival connective tissue. The passage of the drug via the sulcular epithelium may be facilitated by a folic acid deficiency-induced defect in the epithelial barrier.[72] If a compromise in the barrier function was the only important factor, one would expect a significant inflammatory cell infiltrate in the connective tissue, a feature which is usually not observed in humans.[69] Thus, the part played by an underlying folic acid deficiency in DPH-induced gingival overgrowth remains to be clarified. The efficacy of folic acid supplementation in preventing or reversing the gingival hyperplasia is open to question. There is a troublesome issue related to folic acid supplementation in individuals receiving DPH. Apparently, the chronic administration of folate can decrease the effectiveness of phenytoin either by inhibiting drug absorption or by a direct antagonism of the central antiepileptic effects of phenytoin.[73]

Chronic alcohol abuse can eventually lead to, among other disorders, varying degrees of deficiency of not only folic acid, but also that of vitamins A and K, thiamin, and Zn, along with alterations in amino acid metabolism. In a recent symposium, alcohol abuse was incriminated as the major cause among the elderly of deficiencies of both thiamin and folate.[74] The linkage of periodontal pathosis to the deficiency of a particular nutrient is nearly impossible owing to the multiplicity of nutritional problems in the alcoholic. In any event, these nutrients are of importance to periodontal health, and the possible existence of their lack should be borne in mind in the management of oral health problems in alcohol abusers.

### b. Vitamin C

There has been an intense interest in the relationship of ascorbic acid status of individuals and periodontal health and disease. It is very likely that the practice of ascorbic acid supplementation, either self-prescribed or on the recommendations of a member of the health profession, is fairly widespread in the general population. The scientific basis for this practice in the maintenance or for the prevention of the breakdown of periodontal health must be seriously questioned. Although statistically significant improvements in gingival and periodontal health have been reported with ascorbic acid supplementation, other studies have failed to corroborate such findings.[75] Several epidemiologic studies have examined the relationship between ascorbic acid deficiency and periodontal disease status in humans. Fairly extensive studies conducted in eight countries, under the auspices of the Interdepartmental Committee for Nutrition and National Defense, revealed at best a correlation between age and oral hygiene and periodontal health.[76] The Ten-State Nutrition survey reported a "weak" correlation between ascorbic acid deficiency and periodontal disease.[77]

More recently, Ismail et al.[78] analyzed the nutritional and oral health data collected in the course of the first National Health and Nutrition Examination Survey (1971 to 1974) (HANES I). This analysis included data collected from 8609 individuals, aged 25 to 76 years. After an excellent discussion of the limitations in the interpretation of their data, these investigators also concluded that there was only a weak association between periodontal disease and ascorbic acid deficiency.

Given this uncertainty, do we have any clue as to the role of ascorbic acid in the maintenance of periodontal health? Earlier on, attention was drawn to host defense factors that are of importance in the maintenance of oral health. In this context, a series of studies by Mallek[79] are noteworthy. The results of these studies indicate that a direct relationship exists between gingival tissue concentration of ascorbate and the epithelial barrier function of the sulcular epithelium, both of which are independent of leukocyte concentrations (a reliable indicator of ascorbic acid status). In addition, ascorbate supplementation (1 g/day for 4 weeks) of individuals resulted in increased tissue levels of the vitamin, which in turn was associated with increased collagen synthesis and decreased permeability of the gingival sulcular epithelium. All individuals, with one exception, had a mean daily dietary intake of ascorbic acid above the daily recommended requirement. A general conclusion that can be drawn is that ascorbic acid supplementation can lead to a strengthening of host defenses. However, it should be pointed out that periodontal diseases are terribly complex. They are complex in nature owing to the antagonistic and synergistic interactions between periodontal pathogens and the interaction between pathogens and host defense. These interactions may be modulated by factors such as the hormonal status of an individual. In this scheme of things, the weight that can be given to the clinical significance of improving the barrier function (one of several variables) of the sulcular epithelium with ascorbate supplementation remains undetermined.

### 3. Carbohydrates and Proteins

A prominent feature of the diabetic state is altered carbohydrate metabolism. Depending on the type of diabetes, various approaches (diet, insulin, etc.) are employed to obtain good metabolic control. It is widely recognized that diabetics, particularly those exhibiting poor metabolic control, exhibit increased susceptibility to infections, including periodontal disease (for a review on diabetes and periodontal disease see References 80 and 81). Gingivitis is more severe, and the onset of periodontitis occurs earlier in insulin-dependent diabetes mellitus (IDDM). Often, the alveolar bone loss is more pronounced in the first molar incisor regions.[81] The greater severity of periodontal disease in IDDM individuals cannot be attributed to more dental plaque accumulation. Rather, this greater severity may be related to defective neutrophil function.[81] Parenthetically, individuals genetically predisposed to diabetes ("potential abnormality of glucose tolerance") also exhibit a decrease in polymorphonuclear leukocyte chemotactic response in association with severe periodontitis.[82] Yet another interesting feature of IDDM is that the subgingival microorganisms from periodontal lesions differ quantitatively from that associated with other types of periodontal diseases, e.g., adult periodontitis, localized juvenile periodontitis.[83] The subgingival microflora in IDDM is predominated by *Capnocytophaga* species and "anaerobic vibrios". Finally, the increased susceptibility to periodontal disease in IDDM may be related to an increase in capillary basement membrane thickness, which is known to involve a variety of organs, including gingiva.[84] Increase in basement membrane thickness could prevent adequate diffusion of insulin and glucose to host defense cells already outside the small vessels.[85] In the overall management of the diabetic individual, the existence of infections may be an added obstacle in achieving good metabolic control. Failure to recognize and eliminate periodontitis may impede the successful management of the diabetic state.

The relationship between fermentable carbohydrates and incidence of caries is well established. Within the oral cavity these carbohydrates exert a local effect by acting as a substrate for supragingival cariogenic organisms. It is doubtful that a similar cause and effect relationship exists with respect to periodontal disease. Tube feeding in man apparently leads to a decrease in the quantity and altered microbial composition of supragingival plaque.[86,87] Whether or not this is primarily due to the absence of carbohydrates in the oral cavity cannot be stated with certainty. Since it is unlikely that carbohydrates or other foodstuffs, when in the oral cavity, have access to subgingival regions except at sites of food impaction, they probably have very little, if any, effect on the subgingival flora. The Turku studies[88] compared the effect of either sucrose, xylitol, or fructose as the major sweetening agent in the diet on oral health over a 2-year period. These studies failed to show significant clinical differences in periodontal health among the three groups ingesting different sweetening agents.

In developing societies, protein-calorie malnutrition is associated with a higher incidence of acute necrotizing ulcerative gingivitis.[89] More ominous is the observation that in these malnourished individuals, acute necrotizing ulcerative gingivitis extends rapidly into adjacent tissues, producing extensive destruction and necrosis of orofacial tissues.[89] In industrialized societies there is a lack of documented evidence linking dietary protein levels and protein metabolism to periodontal health.

### 4. Minerals

Over a decade ago, Krook et al.[90] suggested that periodontal disease results from low dietary Ca and/or an excess of P. These investigators claimed that nutritional secondary hyperparathyroidism occurs to maintain normal serum Ca levels as a result of altered dietary Ca:P ratio. Ca is reportedly mobilized from all labile bone stores, including the bone-supporting teeth. In addition, Ca supplementation was reported to have a beneficial effect on human periodontal disease.[90] A critical appraisal of these and other related studies, along with the epidemiologic data concerning the relationship of dietary Ca:P ratio, suggests that it is unlikely that nutritional secondary hyperparathyroidism plays a role in the initiation of periodontal disease.[18] Heaney et al.[91] have recently reviewed the relationship of altered dietary Ca:P ratio to circulating parathyroid hormone levels, bone metabolism, and Ca balance. Apparently, the harmful skeletal effect of altered dietary Ca:P ratio is primarily seen in rapidly growing animals. Further, these investigators[91] concluded that "there is very little evidence in man to suggest that even broad variation in phosphorus intake or in dietary Ca/P ratio have any influence on net calcium utilization or on bone balance, and indeed some evidence to the contrary". It is very likely that we have not heard the last on this subject. Thus, for instance, Spencer and Lenden[92] have assessed the adverse effects of Al-containing antacids (a widely used drug) on mineral metabolism. These adverse effects include increased urinary Ca levels, bone resorption, impairment of fluoride absorption, and P depletion. It has been suggested that these changes may contribute to bone disease in the elderly. At the present time, there is no good evidence to support the practice of Ca supplementation in the prevention and management of human periodontal disease.

Freeland et al.[93] examined the relationship of serum Ca, Cu, Fe, Mg, Mn, and Zn to the severity of periodontal disease in a group of 80 dental patients. A direct relationship was noted between the severity of the disease and increased serum Cu levels. This finding may not be meaningful because the index for assessing periodontal disease severity employed by these investigators is not a very sensitive one. However, the Freeland et al.[93] finding should not be dismissed altogether. Etzel et al.[94] have noted increased Cu concentration in gingiva and serum of hamsters following administration of endotoxin from *Bacteroides melaninogenicus,* a dental plaque-associated microorganism.

Mallek[79] investigated the relationship of Fe to sulcular epithelial permeability in 11 individuals with clinically healthy gingiva. Measurement of serum Fe and total Fe binding capacity revealed all subjects to be within normal limits (serum Fe: 35 to 171 $\mu$g/d$\ell$, total Fe binding capacity: 221 to 481 $\mu$g/d$\ell$). No significant correlations were found between serum Fe levels and total Fe binding capacity with either gingival tissue levels of Fe, or permeability of the sulcular epithelium ($p$ <0.01). Average daily Fe intake was determined for 9 of the 11 subjects using daily intake diaries. One male and two females had a mean daily intake below the recommended daily requirement. No correlation was found between daily dietary Fe intake and either tissue levels of Fe or epithelial permeability.

Finally, the reports of the effects of other nutrients (e.g., vitamin E and Zn) on periodontal health are too fragmentary in nature to warrant their inclusion in this chapter.

## IV. SUMMARY AND DIETARY RECOMMENDATIONS

It is clear that periodontal disease is primarily an infectious process involving an inflammatory-immune response to local tissue irritants, particularly by-products of bacterial plaque. Current knowledge suggests that the response can be conditioned by nutritional factors. However, prevention or reversal of this disease by short-term nutritional therapy has not been demonstrated. Thus, the fundamental basis for patient counseling relative to diet, nutrition, and periodontal disease rests with instructing the patient to ingest a nutritionally adequate diet over the long term. Nutritional supplementation of patients with periodontal disease should be based on a complete nutritional assessment or, in some instances, on a carefully controlled clinical trial. If periodontitis responds positively to plaque control and conventional modes of treatment, nutritional supplementation is unwarranted. However, in light of the potential role of nutrition in the etiology and progression of periodontal disease, failure of optimal dental care to return the tissue to health suggests that nutritional intervention (diet counseling and supplementation) should be considered.

The following dietary guidelines are suggested to assist in maintaining optimum host resistance and to alert the consumer to specific dietary factors that may impact on periodontal health.[95]

Individuals should consume a diet that meets the Recommended Dietary Allowances (RDA) for basic nutrients, provides variety, and be in accordance with the U.S. Dietary Guidelines.

The indivdual should be aware of and avoid any nutrition or diet regimen which purports to cure or reverse periodontal disease without optimal dental care, particularly those which apparently act over a short time frame.

The public should be advised that meticulous oral hygiene and regular dental exams remain the foundation for periodontal disease prevention.

## REFERENCES

1. U.S Bureau of the Census, *Projections of the Population of the United States· 1977 to 2050*, No. 704, U.S. Government Printing Office, Washington, D.C., July 1977.
2. National Institutes of Health, *Epidemiology of Aging*, Research Conference, DHEW Publ #NIH 77-71, Department of Health, Education and Welfare, Bethesda, Md., 1972.
3. HHS/NIH, *Challenges for the Eighties, National Institutes of Dental Research Long Range Research Plan*, U.S. Department of Health and Human Services, National Institutes of Health, Bethesda, Md , December 1983.
4. **Douglass, C., Gillings, D., Sollecito, W., and Gammon, M.,** The potential for increase in the periodontal diseases of the aged population, *J. Periodontol* , 54, 721, 1983.
5. **Lindhe, J.,** *Textbook of Clinical Periodontology*, Munksgaard, Copenhagen, 1983, chap. 5.
6. **Socransky, S. S., Haffajee, A. D., and Goodson, J. M., and Lindhe, J.,** New concepts of destructive periodontal disease, *J. Clin. Periodontol.*, 11, 21, 1984.
7. **Page, R. C. and Schroeder, H. E.,** *Periodontitis in Man and Other Animals*, S. Karger, New York, 1982, chap. 4.

8. **Cianciola, L. J., Genco, R. J., Patters, M. R., McKenna, J., and VanOss, C. J.,** Defective poly-morphonuclear leukocyte function in human periodontal disease, *Nature (London),* 265, 455, 1977.

9 **Clark, R. A., Page, R. C., and Wilde, G.,** Defective neutrophil chemotaxis in juvenile periodontitis, *Infect. Immunol.,* 18, 694, 1977.

10. **VanDyke, T. G., Levine, M. J., and Genco, R. J.,** Periodontal diseases and neutrophil abnormalities, in *Host-Parasite Interactions in Periodontal Diseases,* Genco, R. J and Mergenhagen, S. E , Eds., American Society for Microbiology, Washington, D C., 1982, 235

11. **Page, R. C. and Schroeder, H. E.,** Current status of the host response in chronic marginal periodontitis, *J. Periodontol.,* 52, 477, 1981.

12 **Ramfjord, S. P., Kerr, D. A., and Ash, M. M.,** *World Workshop in Periodontics,* University of Michigan Press, Ann Arbor, 1966.

13. **Ranney, R. R.,** Pathogenesis of periodontal disease: a position report and review of the literature, in *Proc. Int. Conf Res. Biol. Periodontal Dis ,* Chicago, 1977, 261

14. **Scrimshaw, N. S., Taylor, C. E., and Gordon, J. E.,** Interactions of nutrition and infection, *WHO Monogr. Ser.,* 57, 1968.

15. **Gontzea, I.,** *Nutrition and Anti-Infectious Defense,* S. Karger, New York, 1974.

16. **Mandel, I. D.,** In defense of the oral cavity, in Saliva and Dental Caries, Kleinberg, I., Ellison, S. A , and Mandel, I. D., Eds., *Microbiol Abstr.,* (Spec. Suppl ), 473, 1979.

17. **Cimasoni, G.,** The crevicular fluid, in *Monographs in Oral Science,* Vol 3, Myers, H., Ed., S. Karger, Basel, 1974.

18. **Alfano, M. C.,** Controversies, perspectives and clinical implications of nutrition in periodontal disease, *Dent Clin N. Am.,* 20, 519, 1976.

19. **Skougaard, M.,** Turnover of the gingival epithelium in marmosets, *Acta Odont. Scand.,* 23, 623, 1965

20. **DePaola, D. P. and Kuftinec, M.,** Nutrition in growth and development of oral tissues, *Dent Clin. N Am.,* 20, 441, 1976.

21. **Winnick, M. and Noble, A.,** Quantitative changes in DNA, RNA and protein during pre-natal and post-natal growth in the rat, *Dev. Biol ,* 12, 451, 1965

22. **Squier, C. A. and Johnson, N. W.,** The permeability of oral mucosa, *Br. Med. Bull ,* 31, 169, 1975

23 **Alfano, M. C., Drummond, J. F., and Miller, S. A.,** Localization of rate limiting barrier to penetration of endotoxin through nonkeratinized oral mucosa *in vitro, J. Dent. Res.,* 54, 1143, 1975.

24 **Schwartz, J., Stinson, F. L., and Parker, R. B.,** The passage of tritiated bacterial endotoxin across intact crevicular epithelium, *J. Periodontol.,* 43, 270, 1972.

25. **Stahl, S. S.,** Response of the periodontium to protein-calorie malnutrition, *J. Oral Med.,* 21, 146, 1966

26. **Stahl, S. S.,** Host resistance and periodontal disease, *J Dent. Res.,* 49(Suppl.), 248, 1970.

27 **Stahl, S. S.,** Nutritional influences on periodontal disease, *World Rev. Nutr. Diet.,* 13, 277, 1971.

28. **Stahl, S. S., Sandler, H. C., and Cahn, L.,** The effects of protein deprivation upon the oral tissues of the rat and particularly upon periodontal structures under irritation, *Oral Surg.,* 8, 760, 1965.

29. **Stein, G. and Ziskin, D. E.,** The effect of a protein free diet on the teeth and periodontium of the albino rat, *J. Dent Res.,* 28, 529, 1949.

30. **Chawla, T. N. and Glickman, I.,** Protein deprivation and the periodontal structures of the albino rat, *Oral Surg.,* 4, 578, 1951.

31. **Carranza, F. A., Cabrini, R. L., Lopez, O. R., and Stahl, S. S.,** Histometric analysis of interradicular bone in protein deficient animals, *J. Periodontol. Res.,* 4, 292, 1969.

32. **Goldman, H. M.,** The effect of dietary protein deprivation and of age on the periodontal tissues of the rat and spider monkey, *J Periodontol ,* 25, 87, 1954.

33. **Squires, B. T.,** Differential staining of buccal epithelial smears as an indicator of poor nutritional status due to protein-calorie deficiency, *J. Pediatr ,* 66, 891, 1965

34. **Dreizen, S., Levy, B., and Bernick, S.,** Studies on the biology of the periodontium of marmosets. VII. The effect of vitamin C deficiency on the marmoset periodontium, *J. Periodontol. Res.,* 4, 274, 1969.

35. **Dreizen, S., Levy, B., and Bernick, S.,** Studies on the biology of the periodontium of marmosets VIII The effect of folic acid deficiency of marmoset oral mucosa, *J Dent. Res.,* 49, 616, 1970

36. **Dreizen, S., Levy, B., and Bernick, S.,** Studies on the biology of the periodontium of marmosets. XIII. Histopathology of niacin deficiency stomatitis in the marmoset, *J. Periodontol ,* 48, 452, 1977.

37. **Glickman, I.,** Acute vitamin C deficiency and periodontal disease. I. The periodontal tissues of the guinea pig in acute vitamin C deficiency, *J. Dent. Res.,* 27, 9, 1948.

38. **Glickman, I. and Stoller, P.,** The periodontal tissues of the albino rat in vitamin A deficiency, *J. Dent. Res.,* 27, 758, 1948

39. **Becks, H., Wainwright, W. W., and Morgan, A. F.,** Comparative study of oral changes in dogs due to pantothenic acid, nicotinic acid and unknowns of B vitamin complex, *Am. J. Orthod.,* 29, 183, 1943.

40. **Topping, N. N. and Fraser, H. F.,** Mouth lesions associated with dietary deficiencies in monkeys, *Public Health Rep.,* 54, 416, 1939.

41. **Ferguson, H. W. and Hertles, R. L.,** The effects of diets deficient in calcium or phosphorus in the presence and absence of supplements of vitamin D on the secondary cementum and alveolar bone of young rats, *Arch Oral Biol*, 9, 647, 1974.

42 **Alfano, M. C., Drummond, J. F., and Miller, S. A.,** Technique for studying the dynamics of oral mucosal permeability *in vitro, J. Dent. Res.*, 54, 194, 1975.

43 **Alfano, M. C., Miller, S. A., and Drummond, J. F.,** Effect of ascorbic acid deficiency on the permeability and collagen biosynthesis of oral mucosal epithelium, *Ann N.Y Acad. Sci.*, 258, 253, 1975.

44 **Alfano, M.,** Nutrition in periodontal diseases, in *New Horizons in Nutrition for the Health Professions,* Slavkin, H C, Ed., University of Southern California Press, Los Angeles, 1981, 160.

45. **Sabin, A. B.,** Constitutional barriers to involvement of the nervous system by certain viruses with special reference to the role of nutrition, *J. Pediatr.*, 19, 596, 1941.

46. **Sperber, R.,** The Effect of Preweaning Iron Deficiency and Subsequent Rehabilitation on Barrier Function of Mucosal Epithelium, Doctoral thesis, Massachusetts Institute of Technology, Cambridge, 1977

47 **Joseph, C. E., Ashrafi, S. H., Steinberg, A. D., and Waterhouse, J. P.,** Zinc deficiency changes in the permeability of rabbit periodontium to $^{14}C$-phenytoin and $^{14}C$-albumin, *J Periodontol.*, 53, 251, 1982.

48. **Ross, C. W. and Knight, R.,** Dietary factors affecting the pathogenicity of Entamoeba histolytica in rats, *Trans. R. Soc. Trop. Med. Hyg.*, 67, 560, 1973.

49 **Worthington, B. S. and Boatman, E. S.,** The influence of protein malnutrition on ileal permeability to macromolecules in the rat, *Am. J Dig. Dis.*, 19, 43, 1974.

50 **Worthington, B. S., Boatman, E. S., and Kenney, G. E.,** Intestinal absorption of intact protein in normal and protein deficient rats, *Am. J. Clin Nutr.*, 27, 276, 1974

51 **Alvares, O. and Siegel, I.,** Permeability of gingival sulcular epithelium in the development of scorbutic gingivitis, *J Oral Pathol.*, 10, 40, 1981.

52. **Barnes, M. J.,** Functions of ascorbic acid in collagen biosynthesis, *Ann N Y Acad Sci*, 258, 264, 1975.

53. **Sulkin, D. F., Sulkin, N. M., and Nusham, H.,** Corneal fine structure in experimental scorbutus, *Invest Ophthalmol*, 2, 633, 1972.

54. **Alvares, O., Altman, L. C., Springmeyer, S., Ensign, W., and Jacobson, K.,** The effect of subclinical ascorbate deficiency on periodontal health in nonhuman primates, *J. Periodontol Res*, 16, 628, 1981.

55 **Alvares, O., Burt, G., and Leik, J.,** The gingival inflammatory infiltrate in subclinically deficient nonhuman primates, *J Dent. Res*, 61, A182, 1982.

56 **Lindhe, J. and Wicen, P. D.,** The effects on the gingiva of chewing fibrous foods, *J. Periodontal. Res*, 4, 193, 1969.

57 **Wade, A. B.,** Effect on dental plaque of chewing apples, *Dent. Pract.*, 21, 194, 1971.

58. **Johnson, D. A.,** Effect of liquid diet on the protein composition of rat parotid saliva, *J. Nutr.*, 112, 175, 1982.

59. **Hall, H. D., Merig, J. J., Jr., and Schneyer, C. A.,** Metrecal-induced changes in human saliva, *Proc. Soc. Exp. Biol Med.*, 124, 532, 1967

60 **Anon.,** Dietary intake and biochemical findings, in *Health and Nutrition Examination Survey, 1971—72,* DHEW Publ. #74-1219-1, U.S. Department of Health, Education, and Welfare, Rockville, Md, 1974

61. **Smith, C. J.,** Marginal nutritional states and conditioned deficiencies, in *Alcohol and Nutrition,* Li, T K, Schenker, S, and Lumeng, L, Eds, Publ. #ADM 79-780, U S Department of Health, Education, and Welfare, 1979.

62 **Roe, D. A.,** *Drug-Induced Nutrition Deficiencies,* VI, Westport, Conn., 1976.

63 **Whitehead, N., Reyner, F., and Lindenbaum, J.,** Megaloblastic changes in cervical epithelium, *JAMA,* 226, 1421, 1973.

64 **Lindenbaum, J., Whitehead, N., and Reyner, F.,** Oral contraceptive hormones, folate metabolism and the cervical epithelium, *Am J. Clin. Nutr*, 28, 346, 1975

65. **Vogel, R., Deasy, M., Alfano, M., and Schneider, L.,** The effect of folic acid on gingival health in women taking oral contraceptives, *J. Prev. Dent*, 6, 221, 1980

66. **Vogel, R. and Deasy, M.,** The effect of folic acid in experimentally induced gingivitis, *J. Prev. Dent.,* 5, 30, 1978.

67 **Thomson, M. and Pack, A.,** Effect of extended systemic and topical folate supplementation on gingivitis and pregnancy, *J. Clin Periodontol.,* 9, 285, 1982.

68 **Halsted, C. H.,** Drug and water-soluble vitamin absorption, in *Nutrition and Drug Interactions,* Hathcock, J H and Coon, J., Eds., Academic Press, New York, 1975.

69. **Mallek, H. M. and Nakamoto, T.,** Dilantin and folic acid status Clinical implications for the periodontist, *J. Periodontol.,* 52, 225, 1981

70. **Klipstein, F. A.,** Subnormal serum folate and macrocytosis associated with anticonvulsant drug therapy, *Blood,* 23, 68, 1964

71 **Steinberg, A. D., Alvarez, J., and Jeffay, H.,** Distribution and metabolism of diphenylhydantoin in oral and non-oral tissues of ferrets, *J. Dent. Res.,* 52, 267

72. **Vogel, R. I.,** Gingival hyperplasia and folic acid deficiency from anticonvulsive drug therapy. A theoretical relationship, *J. Theor. Biol.,* 67, 269, 1977.
73. **Goodman, L. S. and Gilman, A.,** *The Pharmacological Basis of Therapeutics,* Macmillan, New York, 1980, 455.
74. **Rivlin, R. S.,** Summary and concluding statement: evidence relating selected vitamins and minerals to health and disease in the elderly population in the United States, *Am. J. Clin. Nutr.,* 36, 1083, 1982.
75. **Woolfe, S. N., Hume, W. R., and Kenney, E. B.,** Ascorbic acid and periodontal disease A review of the literature, *J. West Soc. Periodontol.,* 82, 44, 1980
76. **Russell, A. L.,** International nutritional surveys: a summary of preliminary dental findings, *J Dent. Res.,* 42, 233, 1963.
77 Center for Disease Control, *Ten State Nutrition Survey, 1968—70,* Publ #(HSM) 72-8131, Department of Health, Education and Welfare. Washington. D.C., 1972, 87.
78. **Ismail, A. I., Burt, B., and Eklund, S.,** Relationship between ascorbic acid intake and periodontal disease in the United States, *J. Am. Dent. Assoc.,* 107, 927, 1983.
79. **Mallek, H.,** An Investigation of the Role of Ascorbic Acid and Iron in the Etiology of Gingivitis in Humans, Ph.D. thesis, Massachusetts Institute of Technology, Cambridge, 1978.
80. **Saadoun, A. P.,** Diabetes and periodontal disease: a review and up-date, *J. West. Soc. Periodontol.,* 28, 116, 1980.
81 **Cianciola, L. J., Park, B. H., Bruck, E., Mosovich, L., and Genco, R. J.,** Prevalence of periodontal disease in insulin-dependent diabetes mellitus (juvenile diabetes), *J. Am. Dent. Assoc.,* 104, 653, 1982.
82. **McMullen, J. A., VanDyke, T. E., Horoszewicz, H., and Genco, R. J.,** Neutrophil chemotaxis in individuals with advanced periodontal disease and a genetic predisposition to diabetes mellitus, *J. Periodontol.,* 52, 167, 1981.
83. **Mashimo, P. A., Yamamoto, Y., Slots, J., Park, B. H., and Genco, R. J.,** The periodontal microflora of juvenile diabetes Culture, immunofluorescence and serum antibody studies, *J. Periodontol.,* 54, 420, 1983.
84. **Frantzis, T. G., Reeve, C. M., and Brown, A. L.,** The ultrastructure of capillary basement membranes in the attached gingiva of diabetic and non-diabetic patients with periodontal disease, *J. Periodontol.,* 42, 406, 1971.
85. **Mowat, A. G. and Baum, J.,** Chemotaxis of polymorphonuclear leukocytes from patients with diabetes mellitus, *N. Engl. J. Med.,* 284, 621, 1971.
86. **Cooke, V., Petropolou, K., Mandel, I., and Ellison, S. A.,** Plaque in tube-fed persons. I. Metabolism and chemistry, *J. Dent. Res.,* 61(Abstr.), 250, 1982.
87. **Ellison, S. A., Cooke, V., and Mandel, I.,** Plaque in tube-fed persons II. Microbiology, *J Dent. Res.,* 61(Abstr.), 250, 1982.
88. **Paunio, K., Makinen, K., Scheinin, A., and Ylitalo, K.,** Turku Sugar Studies. IX. Principal periodontal findings, *Acta Odont. Scand.,* 33(Abstr.), 217, 1975.
89. **Enwonwu, C. O.,** Epidemiological and biochemical studies of necrotizing ulcerative gingivitis and noma (cancrum oris) in Nigerian children, *Arch. Oral. Biol.,* 17, 1357, 1972.
90. **Krook, L., Litwak, L., and Whalen, J.,** Human periodontal disease. Morphology and response to calcium therapy, *Cornell Vet.,* 62, 32, 1972.
91. **Heaney, R. P., Gallagher, J. C., Johnston, C. C., Neer, R., Parfitt, A. M., and Whedon, G. D.,** Calcium, nutrition and bone health in the elderly, *Am. J. Clin Nutr.,* 36(Suppl.), 986, 1982.
92. **Spencer, H. and Lender, M.,** Adverse effects of aluminum-containing antacid on mineral metabolism, *Gastroenterology,* 76, 603, 1979.
93. **Freeland, H. H., Cousins, R. J., and Schwartz, R.,** Relationship of mineral status to periodontal disease, *Am. J Clin. Nutr.,* 29, 745, 1976.
94. **Etzel, K. R., Swerdl, M. R., Swerdl, H. N., and Cousins, R. J.,** Endotoxin-induced changes in copper and zinc metabolism in the Syrian hamster, *J. Nutr ,* 112, 2363, 1982.
95. **Burakoff, R. P. and Goldsmith, J. A.,** A workshop's recommendation: Dental Nutrition Guidelines for the 1980s, *N.Y State J. Dent.,* 47, 447, 1981.
96. **Johnson, D. A.,** Personal communication.

Chapter 8

# HYPERTENSION

**Rudy A. Bernard**

## TABLE OF CONTENTS

## I. INTRODUCTION

Hypertension is a condition in which blood pressure rises above normal levels. Measurements of blood pressure are usually expressed with two values: a higher one, called systolic pressure, and a lower one, called diastolic pressure (e.g., 120/80 mmHg). The systolic value represents the pressure encountered by the heart when it is fully contracted, whereas diastolic pressure is what the heart confronts when it is relaxed, between beats. It is the rapid alternation between these two pressure levels that makes it possible to feel a pulse and to count the heart rate. In general, there are two major ways in which elevation of blood pressure is brought about. One is an increase in the total volume of fluid that the heart has to pump, and the other is a change in the ease with which blood can flow through the arteries and veins. For these reasons, the kidneys and the hormones associated with their function of controlling body fluid volume play an important role in the study of hypertension. Similarly, the blood vessels, whose inner diameter and resistance to flow are controlled by neural and hormonal factors, are another major area of interest in the study of this disease.

Inherent in the definition of hypertension is the concept of normality. Until recently it was considered normal for blood pressure to increase with advancing age. The fact that it does so in our and other modern industrial societies (Figure 1)[1] makes it normal in a statistical sense, but the fact that blood pressure does not increase with age in most of the preindustrial, non-Westernized populations that have been studied[2,3] indicates very strongly that it is not biologically normal for blood pressure to increase with age. Thus, hypertension is not a natural consequence of the aging process as such. This is also evident in the fact that even in modern societies many individuals grow old without becoming hypertensive.

It is also generally agreed that there is a genetic component in hypertension,[1] but studies of migration and acculturation in pre-modern societies indicate that the genetic element is permissive, requiring the environmental and behavioral conditions of modern life for its expression.

The definition of hypertension is an arbitrary one since blood pressure is distributed over a continuous range of values. Thus, there is no natural dividing line between normal blood pressure and hypertension.[4] Any dividing line is an artifact, but out of practical clinical considerations the World Health Organization has defined hypertension as a single sitting or recumbent blood pressure exceeding 160/95. Values below 140/90 are considered to be normotensive. In practice the term *borderline hypertension* is used to describe individuals whose blood pressure falls between these two limits.[5]

By itself, hypertension, unless it is of the rapid onset and malignant variety, is not an immediate danger to health. Also, it is not usually accompanied by symptoms that are noticeable to the individual. Its existence requires measurement by an appropriate device, many varieties of which are widely distributed and easily used at home without special medical training. Because of the absence of symptoms, it is called the "silent killer", for, if left untreated, hypertension leads to cardiovascular diseases and death from stroke, heart failure, etc.

## II. CLASSIFICATION

Hypertension is usually divided into the two categories of essential and secondary. The term *essential* is used to denote that the cause of the pressure elevation is unknown, for which reason some prefer to use the terms *idiopathic* or *primary* instead. The term "secondary hypertension" is used when there is an association with a disease or a factor known to raise blood pressure, such as renal disease or endocrine disturbances.[5] It is widely held that 90 to 95% of hypertension is of the *essential* variety. Genest et al.[6] propose a separate classification for the hypertension found in the elderly, which is characterized by high systolic

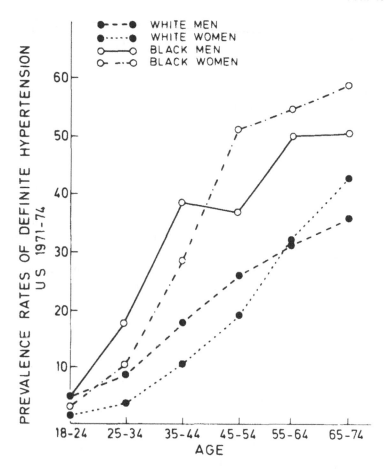

FIGURE 1. The prevalence of hypertension in the U.S., defined as systolic blood pressure of at least 160 mmHg or a diastolic blood pressure of at least 95 mmHg (data from HANES I).[1]

pressure and mildly elevated or normal diastolic pressures. They suggest the term "athero-arteriosclerotic hypertension" for this form of the disease, since it is always associated with signs of atherosclerosis and arteriosclerosis. For similar reasons, Franklin[7] developed the term "geriatric hypertension" with two subsets: (1) isolated systolic hypertension and (2) combined diastolic-systolic hypertension, with a disproportionately elevated systolic component. The importance of a special classification for the hypertension found in the older segment of the population lies in the now-disproven idea that elevated diastolic pressure is a greater risk to the patient than elevated systolic pressure. It is now clear that treatment even of mildly elevated diastolic pressure reduces the risk of cardiovascular disease, especially in the elderly.[8] Both proponents of this special classification suggest that this form of hypertension differs from essential hypertension, which generally develops by the age of 40 or 50, whereas the elderly form begins later in life. This, of course, has important etiological implications and suggests that different nutritional factors may be involved.

## III. CAUSES OF HYPERTENSION

In a recent review of the physiology of hypertension, Folkow[9] mentioned that there are few points on which authorities agreed, leaving the overwhelming majority of them as subjects for disagreement. Part of the problem is due to the inherent complexity of the subject matter as well as to the overwhelming amount of information available. Although there are some important points on which a consensus seems to exist, there are so many apparently contradictory reports that painstaking interpretation is required to avoid confusion.

## A. Role of Sodium

In the search for a nutritional basis of essential hypertension, most of the interest as well as the controversy has focused on salt (NaCl). That Na has some role in hypertension seems to be generally accepted. The controversy is about the extent and importance of its involvement. The average daily NaCl intake by Americans has been estimated to range from 10 to 14.5 g (170 to 250 mEq of Na) per person,[10] whereas the daily amount recommended by the National Research Council ranges from 1 to 3 g (15 to 50 mEq of Na) per person.[11] Critics of the Na theory say that the level of Na intake in the American diet is not a significant health hazard for the majority of the population.[4,12] Yet reduction of Na intake remains a national health goal,[13] promoted by educational campaigns of the Food and Drug Administration and the Department of Agriculture.

The evidence that excessive Na intake is related to essential hypertension comes from three major sources: (1) epidemiological studies, (2) salt deprivation and salt loading studies in humans, and (3) animal experiments.

### 1. Epidemiological Studies

In general, studies across populations have shown that blood pressure is correlated with Na intake, whereas studies within a population have not shown a similar correlation. Particularly striking are the findings in nonindustrialized societies, where hypertension and cardiovascular disease are rare and blood pressure does not increase with age. At least 20 such societies have been found.[14] What these people seem to have in common is a very low Na intake by modern standards (less than 85 mEq/24 hr). Critics of the Na theory point out that such people tend to be lean, do not gain weight with age, are very active, and have diets that are high in K as well as low in Na. Whatever the explanation for the lack of cardiovascular disease in such societies, such data make it clear that hypertension is not a necessary part of the aging process. Although the bodily changes that do occur in the process of aging may predispose or increase susceptibility to cardiovascular disease, the data from pre-modern societies clearly indicate that it is a combination of environmental and behavioral factors that brings about the high rate of hypertension and cardiovascular disease, which is such a prominent feature of modern society.

Nutritional surveys of the U.S. population, such as the National Health and Nutrition Examination Survey I (HANES I), have failed to show a positive correlation between blood pressure and dietary intake of Na. Stanton et al.[15] analyzed a subset of 10,419 adults of the total HANES I population of 20,749. Excluded from analysis were those under 18 years and adults being treated for hypertension, on a low-salt diet, or pregnant. Age and body mass index accounted for 94% of the variance in blood pressure. Diet explained less than 1% of the total variance for whites and less than 5% for nonwhites. Dietary variables such as food Na, food K, Na:K ratio, calories from fat, and percent saturated fat were not significantly correlated to blood pressure. However, food Ca, Na:Ca ratio, food vitamin C, and Ca:P ratio were all significantly correlated to both systolic and diastolic blood pressure. Dietary variables were more strongly correlated with blood pressure in nonwhite males, where the effect of body mass index was substantially less. The authors point out that family history of hypertension, which was not included in this survey, is probably an important variable. Another shortcoming was that the mean Na intake did not include Na used in cooking or added at the table. Other studies have also failed to show a significant difference in Na intake between normotensive and hypertensive persons.[12,16]

Proponents of the Na theory point out the methodological difficulties in such surveys due to numerous sources of variation for measuring Na intake and to the homogeneously high

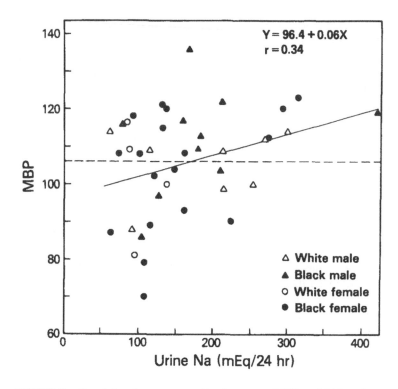

FIGURE 2.   Correlation between mean blood pressure (MBP) and 24-hr urinary sodium excretion in 40 individuals from the Philadelphia area, ranging in age from 22 to 61 years  The correlation is statistically significant ($p = 0.02$), due to the values above 270 mEq/24 hr. (From Bernard, R. A. et al , *Biological and Behavioral Aspects of Salt Intake*, Kare, M. R., Fregly, M J., and Bernard, R. A., Eds , Academic Press, New York, 1980, 397. With permission)

NaCl intake characteristic of Western populations.[17] In the best of such studies,[18] Na intake is estimated by a single 24-hr urine Na sample. In order to reduce to 10% the error in estimating Na output, it is now estimated that 9 or more 24-hr urine samples are required, and 5 or more are needed to adequately separate individuals into high, medium, and low percentiles of Na intake.[18] Other studies have used questionnaire assessments of Na intake, which may be valid for population estimates but are not suitable for individual correlations. In addition to the technical sources of variation, there is the biological variability in salt excretion due to temperature, sweating, and NaCl intake itself. Thus, all these intraindividual variations sum up to approach the level of interindividual variability, leading to weak correlation between the measured variables and disease. The other major factor producing weak correlations is the narrow range of NaCl consumption often found in these studies.

Langford and Watson[19] found a significant correlation between urine constituents and blood pressure. Their methods involved pressure measurements for 8 days and 6 days of urine measures. In another study[20] involving 40 subjects (24 hypertensive, 16 normotensive; approximately 70% black and 30% white) there was a wide enough range of Na excretion (62 to 437 mEq/24 hr) and mean blood pressures (70 to 136 mmHg) to establish a significant correlation. When Na excretion values above 270 mEq/24 hr were excluded, the correlation was very weak and no longer statistically significant (Figure 2).

From the foregoing considerations it appears that with appropriate methodology and a wide enough range of values it is possible to demonstrate a significant correlation between blood pressure and Na intake within a population. However, it remains true that in the middle range of Na intake values, which accounts for the bulk of the U.S. population, there is little correspondence between an individual's Na intake and blood pressure. The most probable explanation for this comes from the idea that essential hypertension is not a single entity, but rather that it is a heterogenous disease expressing itself in many forms and initiated and maintained by many different mechanisms. As Folkow[9] points out, the search for a unitary cause of essential hypertension is the source of much of the confusion surrounding this subject. Based on the low-renin concept, Laragh and Pecker[12] have estimated that approximately 30% of individuals with essential hypertension may have a Na-dependent form of the disease. The concept that only a certain percentage of hypertensive people have a Na-dependent form of the disease represents an important modification of the sodium hypothesis. A corollary to this concept is that NaCl sensitivity is a genetic factor and that many people may have this trait without developing hypertension because of low NaCl intake. Identifying such people then becomes an important public health consideration. Recent experiments[21-24] have shown that individuals with essential hypertension and their normotensive close relatives have abnormalities in red cell Na transport mechanisms. On the assumption that such abnormalities affect other cell types, in particular those of vascular smooth muscle, much interest has focused on the possibility of using this trait as a simple genetic marker. Although there are some conflicting results, part of the confusion may be related to the expectation that all essential hypertensives have this trait, whereas in reality essential hypertension is not a unitary phenomenon and it is misleading to look for a single mechanism that will apply to all manifestations of the disease, even at the cellular level.

### 2. Salt Deprivation and Salt Loading Studies

The second line of evidence for the role of Na arises from the various therapeutic approaches used to treat hypertension. One of the most widely known is Kempner's rice-fruit diet.[25] This diet is low in protein and Na (less than 10 mEq/24 hr), with a correspondingly low Na:K ratio, and its use led to reductions in blood pressure for about 60% of the patients. Others[26-28] showed that the diet resulted in contraction of extracellular fluid volume and that the beneficial effects were due to Na restriction. Laragh and Pecker[12] proposed that other factors besides salt restriction may have played a role, since the diet was not appetizing and weight loss was common. Reisen et al.[29] and Tuck et al.[30] have shown in carefully controlled studies that weight loss has strong effects on lowering blood pressure even when salt intake is not restricted. Such results tend to complicate the interpretation of severe salt restriction experiments when weight loss is not controlled. Several other severe salt restriction studies in the 1940s and early 1950s, reviewed by Laragh and Pecker,[12] reported effectiveness rates between 20 and 40%, whereas studies with less drastic Na restriction did not produce beneficial effects.

More recently, prospective studies of moderate salt restrictions,[31-33] alone or combined with diuretic therapy, yielded encouraging results. These trials did not select individuals who might be more sensitive to Na restriction. Although the reductions in blood pressure achieved by these trials were modest, (averaging 8 mmHg), Luft and Weinberger[34] point out that such a pressure reduction, based on the Framingham data,[35] would produce a change in death rate in the order of 10 to 15%.

The important concept of NaCl-sensitive and non-NaCl-sensitive hypertensive individuals was beautifully illustrated in two very carefully controlled studies from Bartter's laboratory.[36,37] The subjects (19 in the first and 15 in the second) were housed in a metabolic ward and kept on a balance regimen. The basic diet was low in Na (9 mEq/day) to which was added 100 mEq Na per day to make a normal Na diet, and 240 mEq Na per day for the high-Na diet. Starting with the normal-Na diet, 6 days were spent on each diet. Body weight was measured each morning; all urine was collected every day. Blood was collected every 4 hr on the 6th day and 2 more samples were taken on day 7. Blood pressure was measured every 4 hr, and on the 6th day, it was measured every 30 min. In the second group, cardiac output by echocardiography and peripheral resistance were calculated on day 6 of the low- and the high-NaCl diet.

Mean blood pressure dropped 9 mmHg by the 6th day on the low-Na diet and increased by 11 mmHg at the end of the high-Na diet. In 9 of the 19 patients in the first study, and in 8 of the 15 in the second, mean blood pressure rose by 10% or more of that patient's mean value on the low-Na diet. These patients were classified as "NaCl sensitive", the remainder "non-NaCl sensitive". The NaCl-sensitive subjects showed greater Na retention, greater weight gain, and a larger increase in blood pressure than the non-NaCl-sensitive subjects on the high-Na diet. The second study showed that the NaCl-sensitive patients had significantly increased cardiac output and decreased total peripheral resistance. These results suggest that the increase in blood pressure produced by NaCl loading in the NaCl-sensitive patients was brought about by an increase in blood volume with a corresponding increase in venous return to the heart. The authors emphasize that these results apply only to acute NaCl loading and do not indicate whether chronic high-NaCl intake would eventually lead to increased peripheral resistance and normal cardiac output, as is found in the established phase of essential hypertension.

NaCl loading of normotensive individuals produced a considerable heterogeneity in response in experiments conducted by Luft et al.[38-40] Very large intakes of Na (800 mEq/day) were required to raise blood pressure, and some individuals demonstrated no increase in blood pressure, even at 1500 mEq/day Na intake. Black subjects showed a greater increase than whites. Na loading augmented K excretion, leading to a net loss. When negative K balances were avoided, the increases in blood pressure were significantly reduced, thus illustrating the need for considering the interaction between Na and K in studying the role of Na in hypertension. Recent studies referred to below suggest that Ca and other ions may also play a role in this disease.

### 3. Animal Experiments

Experiments in rats have shown a linear correlation between Na intake and blood pressure (Figure 3).[41-43] As in humans, there was a considerable scatter of individual values. In addition, some of these studies[41,42] showed significant differences in survival time with the median duration of life for rats eating 8.4% sodium chloride being 8 months less than the controls. Experiments carried out on older rats produced effects that were less severe, and the effect was less on female rats. Some have questioned the applicability of such high Na levels to the normal human range,[12,44] but theoretical calculations based on differences in food turnover and percentage of nutrients have shown that there was a close approximation between the general human range of NaCl intake and that used in the rat experiments.[45] However, the most convincing experiment on the effect on rats of a human level of NaCl intake was performed by Dahl et al.[46] when they showed that after 6 months on a commercial baby food diet (before their NaCl levels were reduced), 25 test rats had blood pressures of 190 vs. 138 mmHg for the controls. Of the test rats, 12 died and 2 others became seriously ill after 8 months of study, whereas the controls were all alive at 8 months.

In view of the large variability in the response to NaCl, Dahl et al.[43] selectively bred rats over five to seven generations and produced two strains which were either sensitive (S) or resistant (R) to the effect of salt. The R and S strains have been widely studied since then as possible models for the physiological mechanisms underlying the development of NaCl-dependent hypertension.

Another widely studied model of essential hypertension, the spontaneously hypertensive rat (SHR), showed an elevated preference and intake of Na and K salt solutions,[20,47-50] but developed high blood pressure without NaCl in the diet, although added NaCl increased the level that the pressure finally reached.[51,52] Adding K to the diet promoted NaCl excretion and ameliorated the pressure elevation.[51]

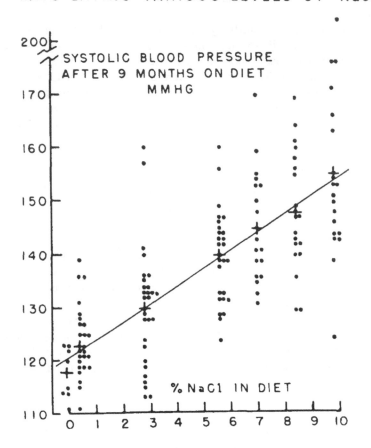

FIGURE 3    Hypertensive action of dietary sodium chloride in male rats. (From Ball, C. O. T. and Meneely, G. R., *J. Am. Dietet. Assoc.*, 33, 366, 1957. With permission.)

Dahl's S rats, on the other hand, showed an aversion to NaCl[53] when compared to the R strain, as did rats made hypertensive by interference with normal kidney function.[54,55]

A new perspective on the relationship between NaCl intake and hypertension was opened by recent work[56-60] showing that chronic exposure to stress or chronic administration of stress-related hormones have a strong stimulatory effect on NaCl intake and blood pressure in rats, rabbits, and sheep. Such findings provide evidence for the idea that the stressful conditions of modern life are one of the most important factors explaining the high incidence of hypertension in industrialized societies and its virtual absence in pre-modern societies. They also provide a good explanation for the positive correlation found between NaCl intake and blood pressure when different populations are compared.

In summary, the evidence obtained from animal experiments provides support for the idea that a high level of NaCl consumption is a very important causal factor in a substantial percentage of the population which already has, or is in the process of developing hypertension.

## B. Role of Other Ions

As indicated in the previous section, K and Ca are other ions which have been implicated in the development of hypertension. This is not surprising in view of the close interrelationship that exists between Na and K in cellular processes and the role of Ca in the mechanism of smooth muscle contraction. In addition, there are some data[61] suggesting that Mg may be involved, along with Ca, and that some of the effects of Na may be dependent on chloride, since other Na salts do not have the same effects.[62] Some studies[63] have also shown that essential trace elements, such as Zn, Fe, and Cu, may be involved because of their participation in enzyme reactions related to the control of blood pressure, and that toxic heavy metals, such as Cd, Pb, Hg, and Tl may cause hypertension[64] by affecting hormone metabolism, vasoconstriction, and renal tubular function. However, there is no evidence that these represent primary causes of the disease in the American population. Most of the evidence for the role of other ions relates to K and Ca, and as for Na it comes from epidemiological and experimental studies in humans and animals.

### 1. Potassium

Some of the best evidence for the role of K in hypertension comes from populations with low salt intake. In addition to having a low Na intake, such populations usually have a correspondingly elevated intake of K, not only in comparison to their low Na intake, but also to the high Na intakes of modern populations. A well-documented example is provided by the Yanomamo Indians of northern Brazil and southern Venezuela.[65] They number 12,000 to 15,000 distributed among 150 villages in an area of approximately 100,000 mi². Since 1966 they have been carefully studied by many multidisciplinary groups. The linguistic and genetic data indicate that this tribe has had a high degree of isolation for at least the past several thousand years.[66] Their diet is very simple, with the major staple consisting of the cooking banana, supplemented by irregular additions of game, fish, insects, and wild vegetable foods. The tribe had no access to NaCl until it was introduced by caucasians. In a representative sample of 506 Indians from 16 different villages, blood pressure average 102/64, with no tendency to increase with age. In a small subgroup from which 24-hr urine samples were obtained, daily excretion averaged 1 mEq of Na and 152 mEq of K, whereas the control values from the members of the expedition averaged 104 mEq of Na and 37 mEq of K. Correspondingly, the Indians had very high levels of aldosterone and plasma renin activity, in the range found in North Americans placed on short-term low-Na diets. Such elevated levels of aldosterone and renin were probably the norm for people during much of human evolution and suggest that the values observed in modern controls are depressed by a biologically excessive NaCl intake in present-day diets. It is important to note that the absence of hypertension in these people cannot be due to low Na or high K intakes alone, since these people are seldom obese and rarely demonstrate weight gain with advance in age. Furthermore, they are physically very active with an aggressive lifestyle very different from our present culture,[67] suggesting that their low blood pressure is not due to a placid and undisturbed lifestyle.

Epidemiologic studies in this country have also shown that even when there was no correlation between Na excretion and blood pressure, significant correlations were found between urinary Na:K ratio and blood pressure.[68,69] Evidence from a study by Grim et al.[70] in Evans County, Ga., showed that, as in almost all American studies, blacks had significantly higher blood pressures than whites, that they ate and excreted similar quantities of Na, but that the dietary K intake of blacks was consistently less than that of whites. In another study in which there was no difference in urinary Na excretion between normotensive and hypertensive individuals, there was a highly significant negative correlation between urinary K excretion and blood pressure.[71,72] Similarly, in an analysis of the HANES I survey, McCarron et al.[73] found that significant decreases in the consumption of Ca, K, and vitamins A and C were the nutritional factors that distinguished hypertensive from normotensive subjects.

Experimental studies in which K intake has been increased or decreased have produced mixed results.[74-76] Many studies have shown that K depletion decreased blood pressure in normal animals and in hypertensive animals and humans, whereas high K intake has also been found to lower blood pressure in hypertensive animals and humans. However, high K intake has not been effective in lowering blood pressure in normal humans, dogs, and rats.

According to Paller and Linas,[75] the mechanism by which elevated K reduces blood pressure in hypertensive animals has never been shown to be dissociated from its effect of increasing Na excretion. The decrease in blood pressure produced by chronic K deficiency may be a direct vasodilatory effect on vascular smooth muscle. In view of these mixed effects, further studies are required to determine how body levels of K could be linked to the high incidence of hypertension in this country and whether dietary or pharmacological control of K levels would be useful in treating hypertension.

### 2. Calcium

Interest in the potential role of Ca in hypertension is of more recent origin. It began with the work of Kobayashi in 1957 (described by Shaper et al.[77]), who showed that the death rate from stroke, which was the leading cause of death in Japan at that time, was lower in regions in which the river water had higher levels of calcium carbonate. Schroeder[78,79] went on to show, both in Japan and in the U.S., that soft- and hard-water areas had considerable differences in cardiovascular death rates, especially with respect to hypertension as an associated factor. Shaper et al.[77] found significantly higher percentages of hypertension in six soft-water towns, than in six matched hard-water towns in England. Although many other factors, such as Cd in soft water, have been implicated, Ca remains the most likely candidate for the beneficial effects of hard water.[80]

More direct evidence for the role of Ca in hypertension comes from the work of McCarron et al.[73,81,82] In a pilot survey McCarron and associates[81] found that subjects with essential hypertension had significantly lower daily Ca ingestion than normotensive controls matched for age, sex, and race. The difference in Ca intake arose primarily from lower ingestion of nonfluid dairy products. In a subgroup of this population[82] he also found that the hypertensives had lower serum concentrations of ionized Ca. Although the difference was statistically significant, it was very small. In an analysis of the HANES I data, referred to in the previous section, McCarron et al.[73] found that a lower Ca intake was the most consistent factor in hypertensive individuals, independent of age, race, sex, body mass index, and alcohol consumption. Strikingly, they also found that diets low in Na were associated with higher blood pressures, while high-Na diets were associated with the lowest blood pressures. The inverse relationship found by these authors between Na and blood pressure flies in the face of a huge amount of other epidemiological and experimental data and serves to add more fuel to the controversial fire over the role of sodium in hypertension.

On the other hand, in an epidemiological survey of blood pressure in 9321 men in the Belgian army, Kesteloot and Geboers[83] found an independent and highly significant correlation between serum Ca and both systolic and diastolic pressure. A significant, but weaker, correlation was also found between 24-hr urinary Ca excretion and blood pressure. In this same study, neither a positive correlation between 24-hr Na excretion nor a negative one between 24-hr K excretion and blood pressure could be demonstrated.

The role of Ca has received support from studies in Puerto Rico[84] involving 7932 men aged 45 to 64 years. A twofold increase in hypertension was found in subgroups who drank no milk compared to those who consumed over 1 quart of milk per day. Similar trends were found when an estimate of total Ca intake from food, principally from milk, was used. In another study,[85] it was shown that in populations with low Ca intake, the incidence of toxemia of pregnancy, which involves hypertension, is higher. Guatemalan women, with a diet low in calories, proteins, and vitamins, but high in Ca, have one of the lowest incidence rates of eclampsia in the world. Poor Guatemalan populations ingest large amounts of Ca as corn tortillas, due to the maize being cooked and soaked in lime water, and the tortilla cooked on a grill coated with calcium hydroxide. Worldwide incidence of eclampsia is inversely related to daily Ca intake. Although low Ca intakes are usually related to low milk intakes and low socioeconomic level, Guatemalans and Ethiopians, who eat a grain high in amino acids, Ca, and Fe, are exceptions to the rule of high incidence of eclampsia in populations of low socioeconomic level.

The role of Ca in hypertension has also received support in experimental animal studies. Ayachi[86] found that doubling the Ca concentration of regular lab chow attenuated the time course of hypertension in SHR rats. Although they became hypertensive, their final blood pressure was lower than the controls on the regular diet. The blood pressure lowering effect was probably due to an osmotic diuresis produced by the high Ca level. In a similar experiment, McCarron[87] found that the blood pressure of mature SHR rats was inversely related to the Ca content of their diet. It was suggested that the bioavailability of Ca is reduced in SHR, and possibly in human, hypertension and that dietary Ca supplementation may reverse this defect.

The role of Ca in hypertension is not yet fully determined and more studies will undoubtedly be performed in the near future. The importance of this topic is underlined by the fact that Ca intake tends to decrease with advancing age, whereas blood pressure increases, at least in modern societies. If the relationship between low Ca intake and hypertension can be firmly established, this may provide an important opportunity for nutritional intervention to forestall the development of hypertension in aging individuals.

### C. Other Nutritional Factors

Except for the effect of advancing age, the most consistent finding of population surveys of hypertension is the correlation of blood pressure with weight.[88-92] According to Pickering,[4] an eminent clinician and researcher in hypertension who died in 1980, no environmental factor of significance other than over-nutrition has been identified in Western societies. For him, and for many others in medicine, the evidence for the role of NaCl in hypertension is not persuasive in terms of the intake levels that occur in everyday life. It remains to be seen whether more recent evidence and concepts, as presented in previous sections, will be more convincing.

Corresponding to the strong relationship between obesity and high blood pressure is the strong effect that weight loss has on reducing blood pressure.[88,92] Although a decreased Na intake associated with weight loss has been proposed as a mechanism for the reduction in blood pressure, Reisin et al.[29] have shown that a significant reduction in blood pressure occurred with weight loss even when Na intake was maintained at a high level. Some of the mechanisms operating in obesity that may have an effect on blood pressure are higher insulin levels, which lead to greater Na retention. However, it is also possible that obesity and hypertension are but two expressions of an abnormality that affects endocrine and neuroendocrine factors and vascular smooth muscle.[89]

Recent studies have shown that dietary fats may play an important role in hypertension. In one set of experiments, blood pressure was significantly lowered by changing the ratio of polyunsaturated to saturated fats in the diet, without producing any effect on body weight and without altering the NaCl level of the diet.[93] The authors proposed that the effects on blood pressure were obtained by an increased synthesis of prostaglandins brought about by the higher level of linoleic acid in the high polyunsaturated fat diet. Similar results were reported in animal experiments.[94] Clearly, further studies are needed to clarify this interesting relationship, particularly with respect to the athero-arteriosclerotic hypertension found in older people.[6]

The role of sugars in hypertension has also been recently explored in animal and human experiments.[95] The evidence from these experiments suggests that sucrose in particular can raise blood pressure in NaCl-sensitive rats and can produce a transient elevation of blood pressure in human subjects. In some experiments the effect of the sugars seem to be related to depression of Na excretion. However, other experiments have shown that the pressure effects are independent of the antinatriuretic activity of sugars.

Finally, recent work has indicated that high-fiber diets have a lowering effect on blood pressure.[96] Such observations are in keeping with the absence of hypertension in nonindustrialized tribal cultures described at the beginning of this review. Although their lack of hypertension is usually attributed to a low level of NaCl intake, it is clear that the diet of such cultures also differs in the higher levels of complex carbohydrate and plant fiber that they contain.

## IV. CONCLUSION

It is clear from the foregoing review that the subject of hypertension is very complex and the role of nutrition in causing or preventing the disease is controversial. It is also apparent that even though much has been learned in the recent past, much more still needs to be learned. Some of the most important concepts to emerge from recent research are that hypertension is not a necessary accompaniment of aging, that it can arise from many different causes, that it is beneficial to treat it even at its earliest stages, and better yet to prevent it if possible. Although controversial, and understandably so because of the complex inter-relationships involved, the role of nutrition will become increasingly important as we learn more about the various mechanisms involved in the control of blood pressure. Even now, it is clear to many that hypertension can be prevented or lessened, if it is already established, by nutritional means. Although much of the emphasis has been placed on NaCl and rightly so in view of the large body of information available, the importance of K and Ca has come to recent attention as has the role of trace minerals and toxic metals. Together with the new information about fats, sugars, and fiber, the conclusion emerges that the long-established principles of good nutrition may provide the fundamental basis for solving one of the most important public health problems of the present day. This is a great challenge to all of us.

## REFERENCES

1. **Kannel, W. B.**, Host and environmental determinants of hypertension, in *Epidemiology of Arterial Blood Pressure*, Kesteloot, H. and Joosens, J. V., Eds , Martinus Nijhoff, The Hague, 1980, 265.
2. **Freis, E. D.**, Salt, volume and the prevention of hypertension, *Circulation*, 53, 589, 1976.
3. **Page, L. B.**, Epidemiological evidence on the etiology of human hypertension and its possible prevention, *Am Heart J.*, 91, 527, 1976.
4. **Pickering, G.**, Position paper: dietary sodium and human hypertension, in *Frontiers in Hypertension Research*, Laragh, J H., Buhler, F. R., and Selden, D. W., Eds., Springer-Verlag, New York, 1981, 37.
5. **Julius, S. and Hansson, L.**, Classification of hypertension, in *Hypertension*, 2nd ed., Genest, J , Kuchel, O., Hamet, P., and Cantin, M., Eds., McGraw-Hill, New York, 1983, 679.
6. **Genest, J., Larochelle, P., Kuchel, O., Hamet, P., and Cantin, M.**, Hypertension in the elderly· atheroarteriosclerotic hypertension, in *Hypertension*, 2nd ed., Genest, J , Kuchel, O., Hamet, P., and Cantin, M., Eds , McGraw-Hill, New York, 1983, 913
7. **Franklin, S. S.**, Geriatric hypertension, *Med. Clin N. Am.*, 67, 395, 1983
8. Hypertension Detection and Follow-up Program Cooperative Group, Five-year findings of the Hypertension Detection and Follow-up Program I. Reduction in mortality of persons with high blood pressure, including mild hypertension, *JAMA*, 242, 2562, 1979
9. **Folkow, B.**, Physiological aspects of primary hypertension, *Physiol. Rev.*, 62, 347, 1982.
10. **Fregly, M. J.**, Estimates of sodium and potassium intake, *Ann. Intern. Med.*, 98, 792, 1983.
11. **Meneely, G. R. and Battarbee, H. D.**, Sodium and potassium, in *Nutrition Reviews' Present Knowledge in Nutrition*, 4th ed., Nutrition Foundation, New York, 1976, 259.
12. **Laragh, J. H. and Pecker, M. S.**, Dietary sodium and essential hypertension: some myths, hopes, and truths, *Ann. Intern. Med.*, 98, 735, 1983
13. **Kolata, G.**, Value of low-sodium diets questioned, *Science*, 216, 38, 1982.
14. **Page, L. B.**, Hypertension and atherosclerosis in primitive and acculturating societies, in *Hypertension Update: Mechanisms, Epidemiology, Evaluation and Management*, Hunt, J. C , Cooper, T , and Frohlich, E. D., Eds., HLS Press, Bloomfield, N J., 1980, 1.
15. **Stanton, J. L., Braitman, L. E., Riley, A. M., Khoo, C. S., and Smith, J. L.**, Demographic, dietary, life style, and anthropometric correlates of blood pressure, *Hypertension*, 4(Suppl. 3), 135, 1982.

16. **Dawber, T. R., Kannel, W. B., Kagan, A., Donabedian, R. K., McNamara, P. M., and Pearson, G.,** Environmental factors in hypertension, in *The Epidemiology of Hypertension*, Stamler, J., Stamler, R., and Pullman, T., Eds., Grune & Stratton, New York, 1967, 255.

17. **Blackburn, H. and Prineas, R.,** Diet and hypertension: anthropology, epidemiology, and public health implications, *Prog. Biochem. Pharmacol.*, 19, 31, 1983.

18. **Liu, K., Cooper, R., McKeever, J., McKeever, P., Byington, R., Soltero, I., Stamler, R., Gosch, F., Stevens, E., and Stamler, J.,** Assessment of the association between habitual salt intake and high blood pressure, *Am. J. Epidemiol*, 110, 219, 1979.

19. **Langford, H. G. and Watson, R. L.,** Electrolytes and hypertension, in *Epidemiology and Control of Hypertension*, Paul, O., Ed., Stratton Intercontinental Medical Book, New York, 1975, 119.

20. **Bernard, R. A., Doty, R. L., Engelman, K., and Weiss, R. A.,** Taste and salt intake in human hypertension, in *Biological and Behavioral Aspects of Salt Intake*, Kare, M. R., Fregly, M. J., and Bernard, R. A., Eds., Academic Press, New York, 1980, 397.

21. **Canessa, M., Adragna, N., Solomon, H. S., Connolly, T. M., and Tosteson, D. C.,** Increased sodium-lithium countertransport in red cells of patients with essential hypertension, *N. Engl. J. Med.*, 302, 772, 1980.

22. **Canessa, M., Bize, I., Adragna, N., and Tosteson, D.,** Cotransport of lithium and potassium in human red cells, *J Gen Physiol.*, 80, 149, 1982.

23. **Garby, R. P., Dagher, G., Pernollet, M. G., Devynck, M. A., and Meyer, P.,** Inherited defect in a $Na^+$, $K^+$-co-transport system in erythrocytes from essential hypertensive patients, *Nature (London)*, 284, 281, 1980.

24. **Garby, R. P., Elghozi, J. L., Dagher, G., and Meyer, P.,** Laboratory distinction between essential and secondary hypertension by measurement of erythrocyte cation fluxes, *N. Engl. J. Med.*, 302, 769, 1980.

25. **Kempner, W.,** Treatment of hypertensive vascular disease with rice diet, *Am. J. Med.*, 4, 545, 1948

26. **Grollman, A., Harrison, T. R., Mason, M. F., Baxter, J., Crampton, J., and Reichsman, F.,** Sodium restriction in diet for hypertension, *JAMA*, 129, 533, 1945.

27. **Murphy, R. J. F.,** The effect of ''rice diet'' on plasma volume and extracellular fluid space in hypertensive patients, *J. Clin. Invest.*, 29, 912, 1950.

28. **Watkin, D. M., Froeb, H. F., Hatch, F. T., and Gutman, A. B.,** Effects of diet in essential hypertension. II. Results with unmodified Kempner rice diet in fifty hospitalized patients, *Am. J. Med.*, 9, 444, 1950.

29. **Reisen, E., Abel, R., Modan, M., Silverberg, D. S., Eliahou, H. E., and Modan, B.,** Effect of weight loss without salt restriction on the reduction of blood pressure in overweight hypertensive patients, *N. Engl. J. Med.*, 298, 1, 1978.

30. **Tuck, M. L., Sowers, J., Dornfeld, L., Kledzik, G., and Maxwell, M.,** The effect of weight reduction on blood pressure, plasma renin activity, and plasma aldosterone levels in obese patients,. *N. Engl. J. Med.*, 304, 930, 1981.

31. **Parijs, J., Joossens, J. V., Van der Linden, L., Verstreken, G., and Amery, A.,** Moderate sodium restriction and diuretics in the treatment of hypertension, *Am. Heart J.*, 85, 22, 1973.

32. **Morgan, T., Gillies, A., Morgan, G., Adam, W., Wilson, M., and Carney S.,** Hypertension treated by salt restriction, *Lancet*, 1, 227, 1978.

33. **MacGregor, G. A., Markandu, N. D., Best, F. E., Elder, D. M., Cam, J. M., Sagnella, G. A., and Squires, M.,** Double-blind randomised crossover trial of moderate sodium restriction in essential hypertension, *Lancet*, 1, 351, 1982.

34. **Luft, F. C. and Wienberger, M. H.,** Sodium intake and essential hypertension, *Hypertension*, 4(Suppl. 3), 14, 1982.

35. Department of Health, Education and Welfare, The Framingham Study, Publ. No. (NIH)74-599, DHEW, Washington, D.C., 1974, Sec. 30.

36. **Kawasaki, T., Delea, C. S., Bartter, F. C., and Smith, H.,** The effect of high-sodium and low-sodium intakes on blood pressure and other related variables in human subjects with idiopathic hypertension, *Am. J. Med.*, 64, 193, 1978.

37. **Bartter, F. C., Fujita, T., Delea, C. S., and Kawasaki, T.,** On the role of sodium in human hypertension, in *Biological and Behavioral Aspects of Salt Intake*, Kare, M R., Fregly, M. J., and Bernard, R. A., Eds., Academic Press, New York, 1980, 341.

38. **Luft, F. C., Rankin, L. I., Block, R., Weyman, A. E., Willis, L. R., Murray, R. H., Grim, C. E., and Weinberger, M. H.,** Cardiovascular and humoral responses to extremes of sodium intake in normal black and white men, *Circulation*, 60, 697, 1979.

39. **Luft, F. C., Weinberger, M. H., and Grim, C. E.,** Sodium sensitivity and resistance in normotensive humans, *Am. J Med.*, 72, 726, 1982.

40. **Luft, F. C., Grim, C. E., Fineberg, N. S., and Weinberger, M. H.,** Effects of volume expansion and contraction in normotensive whites, blacks and subjects of different ages, *Hypertension*, 4, 494, 1982.

41. **Meneely, G. R., Tucker, R. G., Darby, W. J., and Auerbach, S. D.,** Chronic sodium chloride toxicity in albino rat. II. Occurrence of hypertension and of a syndrome of edema and renal failure, *J. Exp. Med.,* 98, 71, 1953.

42. **Ball, C. O. T. and Meneely, G. R.,** Observations on dietary sodium chloride, *J. Am. Dietet. Assoc.,* 33, 366, 1957.

43. **Dahl, L., Heine, M., Tassinari, L.,** Effects of chronic excess salt ingestion evidence that genetic factors play an important role in susceptibility to experimental hypertension, *J. Exp. Med.,* 115, 1173, 1962.

44. **Simpson, F. O.,** Salt and hypertension a skeptical review of the evidence, *Clin. Sci.,* 57, 463s, 1979.

45. **Meneely, G. R. and Dahl, L. K.,** Electrolytes in hypertension: the effects of sodium chloride, *Med. Clin N. Am.,* 45, 271, 1961

46. **Dahl, L. K., Heine, M., Litel, G., and Tassinari, L.,** Hypertension and death from consumption of processed baby foods in rats, *Proc. Soc. Exp. Biol. Med.,* 133, 1405, 1970.

47. **Catalanotto, F., Schechter, P. J., and Henkin, R. I.,** Preference for NaCl in the spontaneously hypertensive rat, *Life Sci.,* 11, 557, 1972

48. **McConnell, S. D. and Henkin, R. I.,** NaCl preference in spontaneously hypertensive rats: age and blood pressure effects, *Am J. Physiol.,* 225, 624, 1973

49. **Forman, S. and Falk, J. L.,** NaCl solution ingestion in genetic (SHR) and aortic ligation hypertension, *Physiol. Behav.,* 22, 371, 1979.

50. **Mogenson, G. J. and Morris, P.,** Increased salt appetite precedes the onset of arterial hypertension in spontaneously hypertensive rats, *Appetite,* 1, 167, 1980.

51. **Louis, W. J., Tabei, R., and Spector, S.,** Effects of sodium intake on inherited hypertension in the rat, *Lancet,* 2, 1283, 1971.

52. **Aoki, K., Yamori, Y., Ooshima, A., and Okamoto, K.,** Effects of high or low sodium intake in spontaneously hypertensive rats, *Jpn. Circ. J.,* 36, 539, 1972

53. **Wolf, G., Dahl, L. K., and Miller, N. E.,** Voluntary sodium chloride intake of two strains of rats with opposite genetic susceptibility to experimental hypertension, *Proc. Soc. Exp. Biol Med.,* 120, 301, 1965.

54. **Abrams, M., DeFriez, A. I. C., Tosteson, D. C., and Landers, E. M.,** Self-selection of salt solutions and water by normal and hypertensive rats, *Am J. Physiol ,* 156, 233, 1949

55. **Fregly, M. J.,** Effect of renal hypertension on the preference threshold of rats for sodium chloride, *Am. J. Physiol.,* 187, 288, 1956.

56. **Bernard, R. A., Priehs, T. W., Fink, G. D., and Nachreiner, R. F.,** Sustained elevation of plasma ACTH increases salt intake and blood pressure in rats, *Physiologist,* 25, 280, 1982

57. **Blaine, E. H., Covelli, M. D., Denton, D. A., Nelson, J. F., Shulkes, A. A.,** The role of ACTH and adrenal glucocorticoids in the salt appetite of wild rabbits *[Oryctolagus cuniculus (L.)],* *Endocrinology,* 97, 793, 1975

58. **Denton, D. A. and Neslon, J. F.,** The influence of reproductive processes on salt appetite, in *Biological and Behavioral Aspects of Salt Intake,* Kare, M. R., Fregly, M. J., and Bernard, R. A., Eds., Academic Press, New York, 1980, 229.

59. **Freeman, R. H., Davis, J. O., and Fullerton, D.,** Chronic ACTH administration and the development of hypertension in rats, *Proc. Soc. Exp. Biol. Med.,* 163, 473, 1980

60. **Weisinger, R. S., Denton, D. A., McKinley, M. J., and Nelson, J. F.,** ACTH induced sodium appetite in the rat, *Pharm. Bioch. Behav.,* 8, 339, 1978

61. **McCarron, D. A.,** Calcium and magnesium nutrition in human hypertension, *Ann. Intern. Med.,* 98, 800, 1983

62. **Kotchen, T. A., Luke, R. G., Ott, C. E., Galla, J. H., and Whitescarver, S.,** Effect of chloride on renin and blood pressure responses to sodium, *Ann. Intern. Med.,* 98, 817, 1983.

63. **Saltman, P.,** Trace elements and blood pressure, *Ann Intern. Med.,* 98, 823, 1983

64. **Perry, H. M. and Perry, E. F.,** Metals and human hypertension, in *Epidemiology and Control of Hypertension,* Paul, O., Ed., Stratton Intercontinental Medical Book, New York, 1975, 147

65. **Oliver, W. J., Cohen, E. L., and Neel, J. V.,** Blood pressure, sodium intake, and sodium related hormones in the Yanomamo Indians, a 'no-salt' culture, *Circulation,* 52, 146, 1975.

66. **Neel, J. V. and Weiss, K. M.,** The genetic structure of a tribal population, the Yanomamo Indians XII. Biodemographic studies, *Am. J. Phys. Anthropol ,* 42, 25, 1975.

67. **Chagnon, N. A.,** *Yanomamo: The Fierce People,* Holt, Rhinehart & Winston, 1968.

68. **Langford, H. G.,** Dietary potassium and hypertension: epidemiologic data, *Ann. Intern. Med ,* 98, 770, 1983.

69. **Watson, R. L., Langford, H. G., Abernethy, J., Barnes, T. Y., and Watson, M. J.,** Urinary electrolytes, body weight, and blood pressure pooled cross-sectional results among four groups of adolescent females, *Hypertension,* 2, 193, 1980.

70. **Grim, C. E., Luft, F. C., Miller, J. Z., Meneely, G. R., Battarbee, H. D., Hames, C. G., and Dahl, L. K.,** Racial differences in blood pressure in Evans County, Georgia: relationship to sodium and potassium intake and plasma renin activity, *J Chronic Dis.,* 33, 87, 1980

71. **Walker, W. G., Whelton, P. K., Saito, H., Russell, R. P., and Hermann, J.**, Relation between blood pressure and renin, renin substrate, angiotensin II, aldosterone and urinary sodium and potassium in 574 ambulatory subjects, *Hypertension*, 1, 287, 1979.

72. **Walker, W. G.**, Relationships between sodium and potassium intake and blood pressure, in *Epidemiology of Arterial Blood Pressure*, Kesteloot, H. and Joossens, J. V , Eds., Martinus Nijhoff, The Hague, 1980, 297.

73. **McCarron, D. A., Morris, C. D., Henry, H. J., and Stanton, J. L.**, Blood pressure and nutrient intake in the United States, *Science*, 224, 1392, 1984.

74. **Tannen, R. L.**, Effects of potassium on blood pressure control, *Ann. Intern. Med.*, 98, 773, 1983

75. **Paller, M. S. and Linas, S. L.**, Hemodynamic effects of alterations in potassium, *Hypertension*, 4(Suppl. 3), 20, 1982

76. **Treasure, J. and Ploth, D.**, Role of dietary potassium in the treatment of hypertension, *Hypertension*, 5, 864, 1983.

77. **Shaper, A. G., Clayton, D. G., and Stanley, F.**, Water hardness and hypertension, in *Epidemiology and Control of Hypertension*, Paul, O , Eds , Stratton Intercontinental Medical Book, New York, 1975, 163.

78. **Schroeder, H. A.**, Degenerative cardiovascular disease in the Orient. II. Hypertension, *J. Chronic Dis* , 8, 312, 1958.

79. **Schroeder, H. A.**, Relation between mortality from cardiovascular disease and treated water supplies. Variations in states and 163 largest municipalities of the United States, *JAMA*, 172, 1902, 1960.

80. **Joossens, J. V.**, Salt and hypertension, water hardness and cardiovascular death rate, *Triangle*, 12, 9, 1973.

81. **McCarron, D. A., Morris, C. D., and Cole, C.**, Dietary calcium in human hypertension, *Science*, 217, 267, 1982.

82. **McCarron, D. A.**, Low serum concentrations of ionized calcium in patients with hypertension, *N. Engl. J. Med.*, 307, 226, 1982

83. **Kesteloot, H. and Geboers, J.**, Calcium and blood pressure, *Lancet*, 1, 813, 1982.

84. **Garcia-Palmieri, M. R., Costas, R., Cruz-Vidal, M., Sorlie, P. D., Tillotson, J., and Havlik, R. J.**, Milk consumption, calcium intake, and decreased hypertension in Puerto Rico, *Hypertension*, 6, 322, 1984.

85. **Belizan, J. M. and Villar, J.**, The relationship between calcium intake and edema-proteinuria- and hypertension-gestosis: an hypothesis, *Am. J. Clin. Nutr.*, 33, 2202, 1980

86. **Ayachi, S.**, Increased dietary calcium lowers blood pressure in the spontaneously hypertensive rat, *Metabolism*, 28, 1234, 1979.

87. **McCarron, D. A.** Calcium, magnesium, and phosphorus balance in human and experimental hypertension, *Hypertension*, 4(Suppl. 3), 27, 1982.

88. **Berchtold, P., Jorgens, V., Kemmer, F. W., and Berger, M.**, Obesity and hypertension: cardiovascular response to weight reduction, *Hypertension*, 4(Suppl. 3), 50, 1982.

89. **Dustan, H. P.**, Mechanisms of hypertension associated with obesity, *Ann. Intern. Med* , 98, 860, 1983.

90. **Havlik, R. J., Hubert, H. B., Fabsitz, R. R., and Feinleib, M.**, Weight and hypertension, *Ann. Intern. Med.*, 98, 855, 1983.

91. **Sims, E. A. H.**, Mechanisms of hypertension in the overweight, *Hypertension*, 4(Suppl. 3), 43, 1982.

92. **Tyroler, H. A., Heyden, S., and Humes, C. G.**, Weight and hypertension: Evans County studies of blacks and whites, in *Epidemiology and Control of Hypertension*, Paul, O., Ed., Stratton Intercontinental Medical Book, New York, 1975, 177.

93 **Iacono, J. M., Dougherty, R. M., and Puska, P.**, Reduction of blood pressure associated with dietary polyunsaturated fat, *Hypertension*, 4(Suppl. 3), 34, 1982.

94. **Smith-Barbaro, P. A. and Pucak, G. J.**, Dietary fat and blood pressure, *Am Intern. Med.*, 98, 828, 1983.

95 **Hodges, R. E. and Rebello, T.**, Carbohydrates and blood pressure, *Am. Intern. Med.*, 98, 838, 1983.

96. **Anderson, J. W.**, Plant fiber and blood pressure, *Ann. Intern Med.*, 98, 842, 1983.

Chapter 9

## DIVERTICULAR DISEASE

**Michael J. Monzel and Thomas P. Almy**

## TABLE OF CONTENTS

## I. INTRODUCTION

In the 19th century diverticula of the colon and their involvement with disease processes were so rare as to be a reportable pathologic entity.[1] Over the first 80 years of the 20th century a dramatic increase in the incidence of diverticular disease has occurred. This incidence consistently increases with age in all populations surveyed to date in which diverticular disease is prevalent.[2] Autopsy series and other epidemiologic studies suggest that approximately 50% of the population in the 7th through 9th decades will be affected by diverticular disease.[3,4] Nonetheless, it is of foremost importance in any discussion of diverticular disease to stress that the vast majority of people with colonic diverticula will remain totally asymptomatic throughout life. Furthermore, it is probable that another larger segment of this population will only develop minimal symptoms of abdominal pain or constipation which will never prompt them to seek medical attention. As close as can be estimated, 20% of patients with colonic diverticula will develop significant symptoms or complications of the disease.[4] A sound prospective survey, geared toward more precisely defining the incidence of disease, is yet to be performed.

As striking as the recent increase in incidence is the occurrence of diverticular disease almost exclusively in industrialized Western-type cultures.[4] This fact, plus the observation that the disease process is appearing in nations undergoing rapid cultural change, such as Japan and South Africa, has led to the conclusion that environmental factors must play a strong role in the evolution of colonic diverticula.[5,6] As this evolution coincided with marked changes in dietary patterns in these cultures, including decreases in consumption of dietary fiber, the hypothesis is now widely entertained that diverticular disease is the result of a nutritional deficiency state.[7] This is not a universally accepted statement, but epidemiologic, pathologic, clinical, and physiologic data continue to support this hypothesis.

## II. EPIDEMIOLOGY

From the late 1880s through the 1940s a progressive increase in case reports in British, German, and American literature is found.[1] By 1930 Mayo and associates[8] reported that 5% of autopsies in people over the age of 40 revealed colonic diverticula. The striking aspect of these clinical reports is that they are derived almost exclusively from the more industrialized Western cultures. Diverticular disease remains an extremely rare disease in areas of Africa, Asia, and South America. Contrasted with the estimated 20% overall prevalence in the U.S., southern Iran and Jordan had prevalence rates of 2.4 and 4.0%, respectively, when radiologic surveys were performed on adult populations.[9,10] In some countries a marked difference in frequency among different ethnic and racial groups exists. In Israel, Ashkenazic Jews have a much higher prevalence of colonic diverticula than Sephardic Jews, while among the Arabs in Israel, the condition remains a rarity.[11] A similar striking discrepancy is found in Johannesburg, when South African blacks are compared with European descendants in the same population.[12] However, the prevalence in the urban South African black is increasing, as it is with Japanese immigrants who have lived a majority of their lifetime in Hawaii.[6,13]

These geographic patterns of prevalence and the emergence of disease over 10- to 30-year time spans, once Western habits are adopted point strongly to an environmental cause. The most consistently reported environmental factor in high-incidence groups is a change in the diet, namely the decline in dietary fiber intake in persons with diverticula vs. age-matched controls.[14] In England it is believed that in the 1970s the crude fiber intake per person was one tenth of that in 1880.[1] From 1909 to 1975 the crude fiber content of the American diet is estimated to have dropped 28%.[15] This historical perspective and evolution of fiber intake will be treated in more detail in a later section.

The hypothesis of a dietary deficiency integrally related to a common disorder of the large intestine is extremely attractive. Changes in the biochemical environment present in the large intestine may create alterations in function and motility. The fact that the death rate from diverticulitis in Great Britain decreased during World War II may be significantly related to a wartime policy to substitute more fruit, grains, and potatoes for meat and more refined carbohydrates.[16] Vegetarians studied in Oxford, England had one third the frequency of diverticula seen in age-matched controls from the same area.[17] Their mean intake of dietary fiber was approximately double that of the control group.

Another important association is the statistically significant increase of certain other diseases in patients with diverticula. There has been a fairly wide acceptance of a dietary basis in some aspects of their pathogenesis. The frequency of cholelithiasis, diabetes mellitus, ischemic heart disease, varicose veins of the legs, and hiatal hernia are increased.[4,18-22] Carcinoma of the colon, which some observers have suggested may also result in part from fiber deficiency, has not been significantly associated in case control studies with diverticulosis.

These numerous associations point strongly to a dietary deficiency as a common etiologic factor. However, it must be remembered that association does not establish cause and effect. A prospective controlled study is needed to fully evaluate the association between diet and prevalence of diverticular disease.

## III. CLINICAL SPECTRUM

### A. Definition

Precise definition of the terms used in describing diverticular disease is essential. It has long been customary to refer to the presence of diverticula as *diverticulosis*. This continues to be accepted.[7] The condition of diverticulosis implies no specific active disease process, but can be either asymptomatic, as usually is the case, or symptomatic. *Diverticulitis,* on the other hand, refers specifically to the pathologic process of necrotizing peridiverticular inflammation, which may involve one or more diverticula, and can be identified by distinct clinical signs and symptoms which will be elaborated on below. Thus, a patient whose diverticula are discovered incidentally on a barium enema roentgenogram has diverticulosis. The labeling of this patient as having diverticulitis, which is not an uncommon error, would be inappropriate. Table 1 summarizes these differences.

In order to simplify discussion, it has become common practice to refer to patients with symptomatic or asymptomatic diverticulosis, as well as those with complications of the process, such as diverticulitis, as having *diverticular disease*. Patients with uncomplicated diverticular disease can be subdivided into *simple massed diverticulosis,* in which the colon is diffusely involved with diverticula, and *spastic colon diverticulosis,* where the sigmoid and descending colon are chiefly involved together with considerable muscle spasm.

As stated previously, the vast majority of colonic diverticula are acquired deformities of the intestine. They are not true diverticula, as the walls do not include the muscle layers, and consist only of mucosa, submucosa, and serosa. Essentially, mucosa and submucosa herniate through the circular muscle layers[23] (Figure 1). This herniation typically occurs where an arterial branch passes through the circular muscle coat.[24] The relationship of mucosa, submucosa, blood vessel, and serosa is very similar to the relationship of the bowel and spermatic cord in an inguinal hernia in a male.[7] Thus, the typical colonic diverticulum is a pseudodiverticulum.

Another important distinction to be made is that of the frequently identified prediverticular state, or what is commonly termed *myochosis*. This term is most commonly used when, on barium enema examination or flexible endoscopic examination of the sigmoid colon, the lumen of the bowel in the sigmoid and descending colon is noted to be narrowed and deeply indented and corrugated. This narrowing is produced by thickened folds of circular muscle.[23]

**Table 1**
**DIVERTICULAR DISEASE AND ITS**
**SUBSETS**

| | |
|---|---|
| Diverticulosis | Presence of diverticula |
| |     Spastic colon diverticulosis |
| |     Simple massed diverticulosis |
| *Myochosis coli* | Spastic colonic segment with or |
| |     without diverticula |
| Single-diverticulum | Acquired or congenital |
| Diverticulitis | Peridiverticular inflammation |

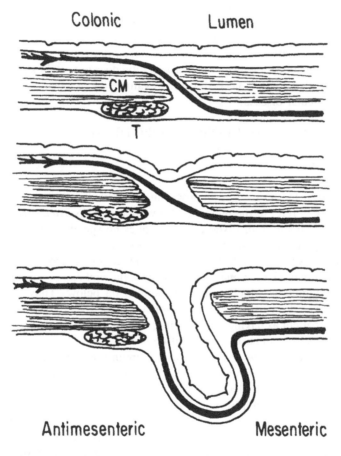

FIGURE 1.   This drawing illustrates the colonic mucosa, circular muscle layer (CM), taenia coli (T), nutrient arteriole, and serosa. (From Meyers, M. A., Alonso, D R., and Gray, G. F., *Gastroenterology*, 71, 577, 1976. With permission.)

Both the circular muscle layer and the longitudinal muscles, the taeniae, are shortened. The majority of diverticula are found in this same region. Over 90% of all patients with diverticula have sigmoid colon involvement, and often the sigmoid and descending colon alone are involved. This is frequently the case when marked myochosis (Figure 2, see glossy insert) is also present in that region.[25] On the other hand, it is almost as common to find diffuse involvement of the entire colon by diverticula.[26] This is especially true of people in the 7th through 9th decades of life. Isolated right colonic diverticula and those occurring in the transverse colon are much rarer.

## B. Clinical Features of Uncomplicated Diverticular Disease

It is estimated that 80% of patients with diverticulosis remain asymptomatic throughout life.[23] When symptoms occur they typically include variable patterns of constipation, diarrhea, and colicky lower abdominal pain.[25] The discomfort is most often felt in the left lower quadrant of the abdomen, and tends to persist for hours to several days. Most affected individuals are able to continue with their daily routine. The pain often worsens after ingestion of a meal, and relief is often associated with a bowel movement. Increase in flatulence and dyspeptic symptoms are not infrequent.

Objective physical findings are usually sparse in uncomplicated symptomatic diverticular disease, and the physical examination may be totally normal. Frequently, a tender loop of sigmoid colon is palpable. Diffuse abdominal distention is also not uncommon. Physical findings of more severe inflammation such as absent bowel sounds, abdominal rigidity, and rebound tenderness are not part of the spectrum of uncomplicated diverticular disease and should alert the observer to a much more devastating process, such as localized bowel perforation or diffuse peritonitis.

A clinical presentation of abdominal pain and tenderness will prompt most physicians to proceed with a barium enema roentgenogram. This is the best available method of demonstrating diverticula and contributes to the exclusion of the other disease states in the colon, which may present with the same symptoms and signs. (A typical example of diverticulosis and myochosis is demonstrated in Figure 2.) The conclusion that the diverticula observed on X-ray produced the patient's symptoms and signs, however, is another matter. This can only be made when conditions such as colon cancer, irritable bowel syndrome, adenomatous polyps, and inflammatory bowel disease have been excluded with a high probability.[4] The decision to perform more testing must be made with knowledge of the high prevalence of diverticula in patients with other lower bowel disease entities. In difficult cases it may be appropriate to examine the bowel by flexible fiberoptic sigmoidoscopy or colonoscopy. This procedure would be indicated, for example, in a patient who relates a history of fairly typical lower abdominal pain, but is found to have occult GI blood loss. In this situation a more aggressive diagnostic approach often would be mandated, for it is common for occult blood loss to be an early sign of colon neoplasm but unusual in uncomplicated diverticular disease. Nonetheless, the vast majority of patients with symptomatic diverticular disease will be sufficiently evaluated by the barium enema studies and will not require more invasive study. It should be stressed that although barium studies are useful in mild disease, those patients who present with a clinical picture in which diverticulitis is suggested should not immediately undergo barium enema study, as this may aggravate the condition or even cause diverticular perforation. There are no other specific laboratory tests of value in uncomplicated diverticular disease.

## C. Complications
### 1. Diverticulitis

Diverticulitis is the most common complication of diverticular disease, and occurs in various series in 10 to 25% of patients with diverticulosis.[4,26] Over approximately a 6-year period in the 1950s, British investigators found a 17% incidence of diverticulitis in an older population of outpatients who previously were known to have had diverticulosis.[27] Furthermore, it is not uncommon for the first manifestation of diverticular disease to be acute diverticulitis.

Pathologically, localized inflammation occurs in the apex of the diverticulum and often is associated with the presence of a trapped fecalith within the diverticulum.[28] The inflammation in symptomatic diverticulitis then extends widely within or through the serosa, and usually remains confined by the adjacent organs or mesentery. The appearance of peridiverticular inflammation presupposes at least a microperforation of a diverticulum. The sigmoid and descending colon are the most common areas of involvement, most likely because of the increased frequency of diverticula in these areas.

Clinically, the picture involves persistent abdominal pain, usually located in the left lower quadrant, often in association with fever and chills.[23] Laboratory evidence of an inflammatory process, in the form of a leukocytosis or elevated sedimentation rate, is not uncommon. Localized tenderness, a palpable and unmovable mass in the left lower quadrant, involuntary guarding, and rebound tenderness are all common diagnostic findings. Many observers feel barium studies are usually contraindicated in acute diverticulitis.[4] It is felt that barium may at times induce a larger perforation and extravasate into the abdominal cavity, compounding the already existing localized peritonitis. When X-rays are taken the typical finding is extension of barium out of the involved diverticulum into the peridiverticular space, thus exhibiting a peridiverticular abscess.[29] The treatment of acute diverticulitis revolves around the principles of nothing by mouth, administration of antibiotics, and fluid replacement. More detailed discussion of the acute treatment of diverticulitis will be found elsewhere.[23]

Life-threatening complications of diverticulitis include massive diverticular perforation, abscess formation, and fistula formation. Most commonly seen are colovesical fistulas; coloenteric fistulae are only slightly less frequent.[30] Large perforations result in general peritonitis and usually require immediate surgical intervention.[31] Partial or total colonic obstruction can occur, and is usually associated with a severe episode of diverticulitis that results in a large peridiverticular and pericolonic mass.[32] A common complication of repetitive episodes of diverticulitis, regardless of their severity, is colonic stricture. This can produce intermittent partial colonic obstruction. On barium enema the appearance of such strictures can be difficult to distinguish from a neoplastic process, and may dictate endoscopic examination. It is essential to stress that the first manifestation of diverticular disease may be one of these complications.[23]

### 2. Diverticular Hemorrhage

Diverticular hemorrhage, which typically manifests itself by the sudden painless passage of a large amount of blood from the rectum, is the second major complication of diverticular disease. Although probably the most common cause of life-threatening lower GI blood loss in the elderly, this complication is estimated to occur in from 3 to 5% of elderly individuals with recognized diverticulosis.[33] The close anatomic association between the nutrient arteriole and the dome and neck of the sac probably is a major factor in the causation of hemorrhage.[34] Interestingly, bleeding tends to occur in uninflamed diverticula: it is uncommon for severe blood loss to accompany acute diverticulitis. Recent studies have shown the site of hemorrhage tends to be on the luminal aspect of the arteriole in the diverticulum.[35] Large volumes of blood are passed at one time — this reflects the capacity of the colon to act as a large reservoir for fluid volumes. Transient abdominal colic may be present just prior to bowel evacuation and the passage of large amounts of blood. This indicates the accumulation of a large volume of material in the colon rather than an unusually rapid rate of bleeding. Some patients will require transfusions, or even emergency surgery, but surprisingly, in the majority of cases, the hemorrhage ceases spontaneously.[4] Of these patients, it is estimated that only 25% will experience a recurrent hemorrhage from diverticula. It must be stressed that other lesions, such as carcinoma, polyps, and arteriovenous malformations, can produce similar bleeding incidents. Nuclear scanning, angiography, and colonoscopy are selective diagnostic techniques which may aid in establishing a correct diagnosis.[36]

### 3. Natural History of Diverticular Disease

As previously stressed, it is important to realize that the vast majority of people with diverticulosis will never develop symptoms of lower bowel dysfunction. Most observers have noted a consistent linear association between number of diverticula and increasing age.[26,37] However, it appears that not all patients with diverticula will have a progressive increase in number with time.[27] This observation suggests that other factors besides age must play a role in progression of disease.[38] It is tempting to consider the possibility that response to symptoms by change in diet may alter the progression of diverticula in some patients, but there is no scientific evidence for this statement to date. One author,[11] however, has predicted a decrease in diverticular disease in Israel over the next 30 years because of changes in fiber consumption.

In patients symptomatic from diverticula, it appears that a large fraction harbor these symptoms for a much shorter period of time than it must have taken for the process to develop.[4] It is further noted that the prediverticular state, or *Myochosis coli*, can produce considerable pain and a change in bowel habits.[39] Of the 10 to 25% of patients who develop the significant complication of diverticulitis, most tend surprisingly not to have recurrent morbidity from the disease, if major complications of diverticulitis were avoided during the initial episode. A retrospective study at Yale University revealed that, over a 9-year period, 3 out of 4 patients who presented with acute diverticulitis had no further symptoms of diverticular disease.[40] It has already been pointed out that 75% of patients with diverticular hemorrhage will not have a recurrence.[4] Thus, it is strongly suggested that the management of the majority of patients with complicated diverticular disease be conservative. Diet is currently felt to play a major role in its management. Whether this can change the incidence of complications over a long term remains unclear. The 25% of patients who do not respond to initial conservative measures and those who have more serious and life-threatening complications must continue to be managed in an aggressive fashion, chiefly by surgery. Diet will have little benefit in the acute situation. The rationale for dietary manipulation for longer-term management can only be considered in light of a fuller appreciation of pathogenesis and the function of fiber in relation to the large intestine.

## IV. PATHOGENESIS

The predominant trend of thought regarding the causation of diverticula revolves around a disorder of intestinal motility interrelated with an abnormality of the colonic wall. It is hypothesized that the formation of diverticula requires the creation of a pressure gradient across the colonic wall and that an area of structural weakness in the wall allows these forces to cause the herniation of bowel layers that compose the typical diverticulum.

Several observations on patients with diverticular disease have pointed toward a primary motility disturbance. Pain has long been presumed to arise from colonic segmentation with pressure increases in these segments.[41] This sustained muscle contraction has been thought to result in the shortening and narrowing of the sigmoid colon, and the striking muscular thickening so often observed, at least in painful diverticular disease.[42] The fact that, in geographic areas where diverticular disease is rare and dietary fiber intake is high, there have been found marked increases in average stool weight and marked decreases in intestinal transit time relative to Western cultures, has been considered to support the concept of a motility disorder.[43] However, when comparison is made between patients with diverticular disease and healthy subjects in the same areas, most studies have found no significant difference in stool weight or intestinal transit time.[41]

In patients with symptomatic "spastic colon" diverticulosis, the concept which its name implies has been supported by observations on motility of the descending and sigmoid colon. Fivefold increases in intraluminal pressure generated by segmentation movements have been recorded,[39] and in some studies these have been correlated with strong contractions observed by cineradiography.[44] Painter[44] observed that in colon affected by myochosis, the contraction of the arcuate ridges of circular muscle often occluded the bowel lumen and isolated certain segments (''little bladders'') creating extremely high pressures in the confined spaces (Figure 3). The predominance of the descending and sigmoid colon as sites of diverticula has been attributed by some to the physical relationship between wall tension and pressure in a contractile hollow cylinder as expressed by a derivative of the Law of Young and Laplace:

$$P = k \ T/R$$

in which P is intraluminal pressure, T = the tension generated by muscular contraction, and R is the radius of the gut. Thus, a given level of contraction of colonic muscle would produce the highest intraluminal pressure in the narrowest portion of the colon, and this pressure would increase as its diameter decreases.

While hypermotility of the distal colon is found in many patients with diverticular disease, especially in those with myochosis and left lower quandrant pain, recent work indicates that it is not a necessary condition for development of diverticula. In comparison with normal controls, patients with asymptomatic diverticulosis exhibit no difference in intracolonic pressure waves, mean stool weight, or mean intestinal transit time.[45-47]

The adequacy of the primary motility disorder as an explanation for the appearance of diverticula in the colon is further weakened by the failure of numerous experiments to produce diverticulosis in experimental animals. Though a few diverticula were found by Carlson and Hoelzel[48] in the colons of aged rats maintained on low-fiber diets, and though Hodgson[49] occasionally produced isolated atypical diverticula by such diets in rabbits, these findings have not been confirmed. More recently, primates fed a low-residue diet have exhibited high intrasigmoidal pressures and altered myoelectrical activity, but no diverticula.[50] In other herbivorous species maintained on low-fiber diets, the sustained contraction of the musculature is evidenced by extreme shortening and narrowing of the colon. The importance of other factors in diverticular formation is apparent.

A second factor in the formation of diverticula is thought to be an abnormality in the colonic wall which exaggerates its weakness at specific points. As previously indicated (Figure 1), the principal sites of mucosal herniation are the gaps in the circular muscle layer through which the nutrient arteries pass. The resistance of these vulnerable areas to intraluminal pressure may be diminished as the result of age-related changes in the structure and function of the connective tissue stroma.[4] The tensile strength of the colonic wall, especially in the distal colon, declines with age, despite increased thickness of the connective tissue between the muscle layers and increased content of collagen, elastin, and reticulum, and despite a 50 to 70% increase in the thickness of the muscle coats. The distal colon of patients with diverticulosis has been shown to have decreased resistance to internal distension, as by a balloon, and presumably would yield more readily to intraluminal pressures generated by its own contractions. These facts implicate changes in viscoelastic properties of the tissue, akin to those occurring in the skin of elderly persons.[41] No isolated form of congenital colonic wall weakness has been described, yet diverticula have been described in young adults with generalized disease of connective tissue, including the Marfan, Ehlers-Danlos, and Williams ''elfin facies'' syndromes.[52-54]

## V. DIETARY FIBER AND THE ETIOLOGY AND TREATMENT OF DIVERTICULAR DISEASE

### A. History

As previously mentioned, the appearance of diverticular disease, linked with increases in other degenerative diseases such as adult onset diabetes mellitus, ischemic heart disease,

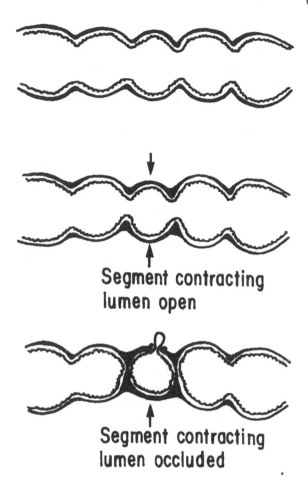

Segment contracting
lumen open

Segment contracting
lumen occluded

FIGURE 3.   This drawing illustrates Painter's theory of segmental muscular contractions along the colon resulting in numerous occluded areas or "little bladders" characterized by extremely high pressure within the segment. (From Painter, N. S., *Ann R Coll. Surg.*, 34, 98, 1964. With permission.)

cholelithiasis, and venous insufficiency, has coincided with major changes in dietary fiber intake. For practical purposes fiber represents all undigested vegetable matter which reaches the large intestine.[55] Historically, in Great Britain and the U.S., the processes of milling cereal grains began to change radically in the late 19th century.[1] In essence, the processing of flour became much more efficient and increased amounts of fiber were removed. That period was furthermore noteworthy for a rapid evolution of transportation networks and therefore allowed for a wider distribution and consumption of these more refined grains and meat products.[20] It was approximately 40 years later that diverticulosis became more prevalent in Western societies.[16] As already mentioned, the fiber intake of the American diet fell 25 to 30% between 1909 and 1975.[15] This decline resulted from lowered consumption of bread, cereal, potatoes, fruits, and dried beans. As these trends have continued the incidence of diverticular disease has increased.

Numerous writers have supported the concept that fiber deficiency is an etiologic factor in colonic disease. Historically, this is a rebirth of an old theory. In the early part of the 20th century diets rich in whole grains, vegetables, and fruits were advocated for constipation and the irritable bowel syndrome. But between 1930 and 1960 it was generally accepted that in diverticular disease "bowel rest", in the form of a bland, low-fiber diet, was therapeutically appropriate.[23] Residue such as seeds and skins of fruits, vegetables, and nuts were often found in the colon at autopsy in cases of diverticulitis, sometimes in the diverticula themselves. An inappropriate conclusion was made from these uncontrolled observations that these food particles were colonic irritants and therefore probably had some causative role in diverticular disease. Professional opinion has now come full circle.[16]

### B. Components of Fiber and their Chemical Properties

Given the concept that increased fiber intake may lead to good health, contemporary society is often saturated with advertisements promoting one product or another because of its content of "crude" fiber, or unprocessed bran or "natural" dietary fiber. It is not infrequent that these terms are misused. Crude fiber is the plant cell residue left after laboratory chemical extraction.[14] Dietary fiber, on the other hand, is the sum total of undigested residue of plant material which reaches the large intestine. Its weight is usually two to five times that of crude fiber content. The determination of the dietary fiber content of food can be made by a variety of assays using numerous methods described elsewhere.[56] Most methods of determination involve fractionating the various components of that fiber through a variety of chemical methods and then adding the weights of these individual components to obtain a sum total.

The components of dietary fiber that are of greatest nutritional interest are celluloses, hemicelluloses, lignins, gums, and pectins.[56] All are polysaccharides with exception of the lignins. The amount and proportion of these substances are related to the age of the plant cell and the species of plant involved. As plants become older, more fibrous tissue, which has a large component of lignin and cellulose, becomes part of the structure of the cell. Younger plants tend to be richer in pectins and hemicellulose. It is essential to realize that laboratory methods which measure fiber usually are assessing the amount of crude fiber present unless otherwise specified.[56] That is, certain proportions of the cellulose, hemicellulose, lignin, and pectin content are removed by the acid or alkali treatment used to extract the fiber.[57] Therefore, if the crude fiber content of a substance is listed, this may actually only represent 20 to 25% of the total fiber content of the substance. Variations in the measurement of amounts of fiber and methods of definition have often led to confusion in the literature. It has been fairly consistently shown that different types of relatively high-fiber foods exert different degrees of physiologic effect on intestinal function, stool weight, transit times, and symptom relief.[58-60] These properties depend on the shape, size, molecular configuration, and texture of the specific type of fiber compound.[61]

### C. Effects of Fiber on Intestinal Function

One of the more reproducible effects of dietary fiber, on increasing its intake in a diet previously fiber deficient, is a significant increase in stool weight, usually accompanied by a decrease in intestinal transit time.[56,60,62] The capacity of various fibers to imbibe water probably adds to the bulk of the stool and eases its passage.[63] In 1978, Cummings[60] discovered that various fibers caused differing degrees of increase in stool weight, which correlated with the amount of pentose-containing polysaccharide in the specific fiber. For instance, bran appeared to increase the fecal weight of healthy volunteers by 127% of that observed on their "typical" Western diet. Cabbage, apples, and guar gum were only one half as effective as bran in increasing stool weight. Of these, bran had the highest content of pentose-containing polysaccharide, which was felt to determine the ability of the particular fiber to hold water. More recent work, however, has shown that other polysaccharides in fiber also hold water and that in fact there may actually be an inverse relationship between pentose-containing compounds and water-holding capacity.[64] Other workers now find there is no direct correlation between water-holding capacity of different types of fiber and resultant increases in fecal weight.[65] Furthermore, the fact that ingested fiber is partially digested before leaving the colon suggests that other mechanisms may be involved in the beneficial effects of bran.

Bacteria play a major role in the digestion of fiber — especially in the colon.[66] They form a large normal component of the stool mass, and as 80% of their weight is water, they add measurably to the stool water.[64] Dietary fiber is a known substrate for bacterial growth which leads to polysaccharide fermentation, with an increase in the colonic content of volatile free fatty acids.[67] These organic anions, in the form of acetate, propionate, and butyrate, contribute to the water content of stool by acting as osmotic agents. Thus, the action of fiber in increasing stool bulk is a complex one, based upon the inherent water-holding capacity of fiber, the bacterial populations, and the osmotic effects of increased amounts of organic anions.

The interrelationship of fiber and bile salts may have profound effects on metabolic processes. The following observations have suggested a role for fiber in another disease state, cholelithiasis. Fiber, especially the lignins, binds both conjugated and deconjugated bile acids, thereby decreasing their enterohepatic circulation and increasing the synthesis of more bile acids from cholesterol. This theoretically has the potential to result in a significant lowering of serum cholesterol.[68] The size and composition of the bile acid pool influences the degree of cholesterol saturation of the bile.[69] In a primate animal model, the bile acid pool size decreases when a fiber-depleted diet is fed.[70] The cholesterol saturation of bile decreases with the oral feeding of chenodeoxycholic acid.[71] In this situation, chenodeoxycholic acid becomes the predominant circulating bile acid, apparently suppressing hepatic cholesterol secretion.[72] Wheat bran, when given to normal subjects for 4 to 6 weeks, brought about an increase in the proportion of chenodeoxycholic acid, and a concurrent decrease in deoxycholic acid, in the bile acid pool.[73] These changes most likely result from a decrease in the absorption of deoxycholate from the intestine and/or an impairment of its production therein. Lignin, for instance, may bind cholic acid, a primary bile acid, and thereby inhibit its conversion to deoxycholate by bacterial action.

The therapeutic and prophylactic implications of this physiologic effect of fiber are not yet clear. It is possible that a return to a diet high in fiber and low in refined sugars and starch will decrease the morbidity from cholesterol gallstones and even from atherosclerotic vascular disease. In general, the populations which consume a high-fiber diet have not only a low prevalence of diverticulosis, but also of cholelithiasis and atherosclerotic disease. Furthermore, a recent 7-year prospective study from Wales suggested that vegetarians, who on average consumed a higher amount of fiber than nonvegetarians, experienced a lower mortality from ischemic heart disease over this time period.[74] The mortality rate of 1088 middle-aged Dutch men surveyed over a 1-year period was noted to be inversely related to their intake of dietary fiber.[75] This relationship was especially prominent for deaths due to cancer.

In addition to its apparently profound effect on bile salt absorption and binding, fiber appears to affect intestinal absorption of other elements. In studies on small numbers of Iranian peasants, who consume a relatively high-fiber diet, Mg, Zn, and Ca deficiencies have been reported.[76] It appears that high-fiber diets do increase the excretion of Ca, Zn, and Mg in some otherwise normal subjects. This could result in a negative balance of these minerals depending on baseline intake and other dietary practices.[76] Animal studies have demonstrated that high fiber intake may delay carbohydrate and amino acid absorption by altering the intestinal mucosa and disrupting active transport processes when compared to animals on fiber-free diets.[77] Disaccharidase levels in rat jejunum are reduced by a diet deficient in fiber.[78] In vitro studies on pancreatic enzyme activity have shown that pectin can reduce trypsin and amylase activity.[79] It remains unclear what is the total clinical significance of any of these experimental data and whether individuals who take in a high-fiber diet may be prone to specific nutritional deficiencies. The only clinical conditions generally attributable to high fiber intake are colonic volvulus, which is a more prevalent disease in developing countries, and intestinal obstruction secondary to impaction by large amounts of undigested fiber in the intestine, usually at the sites of stenosis or stricture.[80]

## Table 2
## FIBER AMOUNTS OF VARIOUS FOOD PRODUCTS

| Source | Serving size | Fiber (g) |
|---|---|---|
| Unprocessed bran | 3 tsp | 10 |
| Whole meal bread | 1 slice | 3 3 |
| Whole wheat bread | 1 slice | 1.3 |
| Rye bread | 1 slice | 0.8 |
| All Bran 100% | ½ cup | 14 |
| Bran Chex | ½ cup | 4 |
| Shredded Wheat | 1 biscuit | 2.8 |
| Wheaties | ½ cup | 1.7 |
| Apple | 1 large | 4 |
| Figs (dried) | 1 | 3 7 |
| Peach | 1 medium | 2.5 |
| Strawberries | 1 cup | 3 1 |
| Brown rice | ½ cup | 2.7 |
| White rice | ½ cup | 0 8 |
| Broccoli | ½ cup | 3.5 |
| Lettuce | 1 cup | 0.8 |
| Spinach (raw) | 1 cup | 0.2 |
| Carrots | ½ cup | 2.4 |
| Potatoes (baked) | 1 medium | 3.8 |
| Green beans | ½ cup | 2.1 |
| Tomatoes | 1 small | 1.5 |
| Zucchini | ½ cup | 2.0 |

Adapted from Anderson, J. W., *Plant Fiber in Foods*, University of Kentucky Medical Center, Lexington, 1980.

## D. Sources of Dietary Fiber

If one accepts the theoretical arguments supporting physiologic benefit of fiber, one must then choose the appropriate type, source, and amount of fiber to be used. Estimates of current dietary fiber intake in Western cultures range from 10 to 30 g/day; the majority of the population consumes less than 20 g/day.[17,81] The quantity which must be taken to produce a therapeutic benefit is difficult to define, and is probably highly variable. It appears that fecal weight must be doubled to produce significant decreases in sigmoid colon resting pressure.[60] To achieve this, 20 g of dietary fiber usually needs to be added to the baseline diet.[82] In practice, nevertheless, the appropriate amount of fiber is usually judged from the patient's clinical response, in terms of relief of pain or constipation, to stepwise increases in intake.[83]

Unprocessed wheat bran (6 tsp/day) appears to be the most efficient way of providing such an amount of fiber;[60,84] however, its taste has been likened to sawdust or compared with a hair shirt.[14,42] Similar amounts of wheat fiber are supplied by five to six slices of wholemeal bread, which is more palatable.[81] Commercial breakfast cereals are also fair sources of bran, but it requires two servings of even the most concentrated bran cereal to provide 20 g of dietary fiber.[85] Vegetables and fruits are much less fiber rich than bran (see Table 2), and must therefore be ingested in much larger quantities. It is noteworthy that in vegetarians who have diverticula, their total dietary fiber intake is usually well above the norm but they typically have a low intake of cereal fiber when compared to those vegetarians who have no diverticula.[86] Aside from truly natural sources of dietary fiber, hydrophilic colloids refined from vegetable sources appear to be effective substitutes.[4] Psyllium preparations, sterculia, and methyl cellulose have been shown to be bulk forming agents which assist in increasing stool weights and decreasing intestinal transit times.[87,88]

### E. Diet in the Treatment of Diverticular Disease

There are at present no data to suggest that a high-fiber diet can result in regression or disappearance of colonic diverticula. There are two controlled trials, however, which indicate benefit from bran in relieving symptoms of pain and constipation in uncomplicated but symptomatic diverticular disease.[89,90] In another report, the benefit of bran in alleviating mild symptoms of diverticular disease has recently been challenged.[91] Motility indexes of the sigmoid colon, intestinal transit times, and stool weights have all been favorably affected by bran over short periods of observation.[92] The long-term effects of increased fiber intake, in terms of eventually lowering the incidence of diverticulitis or through metabolic effects altering the prevalence of other disease states, remain unclear. In the short term, bran supplementation has in various studies lowered oral glucose tolerance curves, increased fecal fat excretion, altered serum lipoproteins, decreased urinary calcium excretion, and decreased serum folate levels.[93] It will take years, if not generations, to assess the impact of such possible dietary changes on complication rates and mortality from diverticular disease or other colonic disorders.[94] The advocates of the fiber hypothesis predict that one result of such dietary changes, if started early in life, would be the primary prevention of colonic diverticula.[95] The previously cited radiographic survey in Oxford vegetarians impressively suggests that the lower prevalence of asymptomatic diverticular disease in vegetarians compared to nonvegetarians is significantly associated with the level of intake of dietary fiber.[86]

When the decision is made to treat a patient with symptomatic diverticular disease it is essential that the patient be carefully instructed about the various sources of fiber. Most often bran will be used. In order to achieve a 10- to 25-g increase in intake, a slow buildup in supplementation is mandatory. Bran tends to produce abdominal distention and increased flatulence, which usually stabilize or diminish over several weeks. Patients who cannot tolerate doses of bran in the 20-g range or other combinations of fiber can be treated with combinations of diet and hydrophilic colloids. Other drugs, such as antispasmodic agents, are not highly effective in uncomplicated but symptomatic diverticular disease.[23] Diverticulitis, its complications, and diverticular hemorrhage are usually managed by first placing the patient on nothing by mouth. The other aspects of their acute management include the use of antibiotics, analgesics, and at times surgery. As the acute phase resolves it is often wise to place the patient on a soft diet for 2 to 3 days before gradually advancing him to a high-fiber diet as tolerated.

## VI. CONCLUSION

As stressed so often in this chapter, present concepts of the pathogenesis and treatment of diverticular disease are supported chiefly by epidemiological and physiological observations which suggest that a deficiency of dietary fiber is a major and remediable factor in the disease process. The results of numerous, retrospective, uncontrolled therapeutic experiments have recently been corroborated by well-designed controlled trials, and some evidence now suggests that primary prevention of colonic diverticula may be afforded by doubling the average intake of vegetable fiber. In the light of negative effects of fiber on balances of essential minerals at marginal levels of mineral intake, the recommendation to the public at large of diets supplemented with fiber-rich foods cannot be made with entire confidence, yet the prevention of diverticulosis, which otherwise may affect 50% of our elderly population and cause significant morbidity in 10 to 20% of those affected, and which is now the 4th ranking digestive disorder leading to hospitalization in this age group, might in itself yield benefits which outweigh the risks. If the potential benefits of a high-fiber diet in the mitigation of obesity, diabetes, atherosclerosis, and gallstones are then considered, the probable gains for the health of the elderly from lifetime consumption of such a diet seem large indeed.

# REFERENCES

1. **Painter, N. S., and Burkitt, D. P.,** Diverticular disease of the colon, a 20th century problem, *Clin. Gastroenterol ,* 4, 3, 1975

2. **Connell, A. M.,** Pathogenesis of diverticular disease of the colon, *Adv. Intern. Med ,* 22, 377, 1977.

3. **Parks, T. G.,** Post-mortem studies of the colon with special references to diverticular disease, *Proc R. Soc. Med ,* 61, 30, 1968.

4. **Almy, T. P. and Howell, D. A.,** Diverticular disease of colon, *N Engl J. Med ,* 302, 324, 1980

5. **Sato, E., Ouchi, A., Sasano, N., and Ishidate, S.,** Polyps and diverticulosis of large bowel in autopsy populations of Akita prefecture compared with Miyagi high risk for colorectal cancer in Japan, *Cancer,* 37, 1316, 1976.

6. **Segal, I. and Walker, A. R. P.,** Diverticular disease in urban Africans in South Africa, *Digestion,* 24, 42, 1982

7. **Painter, N. S.,** Diverticular disease of the colon, *S. Afr. Med. J.,* 61, 1016, 1982

8 **Mayo, W. J.,** Diverticula of the sigmoid, *Ann. Surg ,* 92, 739, 1932

9. **Debestani, A., Aliabadi, P., Shah-Rookh, F. D., and Borhanmanesh, F. A.,** Prevalence of colonic diverticular disease in Southern Iran, *Dis Colon Rectum,* 24, 385, 1981

10 **Fatayer, W. T., A-Khalaf, M. M., Shalan, K. A., Toukan, A. U., Daker, M. R., and Arnaout, M. A.,** Diverticular disease of the colon in Jordan, *Dis Colon Rectum,* 26, 247, 1983

11. **Levy, N., Luboshitzki, R., Shiratzk, Y., and Ghirarello, M.,** Diverticulosis of the colon in Israel, *Dis Colon Rectum,* 20, 477, 1977.

12 **Segal, I., Solomon, A., and Hunt, J. A.,** Emergence of diverticular disease in the urban South African Black, *Gastroenterology,* 72, 215, 1977

13 **Stemmermann, G. N. and Yatini, R.,** Diverticulosis and polyps of the large intestine a necropsy study of Hawaii Japanese, *Cancer,* 31, 1260, 1973

14 **Cello, J. P.,** Diverticular disease of the colon, *West J Med ,* 134, 515, 1981

15 **Heller, S. N. and Hackler, L. R.,** Changes in the crude fiber content of the American diet, *Am. J Clin Nutr.,* 31, 1510, 1978.

16. **Painter, N. S. and Burkitt, D. P.,** Diverticular disease of the colon: a deficiency disease of Western civilization, *Br Med. J ,* 2, 450, 1971.

17 **Gear, J. S. S., Ware, A., and Fursdon, P.,** Symptomless diverticular disease and intake of dietary fiber, *Lancet,* 1, 511, 1979.

18. **Capron, J. P., Piperaud, R., Dupas, J. L., Delamarre, J., and Lorriaux, A.,** Evidence for an association between cholelithiasis and diverticular disease of the colon, *Dig. Dis. Sci ,* 26, 523, 1981

19. **Castleden, W. M., Douss, T. W., Jennings, K. P., and Leighton, M.,** Gallstones, carcinoma of the colon, and diverticular disease, *Clin. Oncol.,* 4, 139, 1978.

20. **Cleave, T. L., Campbell, G. D., and Painter, N. S.,** *Diabetes, Coronary Thrombosis, and the Saccharine Disease,* John Wright & Sons, Bristol, 1969

21. **Foster, K. J., Holdstock, G., Wright, R., Whorwell, P. J., Guyer, P., and Wright, R.,** Prevalence of diverticular disease of the colon in patients with ischemic heart disease, *Gut,* 19, 1054, 1978

22. **Latto, C.,** Diverticular disease and varicose veins, *Am Heart J ,* 90, 274, 1975.

23 **Almy, T. P. and Naitove, A.,** Diverticular disease of the colon, in *Gastrointestinal Disease,* Sleisenger, M H and Fordtran, J S., Eds., W B. Saunders, Philadelphia, 1983, chap. 55.

24 **Noer, R. J.,** Hemorrhage as a complication of diverticulitis, *Ann Surg.,* 141, 674, 1955

25. **Ryan, P.,** Changing concepts in diverticular disease, *Dis. Colon Rectum,* 26, 12, 1983

26 **Parks, T. G.,** Natural history of diverticular disease of the colon, *Clin Gastroenterol ,* 4, 53, 1975.

27. **Horner, J. L.,** Natural history of diverticulosis of the colon, *Am. J Dig. Dis ,* 3, 343, 1958

28. **Morson, B. C.,** Pathology of diverticular disease of the colon, *Clin. Gastroenterol ,* 4, 37, 1975

29. **Fleischner, F. G.,** The question of barium enema as a cause of perforation in diverticulitis, *Gastroenterology,* 51, 290, 1966.

30. **Small, W. P. and Smith, A. N.,** Fistula and conditions associated with diverticular disease of the colon, *Clin. Gastroenterol.,* 4, 171, 1975.

31. **Linhardt, G. E., Moore, R. C., and Mason, G. R.,** Prognostic indices in the treatment of acute diverticulitis: a retrospective study, *Am Surg.,* 48, 217, 1982.

32 **Hughes, L. E.,** Complications of diverticular disease: inflammation, obstruction, and hemorrhage, *Clin. Gastroenterol ,* 4, 147, 1975

33 **Noer, R. J., Hamilton, J. E., Williams, D. G., and Broughton, D. S.,** Rectal hemorrhage: moderate and severe, *Ann Surg ,* 155, 794, 1962

34. **Meyers, M. A., Volberg, F., Katzen, B., Alonso, D., and Abbot, G.,** The angio-architecture of colonic diverticula: significance in bleeding diverticulosis, *Radiology,* 108, 249, 1973

35. **Meyers, M. A., Alonso, D. R., and Baer, J. W.,** Pathogenesis of massively bleeding colonic diverticulosis: new observations, *Am J. Radiol.,* 127, 901, 1976.

36 **Colacchio, T. A., Forde, K. A., Patsos, T. J., and Nunez, D.,** Impact of modern diagnostic methods on the management of active rectal bleeding, *Am. J. Surg.*, 143, 607, 1982.

37 **Parks, T. G.,** Natural history of diverticular disease of the colon: a review of 521 cases, *Br Med J.*, 4, 639, 1969.

38 **Hughes, L. E.,** Post-mortem survey of diverticular disease I. Diverticulosis and diverticulitis, *Gut,* 10, 336, 1969.

39 **Arfwidsson, S. and Dock, N. G.,** Pathogenesis of multiple diverticula of the sigmoid colon in diverticular disease, *Acta Chir. Scand. (Suppl ),* 342, 5, 1964.

40. **Larson, D. M., Masters, S. S., and Spiro, H. M.,** Medical and surgical therapy in diverticular disease. A comparative study, *Gastroenterology,* 71, 734, 1976.

41 **Eastwood, M. A., Watters, D. A., and Smith, A. N.,** Diverticular disease — is it a motility disorder? *Clin Gastroenterol.,* 11, 545, 1982

42. **Eastwood, M. A.,** Medical and dietary management. *Clin. Gastroenterol.,* 4, 85, 1975

43. **Burkitt, D. P. and Trowell, H. C.,** *Refined Carbohydrate Foods and Disease,* Academic Press, London, 1975.

44. **Painter, N. S., Truelove, S. C., Ardran, G. M., and Tuckey, M.,** Segmentation and the localization of intraluminal pressures in the human colon, with special reference to the pathogenesis of colonic diverticula. *Gastroenterology,* 49, 169, 1965.

45. **Weinreich, J. and Anderson, D.,** Intraluminal pressure in the sigmoid colon. I. Methods and results in normal subjects, *Scand. J Gastroenterol.,* 11, 557, 1976.

46 **Eastwood, M., Brydon, N. G., Smith, A. N., and Pritchard, J.,** Colonic function in patients with diverticular disease, *Lancet,* 1, 1181, 1978.

47 **Howell, D. A., Crow, H. C., Almy, T. P., and Ramsey, W. H.,** A controlled double-blind study of sigmoid motility using psyllium mucilloid in diverticular disease, *Gastroenterology,* 74(Abstr ), 1046, 1978.

48 **Carlson, A. J. and Hoelzel, F.,** Relation of diet to diverticulosis of the colon in rats, *Gastroenterology,* 12, 108, 1949

49. **Hodgson, J.,** Animal models in the study of diverticula. I Aetiology and treatment, *Clin Gastroenterol.,* 77, 70, 1975

50. **Brodribb, A. J. M., Condon, R. E., Cowles, V., and DeCoss, J. J.,** Effect of dietary fiber on intraluminal pressure and myeloelectric activity of the left colon in monkeys, *Gastroenterology,* 77, 70, 1979

51 **Bornstein, P.,** Disorders of connective tissue function and the ageing process, a synthesis and review of current concepts and findings, *Mech. Ageing Dev ,* 5, 305, 1976

52 **Brighton, P. H., Murdoch, J. L., and Votteler, T.,** Gastrointestinal complications of the Ehlers-Danlos Syndrome, *Gut,* 10, 1004, 1969.

53. **Cook, J. M.,** Spontaneous perforations of the colon: report of two cases in a family exhibiting Marfan stigmata, *Ohio Med. J.,* 64, 73, 1968.

54. **Pleatman, S. I. and Dunbar, J. S.,** Colon diverticula in Williams "elfinfacies" syndrome, *Radiology,* 137, 869, 1980.

55 **Cummings, J. H.,** What is fiber?, in *Fiber in Human Nutrition,* Spiller, G. A. and Amen, R J., Eds , Plenum Press, New York, 1976, chap. 1.

56. **Kelsey, J. L.,** A review of research on effects of fiber intake on man, *Am. J. Clin Nutr ,* 31, 142, 1978.

57. **Van Soest, P. J. and McQueen, P. W.,** The chemistry and estimation of fiber, *Proc Nutr Soc.,* 32, 123, 1973.

58. **Eastwood, M. A., Smith, A. N., Brydon, W. G., and Pritchard, J.,** Comparison of bran, ispaghula, and lactulose on colon function in diverticular disease, *Gut,* 19, 1144, 1978.

59. **Smith, A. N., Drummond, E., and Eastwood, M. A.,** The effect of coarse and fine Canadian Red Spring wheat and French soft wheat bran on colonic motility in patients with diverticular disease, *Am. J. Clin. Nutr.,* 34, 2460, 1981.

60. **Cummings, J. H., Branch, W., Jenkins, D. J. A., Southgate, D. A. T., Houston, H., and James, W. P. T.,** Colonic response to dietary fibre from carrot, cabbage, apple, bran and guar gum, *Lancet,* 1, 5, 1978

61. **Kirwin, W. O., Smith, A. N., and McConnell, A. A.,** Action of different bran preparations on colonic function, *Br Med. J.,* 4, 187, 1974.

62 **Wrick, K. L., Robertson, J. B., Van-Soest, P. J., Lewis, B. A., Rivers, J. M., Roe, D. A., and Hackler, L. R.,** The influence of dietary fiber source on human intestinal transit and stool output, *J Nutr.,* 113, 1464, 1983

63 **Handler, S.,** Dietary fiber: can it prevent certain colonic diseases? *Postgrad. Med.,* 73, 301, 1983.

64 **Cummings, H. and Stephen, A. M.,** The role of dietary fiber can it prevent certain colonic diseases? *Can. Med Assoc. J.,* 123, 1109, 1980.

65 **Stephen, A. and Cummings, J. H.,** Water holding by dietary fiber in vitro and its relationship to faecal output in man, *Gut,* 20, 722, 1979.

66  **Bryant, M. P.,** Cellulose digesting bacteria from human feces, *Am. J Clin. Nutr.*, 31(Suppl.), S113, 1978.

67  **Fleming, S. E. and Rodriguez, M. A.,** Influence of dietary fiber on fecal excretion of volatile fatty acids by human adults, *J. Nutr.*, 113, 1613, 1983.

68  **Heaton, K. W.,** Gallstones and cholecystitis, in *Refined Carbohydrate Foods and Disease. . . Some Implications of Dietary Fiber,* Burkitt, D. P and Trowell, H. C., Eds , Academic Press, London, 1975, 173

69. **Bell, C. C., Vlahcevic, Z. R., Prazich, J., and Suelle, L.,** Evidence that a diminished bile acid pool precedes the formation of cholesterol gallstones in man, *Surg Gynecol. Obstet.*, 136, 961, 1973

70. **Redinger, R. N., Hermann, A. H., and Small, D. M.,** Primate biliary physiology. X Effects of diet and fasting on biliary lipid secretion and relative composition and bile salt metabolism in the rhesus monkey, *Gastroenterology,* 64, 610, 1973.

71. **Danzinger, R. G., Hofmann, A. F., Schoenfield, L. I., and Thistle, J. L.,** Dissolution of cholesterol gallstones by chenodeoxycholic acid, *N Engl. J Med.,* 286, 1, 1972.

72. **Adler, R., Duane, W., Bennion, L., and Grundy, S.,** Effects of bile acid feeding on billiary cholesterol secretion, *Gastroenterology,* 68, 326, 1975.

73. **Pomare, E. W., Heaton, K. W., Low-Beer, T. S., and Espiner, H. J.,** The effect of wheat bran upon bile salt metabolism and upon lipid composition of bile in gallstone patients, *Dig. Dis Sci.,* 21, 521, 1976.

74. **Burr, M. C. and Sweernan, P. M.,** Vegetarianism, dietary fiber, and mortality, *Am. J. Clin. Nutr.,* 36, 873, 1982

75. **Kromhout, D., Bosschieter, E. B., and DeLezenne Coulander, C.,** Dietary fiber and 10-year mortality from coronary heart disease, cancer, and all causes  The Zutphen Study, *Lancet,* 2, 518, 1982.

76  **Ismail-Beigi, F., Reinhold, J. G., Forasi, B., and Abudi, P.,** Effects of cellulose added to diets of low and high fiber content upon the metabolism of calcium, magnesium, zinc, and phosphorous by man, *J. Nutr.,* 107, 510, 1977

77  **Schwartz, S. E., Levine, G. D., and Starr, C. M.,** Effects of dietary fiber on intestinal ion fluxes in rats, *Am. J. Clin Nutr.,* 36, 1102, 1982.

78. **Thomsen, L. I. and Tasman-Jones, C.,** Disaccharidase levels of the rat jejunum are altered by dietary fiber, *Digestion,* 23, 253, 1982

79  **Isaksson, G., Lundquist, I., and Ihse, I.,** In vitro inhibition of pancreatic enzyme activities by dietary fiber, *Digestion,* 24, 54, 1982.

80. **Allen-Meush, T. and DeJode, L. R.,** Is bran useful in diverticular disease? (letter), *Br. Med J ,* 1, 283, 1982

81. **Bingham, S., Cummings, J. H., and McNeil, N. I.,** Intakes and sources of dietary fiber in the British population, *Am. J. Clin. Nutr.,* 32, 1313, 1979.

82. **Heaton, W.,** Is bran useful in diverticular disease? *Br. Med. J.,* 4, 1523, 1981.

83. **Painter, N. S.,** Are fiber supplements really necessary in diverticular disease? *Br. Med. J.,* 3, 140, 1981.

84. **Hyland, J. M. P. and Taylor, I.,** Does a high fiber diet prevent complications of diverticular disease? *Br. J. Surg.,* 67, 77, 1980.

85. **Anon.,** The high fiber diet: its effect on the bowel, *Med Lett.,* 17, 93, 1975.

86. **Gear, J. S., Fursdon, P., Nolan, D. J., Ware, A., Mann, J. I., and Brodribb, A. J. M.,** Symptomless diverticular disease and intake of dietary fibre, *Lancet,* 1, 511, 1979.

87. **Srivastava, G. S., Smith, A. N., and Painter, N. S.,** Sterculia bulk-forming agents with smooth muscle relaxant versus bran in diverticular disease, *Br. Med. J.,* 1, 315, 1976.

88. **Hodgson, J.,** Effect of methyl cellulose on rectal and colonic pressures in treatment of diverticular disease, *Br. Med. J ,* 3, 729, 1972.

89. **Brodribb, A. J. M. and Humphreys, D. M.,** Diverticular disease. II. Treatment with bran, *Br. Med. J.,* 1, 425, 1976.

90. **Painter, N. S., Almeida, A. Z., and Colebourne, K. W.,** Unprocessed bran in treatment of diverticular disease of the colon, *Br. Med. J.,* 2, 137, 1972

91. **Ornstein, M. H., Littlewood, E. R., McClean, B. Fowley, J., Mouth, W. R. S., and Cox, A. G.,** Are fiber supplements really necessary in diverticular disease of the colon? A controlled clinical trial, *Br. Med. J.,* 2, 1353, 1981.

92. **Taylor, I. and Duthie, H. C.,** Bran tablets and diverticular disease, *Br Med. J ,* 1, 988, 1976.

93. **Brodribb, A. J. and Humphreys, D. M.,** Diverticular disease, III  Metabolic effect of bran in patients with diverticular disease, *Br. Med. J.,* 1, 428, 1976.

94. **Talbot, J. M.,** Role of dietary fiber in diverticular disease and colon cancer, *Fed. Proc Fed. Am. Soc. Exp. Biol.,* 40, 2337, 1981.

95. **Burkitt, R.,** Are fiber supplements really necessary in diverticular disease of the colon? *Br. Med J.,* 282, 1546, 1981.

Chapter 10

# NUTRITIONAL ANEMIAS: IRON, FOLACIN, VITAMIN B$_{12}$

## Lynn B. Bailey

## TABLE OF CONTENTS

# I. INTRODUCTION

Anemia may be defined as a reduction in hemoglobin concentration below a specified norm. The occurrence of this condition in the elderly can be particularly debilitating due to the physical problems associated with a reduced oxygen-carrying capacity of the blood. The anemic elderly individual is often described as weak, breathless, and lethargic, with a general inability to perform work-related tasks. These symptoms may lead to a further degeneration in health status as a result of apathy toward diet and physical activity.

It should be recognized that anemia can result from a deficiency of specific essential nutrients as well as from chronic diseases, blood loss, neoplasms, and other causes. The emphasis in this chapter is on nutritional anemias involving deficiencies of Fe, folacin, or vitamin $B_{12}$. The prevalence of anemia resulting from deficiencies of other nutrients (vitamin E, vitamin $B_6$, protein, vitamin C, riboflavin) is low and usually related to some metabolic aberration or interaction with the three major nutrients to be discussed in detail. Since anemia is a late finding in the development of Fe, folacin, and vitamin $B_{12}$ deficiencies, the progressive changes that precede anemia will be covered. Commonly held assumptions related to the magnitude and etiology of Fe, folate, and vitamin $B_{12}$ deficiencies in the elderly age group are questioned. This chapter also includes information on these nutrients in relation to their chemistry and form in food, metabolism and requirement, and methods of status assessment. This is followed by a discussion of the prevalence and diagnosis of anemia.

# II. IRON

## A. Iron in Food

A review of Fe interactions, measurements, and bioavailability in foods has been published.[1] Fe is present in food as either inorganic or heme Fe. The availability of dietary Fe for absorption is as important a factor as overall total Fe content in determining nutritional adequacy.[2] The availability of non-heme Fe is greatly enhanced by the presence of ascorbic acid and meat, and may be inhibited by factors including tea, coffee, bran, and fiber.[1] Meat contains heme Fe, which is more available for absorption than non-heme Fe and is not influenced by inhibitors.[2] Factors that may enhance or inhibit bioavailability should be taken into account when evaluating the overall adequacy of the dietary Fe intake of the elderly.

## B. Metabolism

### 1. Absorption and Malabsorption

Lynch et al.[3] recently reviewed Fe metabolism and requirements in the elderly. The main control of Fe status is at the site of intestinal absorption. There is no physiological regulation of Fe metabolism through increased or decreased excretion. The absorption of Fe is a complex process that is influenced by the intestinal mucosa, the amount and chemical nature of Fe in the ingested food, and a variety of dietary factors that increase or decrease the availability of Fe for absorption.[4-6] Fe absorption in man is modified by the intestinal mucosa as an inverse function of body Fe stores.[4] Bonnet et al.[7] and others have reported decreased Fe absorption in the elderly.[8,9] These data must be interpreted with caution since Fe status and GI diseases are complicating factors. Jacobs and Owens[9] attributed a progressive fall in non-heme Fe absorption with age to hypochlorhydria. Serum ferritin levels were not determined in this study; therefore, the reduction in absorption with age may have been the result of increased body Fe stores. Marx[10] found no effect of age on the rate of non-heme Fe (ferrous ammonium sulfate) absorption. Since hydrochloric acid is relatively unimportant for adequate absorption of soluble ferrous salts, the suggestion of Jacobs and Owen[3] that hypochlorhydria may lead to impaired non-heme food Fe absorption may be valid.

## 2. Transport, Storage, and Excretion

Fe is transported in plasma bound to transferrin, a glycoprotein which accepts Fe from the intestinal tract or from sites of storage or hemoglobin destruction. Transferrin then delivers Fe to the bone marrow for hemoglobin synthesis, to reticuloendothelial cells for storage, and to all cells for Fe-containing enzymes.

Fe in excess of need is stored intracellularly as ferritin and as hemosiderin. The major sites of storage are the hepatic parenchymal cells and reticuloendothelial cells of the bone marrow, liver, and spleen.

The effect of age on the quantity of storage Fe has been investigated and the findings suggest that Fe requirements decrease with age. Hepatic non-heme Fe concentration has been reported for various ages using a liver necropsy technique in Malaysians and Singaporeans.[11] After the neonatal period, liver Fe concentrations in males remained low until puberty, followed by a progressive rise. In women, the low levels persisted until after menopause, followed by a gradual increase until levels similar to those in the men were reached. The highest level of total liver Fe content was found in subjects of both sexes between the age of 60 and 79 years. Similar data have been observed in other populations.[12-17] Stainable bone marrow Fe and serum ferritin concentrations have also been found to increase with age.[18-25]

Garry et al.[25] compared the serum ferritin levels of a group (n = 280) of free-living healthy elderly (60 to 93 years) men and women with levels in a group (n = 271) of young (20 to 39 years) subjects. Ferritin levels were significantly higher in the elderly group compared to the younger subjects. The geometric means for ferritin levels in the elderly men and women were 145 and 104 ng/m$\ell$, respectively, compared to 85 and 37 ng/m$\ell$ for the young men and women. Only 1% of the young group had ferritin levels higher than 200 ng/m$\ell$ compared to 35 and 15% of the elderly men and women, respectively.

The rate of excretion primarily determines Fe requirements in the elderly since growth, pregnancy, and menstrual losses are no longer influencing factors. Finch[26] used a radioisotopic method ($^{55}$Fe) to measure the change in specific activity in the blood over a prolonged time in men and women of different ages. He reported a mean Fe exchange rate of 0.64 mg in elderly (59 to 77 years) nonmenstruating women compared to 1.22 mg in younger (32 to 44 years) menstruating women. In elderly (57 to 84 years) men, the Fe exchange rate was found to be 0.61 mg compared to 0.95, 0.90, and 1.02 mg for three groups of younger men.[26,27] These data suggest that Fe loss in older women and men is no greater than in younger individuals and may be less.

## 3. Function

The major function of Fe is to regulate oxidative processes. Approximately 60 to 70% of the Fe of an adult male is found in hemoglobin and about 3% in myoglobin. Fe associated with various enzyme systems accounts for less than 1% but it is very important to life. The Fe enzymes may be classified as heme-protein, Fe-flavoproteins, and enzymes requiring Fe as a co-factor.[28]

## 4. Requirement and Allowance

The Recommended Dietary Allowance (RDA) for adult men and post-menopausal women is 10 mg. This is based on an estimated daily loss of 1 mg adjusted for 10% efficiency of absorption.[29]

## C. Iron Status

### 1. Dietary Intake

Dietary intake data from both national and regional surveys indicate that the dietary Fe intake for the aged individual is adequate.[3,25,30-36] The most recent comprehensive national

## Table 1
### MEAN ENERGY AND IRON INTAKES FOR MALES AND FEMALES REPORTED IN HANES II (1976—1980)[3]

| Males | | | | Females | | | |
|---|---|---|---|---|---|---|---|
| | | Iron | | | | Iron | |
| Age (year) | Energy (kcal) | (mg) | (mg/1000 kcal) | Age (year) | Energy (kcal) | (mg) | (mg/1000 kcal) |
| 6—11 mo. | 1001 | 12.8 | 12.8 | 6—11 mo. | 991 | 12 9 | 13 0 |
| 1—2 | 1311 | 8.7 | 6.6 | 1—2 | 1262 | 8.5 | 6 7 |
| 3—5 | 1628 | 10 5 | 6.5 | 3—5 | 1508 | 9.5 | 6 3 |
| 6—8 | 1981 | 12.5 | 6.3 | 6—8 | 1807 | 10.8 | 6.0 |
| 9—11 | 2183 | 14.4 | 6.6 | 9—11 | 1857 | 11.4 | 6.2 |
| 12—14 | 2430 | 15.9 | 6.5 | 12—14 | 1813 | 10.8 | 5 9 |
| 15—17 | 2817 | 17.4 | 6 2 | 15—17 | 1731 | 9 9 | 5 7 |
| 18—24 | 3040 | 17.8 | 5 8 | 18—24 | 1687 | 10 6 | 6.3 |
| 25—34 | 2734 | 17 3 | 6.3 | 25—34 | 1643 | 10.9 | 6.6 |
| 35—44 | 2424 | 16.1 | 6.6 | 35—44 | 1579 | 11.2 | 7.1 |
| 45—54 | 2361 | 16.2 | 6 9 | 45—54 | 1439 | 10.4 | 7.3 |
| 55—64 | 2071 | 14 8 | 7 1 | 55—64 | 1401 | 10 7 | 7 6 |
| 65—74 | 1828 | 14.1 | 7 7 | 65—74 | 1295 | 10.2 | 7 9 |

survey is from the U.S. Department of Health, Education and Welfare and Nutrition Examination Survey (HANES II).[3] The mean dietary Fe intake in this survey for both men and women exceeded the 1980 RDA of 10 mg (Table 1).[29]

A review of data in Table 1 indicates that Fe density in the diet increases with age to intakes that exceed 7 mg/1000 kcal. This provides an explanation for maintenance of adequate Fe intakes when caloric intakes are reduced. Bailey et al.[37] found that the average Fe intake of a group of elderly in Florida met the RDA even when calorie intake was low. The mean daily dietary Fe intake of institutionalized elderly men and women was reported to be 16.9 and 12.7 mg, respectively, by Gregor.[38] Garry et al.[25] found average dietary Fe intakes of 15.2 and 12.4 mg for free-living elderly men and women, respectively, in Albuquerque. Fe density was 7.2 and 7.7 mg/1000 kcal for men and women, respectively, which agrees with data from the HANES II survey.[3]

Data from a U.S. Department of Agriculture Food Consumption Survey carried out in 48 states were based on the average of 3 days intake.[3] Approximately 90% of men and women over the age of 50 consumed 70% or more of the RDA for Fe. These data suggest that the overall average intake of dietary Fe in the older population is adequate.

Lynch et al.[3] have presented data relevant to possible changes in the bioavailability of Fe in the diets of the elderly. Sources of dietary Fe intake in the elderly surveyed in the HANES II survey have been summarized and compared to data for younger age groups. As these authors point out, men and women aged 35 to 44 derived 26.4 and 21.3%, respectively, of their Fe from meat, whereas this was the source of only 18.7 and 14.3% in the elderly (65 to 74 years) men and women, respectively. A concomitant rise in the proportion of dietary Fe supplied by breakfast cereals occurs: from 5.6 to 15.9% in men and 7.3 to 14.8% in women. Mean daily heme Fe consumption falls from 2.5 to 1.8 mg in men and from 1.6 to 1.2 mg in women. Ascorbic acid intake was unchanged in the older vs. younger age groups. Garry et al.[25] point out that the elderly group of men and women surveyed in New Mexico consumed substantial quantities of both meat and ascorbic acid. The significance of these dietary changes to food Fe bioavailability in the elderly has not been evaluated.

### 2. Assessment of Iron Status

The progressive stages of Fe deficiency are well documented and it is recognized that anemia is a late finding. The reduction in hemoglobin concentration is first preceded by depletion of Fe stores, which can be measured by determining bone marrow Fe and serum ferritin concentration.[39] Evidence of more severe Fe deficiency includes a rise in red cell-free erythrocyte protoporphyrin, a fall in transferrin saturation, and a decrease in mean red cell volume.[39] Once Fe deficiency is sufficiently severe, a fall in hemoglobin or hematocrit occurs.[39]

The question of an age effect on serum Fe concentration arose following the report of Perrie[40] that there were lower Fe levels in elderly vs. young adults. However, these findings have not been confirmed by other authors.[25] In fact, Garry et al.[25] found higher serum Fe and serum ferritin levels in elderly vs. young subjects.

Serum ferritin levels increase if inflammation is present and may not accurately reflect body Fe stores.[41] This fact emphasizes the point that health status should be evaluated when measuring Fe parameters in the elderly.

Numerous problems arise when it is assumed that measuring hemoglobin concentration is also the best method for detecting nutritional Fe deficiency and determining its prevalence. Monitoring hemoglobin concentration is the most practical way of assessing the response to Fe therapy once deficiency is established. Inherent problems associated with interpreting hemoglobin standards are discussed later in this chapter.

The problems in data interpretation that arise when the prevalence of Fe deficiency is equated with anemia are illustrated in numerous surveys, as has been previously discussed by Lynch et al.[3] For example, in the Ten-State Nutrition Survey (TSNS), "abnormal" hemoglobin levels were much more common in men than women.[30] It is unlikely that Fe deficiency was responsible for a prevalence rate that was so much higher in men than in women. This conclusion is supported by the low level of correlation between both serum Fe and percent transferrin saturation and hemoglobin concentration. The HANES I survey found approximately three times the prevalence of low hemoglobin values in blacks over the age of 60 years than for whites.[42] However, less than 5% of blacks had low serum Fe concentrations or percentage transferrin saturation levels, and in general the percentage of abnormal values was higher in whites.

Data from smaller studies in various parts of the U.S. illustrate the fact that Fe deficiency appears to be uncommon in the elderly. Bailey et al.[43] found a 14% incidence of anemia (hemoglobin [Hb] $<12$ g/d$\ell$) in an elderly, predominately black, population from low-income households in Florida. However, no subject had a serum iron level below 50 µg/d$\ell$ or a transferrin saturation of less than 15%, which led to the conclusion that there was no evidence of Fe deficiency in these elderly subjects. In a study conducted with the elderly in rural Utah, 14% of men and 3% of women had low Hb levels and none were in the deficient range based on TSNS survey norms.[30,44] Only a small proportion had low serum Fe levels. In Corvallis, Ore., an assessment study of elderly men and women showed that while 20% of men had Hb values below 14 g/d$\ell$, only 1% of women had levels below 12 g/d$\ell$.[45] Serum Fe values were low in 4% of the men and none of the women, and transferrin saturations were low in only 1% of the women and none of the men — again suggesting that Fe deficiency was rare. The effect of oral Fe supplementation and food Fe fortification in anemic elderly subjects in the Boston area was reported by Gershoff et al.[46] Following the fortification trial, both control (no fortification) and fortification groups showed a 0.4 g/d$\ell$ rise in Hb. The administration of ferrous sulfate for 3 months to those whose Hb levels remained below 13 g/d$\ell$ was without effect.

The Fe status of a group (n = 280) of healthy elderly from Albuquerque was assessed and compared to that of younger adults (n = 271). Only 2 elderly men and 4 elderly women had low (16%) transferrin saturation levels. Ferritin levels were normal ($>12$ ng/m$\ell$) in all elderly individuals and higher than that of the healthy young adults. Extrapolation of data from surveys of hospitalized elderly has resulted in the erroneous conclusion that Fe deficiency anemia is a major problem in the elderly.[47,48] In these elderly patients, infection,

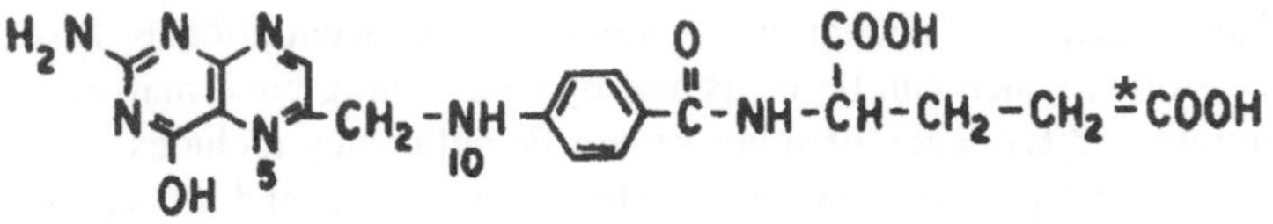

Pteridine     Para-amino     Glutamic
              benzoic Acid   Acid

*Site of attachment of extra glutamate residues

FIGURE 1.    Structural formula of folic acid (pteroylglutamic acid).

chronic diseases, and neoplasms may be the most frequent cause of anemia. Care should be taken in the clinical setting to correctly diagnose the cause of the anemia.

The widespread use of Fe supplements by the elderly results from the misconception that Fe deficiency is prevalent and that supplemental Fe will improve one's health and sense of well being.[49] Studies carried out in both the U.S. and Sweden suggest that some elderly take large quantities of supplemental Fe.[50,51] The potential toxic side effects of large doses of supplemental Fe should be recognized.[52,53] Arbitrary Fe supplementation by the average Fe-repleted elderly individual is at the very least wasteful and unnecessary.

## III. FOLACIN

### A. Chemistry and Form in Food

Folacin is a generic term for this water-soluble vitamin that exists in many different forms. Pteroylglutamic acid (PteGlu) $C_{19}H_{19}N_7O_6$ (441.40 mol wt) is the synthetic pharmaceutical form of the vitamin. The major subunits of the molecule are the pteridine moiety linked by a methylene bridge to *para*-aminobenzoic acid, which is joined by peptide linkage to glutamic acid (Figure 1). Pteroylglutamic acid becomes biochemically active after it has undergone reduction to become tetrahydrofolate. In addition to being reduced, various one-carbon adducts may be linked to tetrahydrofolate at the N-5, N-10, or 5,10 position, and the number of glutamate residues varies. In almost all forms of dietary folacin, the double bonds on the pyrazine ring of the pteridine moiety are reduced, and a methyl or formyl group is attached at the N-5 or N-10 position.[54]

Our knowledge of folacin content of food is incomplete. Early reports of food folacin composition are considered invalid since the essentiality of adding ascorbic acid to the assay was not recognized. More recently, Perloff and Butrum[55] reported the folacin composition of a wide variety of foodstuffs. There is a current research emphasis on developing new and better analytical techniques to more accurately quantitate food folacin. However, much of the current inconsistency observed between laboratories may reflect inherent variability in the concentration of folacin in the same food source. In foods that have been well characterized in relation to their folacin content, most of the vitamin ($\approx 90\%$) is present as polyglutamates with five or more glutamate residues added in $\gamma$-carboxyl linkage to the single glutamic acid found in PteGlu.[56,57] Concentrated food sources include green leafy vegetables and liver. Legumes may also contribute significant amounts of folacin to the diet. Except for liver, animal foods are poor sources of folacin.

### B. Metabolism

#### 1. Absorption and Malabsorption

The process of folacin absorption requires a prior step of deconjugation to the mono-glutamate form by intestinal mucosal $\gamma$-carboxypeptidase (folacin conjugase).[56] Studies have defined two separate folacin conjugase activities in human jejunal mucosa: one located on

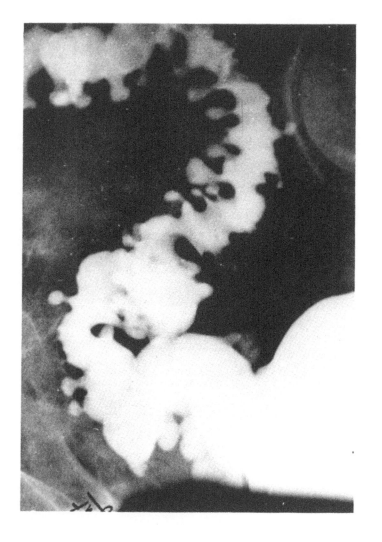

CHAPTER 9, FIGURE 2. An example of a barium enema with numerous diverticula in a narrowed area of spastic sigmoid colon is illustrated (From Sleisenger, M. H. and Fordtran, J. S., Eds., *Gastrointestinal Disease*, W. B. Saunders, Philadelphia, 1983. With permission.)

the brush border with a neutral pH optimum and the other located within the cell with an acid pH optimum.[58]

The absorption of PteGlu is maximal in the jejunum but negligible in the ileum. PteGlu is absorbed in physiological concentrations by an active carrier-mediated transport mechanism. This mechanism is pH dependent with an optimum at pH 6.[59] Drugs or mucosal disease may induce folacin malabsorption by decreasing the hydrolysis of polyglutamyl folacin or the transport of PteGlu. Several reports emphasize the effect of altered intraluminal pH on PteGlu absorption.[60,61] The widespread use of antacids and Cimetidine pose a potential risk to the aged population.[62] By elevating intestinal pH, these agents may adversely affect the pH-sensitive folacin transport system.[60,61]

The effect of alcohol on folacin absorption and metabolism is of concern in the elderly since it is now recognized that 2 to 10% of individuals over the age of 60 in the U.S. are alcoholics.[63,64] Between 15 and 20% of elderly patients admitted to general hospital wards or nursing homes have been found to be chronic alcoholics.[63] In a series of 121 consecutive patients admitted to two New York City hospitals over a 10-year period with megaloblastic anemia due to folacin deficiency associated with alcoholism, 17.4% were over the age of 60 and 9.2% were over 65.[62] Potential causes of folacin deficiency in alcoholism include poor diet, intestinal malabsorption, and defective metabolism due to associated liver disease. Malabsorption of radiolabeled folacin has been documented in two studies of malnourished alcoholics.[65,66] Data from one study suggest that PteGlu malabsorption in chronic alcoholism results from a synergistic effect of folacin deprivation and ethanol toxicity on transport processes in the mucosa.[67]

Drugs other than alcohol interfere with folacin absorption or utilization. The use of anticonvulsant drugs (phenytoin Dilantin®, barbiturates or primidone/Mysoline®) may lead to the development of folacin deficiency in individuals exposed to their action over a long period of time.[68] Sulfasalazine, which is used for the treatment of inflammatory bowel disease, impairs folacin absorption and may cause megaloblastic anemia.[69] The potential effect of these and other drugs on folacin absorption and metabolism in the aged is not clear.

Several groups have investigated the effect of aging per se on folacin absorption. Bhamthumnavin et al.[70] found no difference in the intestinal transport of radiolabeled monoglutamyl folacin in everted gut sacs prepared from rats of various ages. Since the majority of dietary folacin is in the polyglutamate form, the relevance of these data may be questioned. Hurdle and Williams[71] found no difference in the increase in serum folacin following an oral dose of monoglutamyl folacin in aged and young control human subjects. The data are difficult to interpret since tissues were not presaturated and no flushing dose was administered. Kasavan and Noronka[72] investigated the effect of age on mucosal conjugase activity in rats using an everted-sac technique. Conjugase enzyme activity was significantly lower in the aged animals and polyglutamyl folacin was absorbed less efficiently in the older animals. Baker et al.[73] reported that the rise in serum folacin following the oral ingestion of polyglutamyl folacin was significantly less in aged human subjects than the rise observed in young controls. In contrast, absorption of monoglutamyl folacin was comparable in both age groups as was the absorption of other water-soluble vitamins.

Bailey et al.[74] recently compared the absorption of both radiolabeled mono- and polyglutamyl folacin between healthy young and aged subjects using an in vivo triple-lumen perfusion technique. Conjugase enzyme activities were compared following jejunal biopsy in both age groups. No difference was found in the luminal disappearance, urinary recovery, or conjugase enzyme activities between age groups. It was concluded that age per se had no effect on folacin absorption.

Chronic folacin deficiency results in changes in the intestinal epithelial cells similar to the megaloblastic changes in the bone marrow.[75,76] Atrophy of the intestinal villi and impairment of the enzymatic functions of the cells have also been reported.[77] Once these morphological and functional changes in the intestinal epithelium have occurred, the ability of the intestine to absorb folacin is impaired resulting in further depletion of the vitamin. Elsborg[78] demonstrated in elderly subjects that folacin malabsorption was reversible following folate supplementation in folacin-deficient elderly subjects.

## 2. Transport, Storage, and Excretion

Folacin circulates in plasma as free folacin or folacin bound to either high- or low affinity binders. Low-affinity binding does not follow saturation kinetics and is a nonspecific property of many different plasma proteins. The second binder has a high affinity ($\approx 67\%$ saturation), but transports only about 5% of serum folacin.[79] These high-affinity binders are thought to be glycoproteins with a molecular weight of approximately 40,000, and are probably granulocyte derived, but modified by liver and kidney. Folacin is taken up by the body cells in a manner suggestive of active transport. Total body folacin stores are in the range of 5 to 10 mg, of which approximately one half is in the liver. The folacin content of bile is approximately five times that of serum. Enterohepatic recirculation is an important factor in conserving the body pool of folacin.

Results of an autopsy survey of liver folacin concentrations in Canada suggest that liver folacin concentrations are not lower in the elderly.[80] The prevalence of low ($<5$ $\mu g/g$) hepatic folacin concentrations in the individuals above the age of 70 was not different from that of subjects aged 20 to 60 years.

Normal urine in humans contains reduced folacin compounds which have been identified tentatively as 10-formyl-tetrahydrofolate (THFA), 5,10-methenyl THFA, and folinic acid (5-formyl THFA).[81] Folate breakdown products in the urine include a pteridine, pterin-6-aldehyde, and *p*-acetamidobenzolglutamate, which are products of normal folate catabolism.[81] Due to large quantities of folacin synthesized by bacteria in the colon, fecal excretion is not considered a reliable index of folacin metabolism.

## 3. Function

Folacin coenzymes are involved in numerous reactions that involve the transfer of one-carbon units.[79] The reactions include (1) pyrimidine nucleotide biosynthesis (methylation of deoxyuridylic acid to thymidylic acid), (2) purine synthesis, (3) amino acid interconversions including the catabolism of histidine to glutamic acid, interconversion of serine and glycine, and conversion of homocysteine to methionine, (4) generation of formate into formate pool and formate utilization, and (5) methylation of small amounts of transfer RNA. The metabolic interrelationship between folacin and vitamin $B_{12}$ may explain why the hematologic changes that result from a single deficiency of either vitamin are indistinguishable. Both folate and vitamin $B_{12}$ are required for the formation of 5,10-methylene THFA and involved in thymidylate synthesis by way of a vitamin $B_{12}$-containing enzyme. This reaction involves the removal of a methyl group from methyl folate and the delivery of it to homocysteine for the synthesis of methionine. The megaloblastic changes occurring in the bone marrow cells as well as other replicating cells in the body in either a folacin or $B_{12}$ deficiency result from lack of adequate 5,10-methylene THFA. The 5,10-methylene THFA delivers its methyl group to deoxyridylate to convert it to thymidylate for incorporation into DNA.

## 4. Requirement and Allowance

Herbert[82] has determined the requirement of folacin to be about 50 $\mu g$. This was the quantity of synthetic folic acid (PteGlu) that, when given orally or parenterally to patients with folacin deficiency uncomplicated by other diseases, will produce a return to hematologic normality. In a folacin deficiency the estimated daily loss of liver folacin is 48 to 158 $\mu g$/day. A similar rate of loss may occur in other tissues.[81] The folacin requirement is considered by the National Academy of Sciences as the amount required to sustain the tissue

**Table 2**
**FOLACIN INTAKE DATA FOR ELDERLY FLORIDA AND**
**NEW MEXICO RESIDENTS**

| | Daily intake[a] | | % With dietary intake 200 µg | Ref. |
|---|---|---|---|---|
| | No. | µg/day | | |
| Urban low income (Dade County, Fla.) | 128 | 192 | 65 | 43 |
| Middle and high income (retirement community, North Fla.) | 40 | 217 | 45 | U[b] |
| Urban middle and high income (Gainesville, Fla.) | 30 | 250 | 37 | U[b] |
| Middle and high income (Albuquerque, N.M.) | | | | |
| Men | 125 | 240 | 37 | 50 |
| Women | 145 | 210 | 43 | 50 |
| Rural and urban low income (Alachua County, Fla.) | 173 | 184 | 61 | 37 |

[a]  Values represent mean values for Florida studies and median values for Albuquerque study.
[b]  U = Unpublished.

reserves.[29] The RDA is set at 400 µg for adults of all ages to compensate for bioavailability (25 to 50% of dietary folacin estimated to be available).[29]

## C. Folacin Status
### 1. Dietary Intake

There are not extensive data available on the dietary intake of folacin in the elderly. Major national food intake surveys, including those performed by the USDA and both the TSNS and HANES surveys, have not included calculated dietary folacin intakes. In Canada, the daily folacin intake of elderly (>65 years) men and women averaged 151 and 130 µg, respectively.[83] In the U.S., data from several recent regional surveys are of interest.[37,50] Bailey et al. found that the average folacin intake was higher in groups of elderly from middle- and upper-income households when compared to groups from low-income households (Table 2). Approximately 65% of a group of elderly from a densely populated low-income area in Dade County, Fla. consumed less than 200 µg of folacin each day. In contrast, approximately 40% of middle to upper-income groups in Florida and Albuquerque consumed less than 200 µg.[37,50]

Socioeconomic factors such as income, educational level, social interaction, shopping frequency, cooking, refrigeration, and storage facilities may particularly affect nutritional status, and in this case folacin intake. For example, the shopping frequency for low-income elderly living in Dade County, Fla. was less than twice per month. As determined by 24-hr food recall, only 17% of this group indicated that they consumed fresh vegetables and only 30% consumed citrus fruit.[43] Food preparation habits included boiling vegetables for several hours and discarding the cooking water. Folacin supplements were not taken by any subjects in the Florida studies in contrast to 30% of subjects in Albuquerque. Those taking supplements in the Albuquerque study had an average of 48% higher plasma folacin levels and 49% higher red blood cell folacin levels.

Estimates of dietary folacin intakes are low compared to the RDA regardless of socioeconomic status. However, intakes below the RDA standard in the middle to upper income groups were associated with blood folacin levels above the accepted norm. One explanation for this discrepancy is that folacin values in food-nutrient data bases is incomplete. This of course may result in a substantial underestimation of dietary intakes of folacin. Also, actual bioavailability of the vitamin from the diet may be somewhat higher than the percentage assumed in setting the RDA of 400 μg. Another possibility is that the current folacin RDA may be unrealistically high.

### 2. Assessment of Folacin Status

The progressive stages in the development of a folacin deficiency have been documented in the folacin depletion study of Herbert.[84] A reduction in serum folacin level is an early sign in the progression of a folacin deficiency, and levels less than 3 ng/mℓ are considered deficient. The normal range of serum folacin is reported to be 6 to 25 ng/mℓ and is easily influenced by recent dietary intake. A reduction in serum folacin is an early sign in the progressive stages of folacin deficiency which precedes the reduction in red cell folacin concentration and anemia. Red blood cell folacin concentrations are considered better indicators of folacin status since the concentration is reported to parallel that of liver stores of the vitamin. The concentration in red cells is considerably higher than in serum with a wide normal range (160 to 800 ng/mℓ). The standard cut-off used to define folacin deficiency using this parameter is 140 ng/mℓ.[86] Reduction in red cell folacin concentration precedes the development of megaloblastic anemia, which is a late finding in the progression of a folacin deficiency.[84]

The functional consequences of "low" blood levels of folacin and reduced body stores in the absence of anemia are unclear. Low blood folacin levels have been associated with poorer performance on two standardized mental tests. Goodwin et al.[87] reported that elderly with low blood folacin levels (bottom 5% of group) scored significantly lower than the rest of the population on the Halstead-Reitan categories test, a nonverbal test of abstract thinking ability, as well as on the Weschler memory test.[87] While not conclusive, such findings indicate a need for further study into the possible consequences of subclinical folacin deficiency.

There is very little data available at the present time regarding folacin status of the elderly in the U.S. In the TSNS, the percent of the population with serum and red cell levels below accepted norms was not reported.[30] In the HANES I survey, only folacin in serum was measured.[88] Of the 369 persons aged 65 to 74 years, approximately 6% had serum folacin levels below 3.0 ng/mℓ. Several smaller folacin status assessment studies have been conducted on selected elderly population subgroups in the U.S.[37,50 57,89] The biochemical data substantiate the conclusion that elderly from low-socioeconomic households are at particular risk of developing a folacin deficiency. In a densely populated urban poverty area of Florida (Miami), 60% of a predominately black group of elderly had red cell folacin concentration less than 140 ng/mℓ.[57] In a separate study of elderly from low-income households in Alachua County, Fla., Bailey and Wagner[37] found that 31% had deficient (<140 ng/mℓ) red cell levels. In contrast, Wagner and Bailey[89] reported that only 6% of a white elderly group living in a middle- and upper-income retirement community in Florida had deficient (<140 ng/mℓ) red cell folacin levels. A similar prevalence (6% of group with red blood cell level <140 ng/mℓ) was found in a highly educated, nutrition-conscious white elderly group in Albuquerque, N.M.[183]

Institutionalized elderly individuals may be at risk of folacin deficiency due to a number of factors including chronic diseases, multiple medications, and poor dietary intake. Baker et al.[90] reported a higher prevalence of low (<5 ng/mℓ) serum folacin levels in institutionalized vs. noninstitutionalized elderly (19.5 and 9%, respectively). Similar findings have been reported by numerous other investigators in countries other than the U.S.[91-95] Magnus et al.[96] reported that institutionalized elderly in Norway had significantly lower serum and red cell folacin levels than younger healthy persons. This finding was also reported by other investigators.[97,98] Numerous factors may account for this, including poor diet, drug use, and

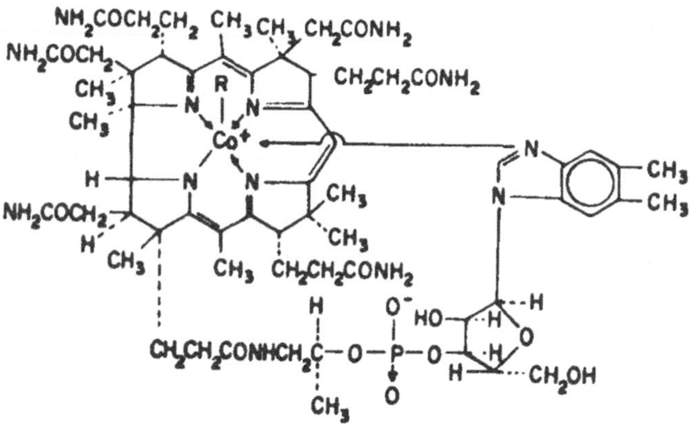

## Vitamin B₁₂

| R | Permissive Name |
|---|---|
| 5'-deoxyadenosyl | Adenosylcobalamin |
| $-CH_3$ | Methylcobalamin |
| $-OH$. | Hydroxocobalamin |
| $-CN$ | Cyanocobalamin |

FIGURE 2.   Structural formula of vitamin $B_{12}$.

chronic disease. Also, the presence of antibiotics in blood may interfere with the growth of *Lactobacillus casei* used in the microbiological assay of blood folacin and result in a false diagnosis of folacin deficiency.[97]

## IV. VITAMIN $B_{12}$

### A. Chemistry and Form in Food

Vitamin $B_{12}$ is a generic term for a series of cobalamin compounds (Figure 2). The principal forms of cobalamin in food are adenosylcobalamin and methylcobalamin, although some cobalamin is converted to hydroxycobalamin during the preparation of food.[99] Food sources of vitamin $B_{12}$ are almost exclusively of animal origin and the vitamin is ingested bound to dietary protein. The common pharmaceutical form of vitamin $B_{12}$ is cyanocobalamin. It is the most stable form and therefore the chemical form in which the vitamin is produced commercially. This form is converted to the nutritionally active forms in humans when given either orally or parenterally. There are three main cobalamins in humans: hydroxycobalamin, adenosylcobalamin, and methylcobalamin. The majority of cobalamin in the blood is methylcobalamin, whereas in the liver, adenosylcobalamin and hydroxycobalamin predominate.

### B. Metabolism

#### 1. Absorption and Malabsorption

In the stomach, acid and pepsin digest cobalamin from the protein with conflicting data on the release of cobalamin from protein in the presence of acid and pepsin.[99] Toskes[99] proposes that after cobalamin is released from its protein bond in the stomach, that it binds to R binders, which are nonintrinsic factor cobalamin-binding proteins found in many body secretions, including gastric. In the duodenum, pancreatic proteases partially degrade the R binders, releasing cobalamin, which then binds to intrinsic factor. The intrinsic factor-cobalamin complex attaches to a specific receptor in the ileum in the presence of Ca and a pH greater than 5.6. Passive absorption of massive doses of vitamin $B_{12}$ also takes place throughout the small intestine.

One physiologic change that may be associated with aging is a reduced ability to absorb vitamin $B_{12}$.[100,101] Some investigators[102,103] have reported diminished $B_{12}$ absorption with advancing age, while others[104,105] have found no decrease. The divergence of opinion appears to result from not screening the subjects for health status, therefore not separating the effect of aging per se from the effect of illness. McEvoy et al.[106] used a whole body retention method to assess the effect of age on $B_{12}$ absorption from "screened" normal elderly individuals. It was concluded that age per se had no effect on $B_{12}$ absorption.

Achlorhydria may be a problem in the aged. This reduction of stomach acid may lead to malabsorption of protein-bound cobalamin due to an inability to cleave cobalamin from its protein bond.[99] King et al.[107] have demonstrated a marked difference in absorption of protein-bound vitamin $B_{12}$ in achlorhydric patients compared to controls. Studies designed to determine the effect of age on vitamin $B_{12}$ absorption have used crystalline unbound cobalamin, which is absorbed normally in achlorhydric individuals.[102,104,106,108,109] These studies found no differences in vitamin $B_{12}$ absorption in the elderly vs. young controls. Further investigations in which absorption of the protein-bound vitamin is compared in different age groups are needed.

The most common cause of vitamin $B_{12}$ malabsorption and deficiency is a lack of intrinsic factor. This may be secondary to pernicious anemia or to surgical removal of the parietal cell containing part of the stomach. Two kinds of antibodies to intrinsic factor have been found in the blood and GI secretions of patients with pernicious anemia. Type I prevents the attachment of cobalamin to intrinsic factor and type II prevents the attachment of the intrinsic factor-cobalamin complex to the ileal receptor. The prevalence of pernicious anemia increases with age.[110]

Diseases and clinical conditions that adversely affect vitamin $B_{12}$ absorption may occur more frequently in the elderly.[110] In addition to total and partial gastrectomy and pernicious anemia, these conditions include disorders of the ileum such as gluten-induced enteropathy, regional ileitis, Crohn's disease, pancreatic insufficiency, intestinal resections, and bacterial overgrowth.[111-113]

Another factor that increases the risk for vitamin $B_{12}$ malabsorption in the elderly is the chronic use of drugs that interfere with $B_{12}$ absorption. para-Aminosalicylic acid (PAS), oral hypoglycemic agents of the biguanide group (phenformin, metformin, and buformin), neomycin, colchicine, and alcohol are the drugs which have been shown to interfere with cobalamin absorption.[99,114,115]

### 2. Transport, Storage, and Excretion

Vitamin $B_{12}$ is taken up by the ileal mucosa cell following attachment of the intrinsic factor-cobalamin complex. Transcobalamin II (TCII), a protein whose function is to transport vitamin $B_{12}$ in the circulation to various tissues of the body, binds to $B_{12}$ inside the ileal cell.[116] TCII is required for uptake of $B_{12}$ by virtually all cells in the body. Other $B_{12}$-binding proteins are collectively called R binders. Since two R binders in plasma are readily separable, the terms "TCI" and "TCIII" are used. A substantial amount of the cobalamin excreted in bile is reabsorbed, thus constituting an enterohepatic circulation for the vitamin.[116]

The total body content of vitamin $B_{12}$ is estimated to be 2 to 2.5 mg, the bulk of which is in the liver.[117] Hsu et al.[118] reported an upward trend in hepatic vitamin $B_{12}$ concentration associated with increased age of the subjects studied. This finding is not in agreement with the work of Swendseid et al.[82] and Halstead et al.,[119] who found no differences in vitamin $B_{12}$ liver stores in the various age groups.

Once plasma binders are saturated with vitamin $B_{12}$, the free vitamin $B_{12}$ in plasma is excreted by the kidney in the same way as insulin (i.e., it is filtered by the glomerulus and is neither excreted or absorbed by the renal tubules). The plasma clearance of $B_{12}$ may be used to measure the glomerular filtration rate. The average clearance of free vitamin $B_{12}$ is 80 to 132 m$\ell$/min.[120]

### 3. Function

Vitamin $B_{12}$ is required for the synthesis of thymidylate and therefore DNA synthesis and cell division are $B_{12}$-dependent.[79] An enzyme containing vitamin $B_{12}$ removes a methyl group from methyl folate and delivers it to homocysteine, which converts homocysteine to methionine and regenerated THFA, from which the 5,10-methylene THFA involved in thymidylate synthesis is made. Cobalamin acts as an intermediary carrier of the methyl group, i.e., methylcobalamin is a transient intermediate between methylfolate and homocysteine.

Adenosylcobalamin is required for the hydrogen transfer and isomerization whereby methylmalonate is converted to succinate.[79] Methylmalonic acid arises in man from propionic acid and from valine, isoleucine, and threonine. The reversible conversion of methylmalonyl-CoA to succinyl CoA involves the intramolecular migration of COSCOA. Vitamin $B_{12}$, therefore, is involved in both fat and carbohydrate metabolism. This fact has lead to speculation that one explanation for neurological damage of patients with vitamin $B_{12}$ deficiency is the inability to synthesize or maintain the lipoprotein myelin sheath.

Vitamin $B_{12}$ is involved in protein synthesis since it is required for methionine synthesis.[79] Methionine is essential for the metabolism of lipotropic substances such as choline and betaine; therefore, this may represent another role in lipid metabolism.

### 4. Requirement and Allowance

The Recommended Dietary Allowance (RDA) for vitamin $B_{12}$ is 3.0 μg and is the same for elderly as for young adults.[29] This is based on an estimated total body cobalamin pool size of 2.5 mg and a turnover time of 960 days (based on a half-life of 480 days) resulting in a daily loss of 2.6 μg.

## C. Vitamin $B_{12}$ Status Assessment
### 1. Dietary Intake

Vitamin $B_{12}$ deficiency from inadequate dietary intake of cobalamin is not a common problem in the U.S. Individuals at risk of inadequate dietary intake include complete vegetarians; however, the risk is reduced by the fact that cobalamin is efficiently reabsorbed from bile and the large body stores available.

The typical mixed diet in the U.S. probably supplies between 5 and 15 μg/day of vitamin $B_{12}$.[29,121] Since noninstitutionalized healthy elderly generally consume at least some animal products, it is unlikely that a dietary deficiency of vitamin $B_{12}$ would become a problem in this age group.

### 2. Assessment of Vitamin $B_{12}$ Status

It is important to recognize the distinction between the assay of cobalamin blood levels and cobalamin absorption tests. Cobalamin blood levels reflect the state of body cobalamin stores. The liver contains 2 to 5 mg of cobalamin and malabsorption must persist for 4 to 6 years before depletion of cobalamin is observed.[122] Vitamin $B_{12}$ absorption tests reflect what is happening now to cobalamin absorption. The Schilling test is the standard method used to evaluate cobalamin absorption.[123]

Serum vitamin $B_{12}$ levels can be measured radioisotopically and with microbiological assay. The normal serum vitamin $B_{12}$ level is very much a function of the method used in its measurement. The mean level is of the order of 450 pg/m$\ell$; the upper limit is of the order 1000 pg/m$\ell$. The definition of a lower limit is not clear-cut; however, the value of

200 pg/m$\ell$ is commonly used.[124] A population survey was recently conducted in Denmark to determine if age influenced the reference interval for plasma cobalamins. The data justifies the use of 200 pg/m$\ell$ as the lower limit for normal in the elderly.[125]

Mollin and Ross[126] reported that a group of elderly patients had a significantly lower mean B$_{12}$ level than the healthy hospital staff under 40 years of age. This finding is not surprising since the elderly patients were suffering from various disease conditions, including megaloblastic anemia. Some investigators have described a progressive fall in serum B$_{12}$ with advancing age.[125,127-129] Nexo[125] observed a decrease in the mean and a broadening of the reference interval for vitamin B$_{12}$ with age (age interval 20 to 80 years). In contrast, other researchers have not found any reduction in serum vitamin B$_{12}$ with age.[130-132] In fact, Meindock and Dvorsky[132] reported higher serum levels of vitamin B$_{12}$ in older age groups compared with young controls.

Vitamin B$_{12}$ tolerance tests involving measuring vitamin B$_{12}$ in the urine at specific intervals after parenteral administration of various doses of the vitamin have been conducted in the elderly.[133] Watkin et al.[133] reported that the elderly showed significantly less excretion or greater retention of the administered dose than young controls. These findings were not explained by differences in dietary intake or renal clearance. The meaning of these tolerance tests is unclear.

As discussed previously, studies have predominately shown that there is no significant variation of vitamin B$_{12}$ absorption with age.[102,104,108,109] In addition, the work of Swenseid et al.[130] and that of Halsted et al.[119] show that the liver content of B$_{12}$ is unaffected by age. Since the liver contains 90% of the vitamin B$_{12}$ in the body, it would appear that the "normal" elderly do not in fact have depleted B$_{12}$ stores. This conclusion agrees with the report by Adams and Boddy[134] that age had no effect on B$_{12}$ turnover.

In nutritional assessment surveys of the elderly, some assessment studies of elderly persons in Europe show a 15 to 20% incidence of low serum vitamin B$_{12}$ levels when either 150 or 200 pg/m$\ell$ were used as the criteria for low.[134-138] In contrast, other investigations have not observed decreased B$_{12}$ blood levels in the elderly.[139,140] Nutritional B$_{12}$ deficiencies in the healthy elderly in Europe, in the absence of strict vegetarianism, are considered rare.[141] There is very little data related to the vitamin B$_{12}$ status of elderly in the U.S. Bailey et al.[142] recently evaluated the vitamin B$_{12}$ status of 111 noninstitutionalized elderly persons (age range 60 to 87 years) living in an urban poverty area. The sample was predominately black (90 subjects); the rest were Spanish Americans. Serum vitamin B$_{12}$ levels were all normal (>200 pg/m$\ell$) and ranged from 226 to 1200 pg/m$\ell$. The findings indicate that vitamin B$_{12}$ deficiency was not a problem in this elderly population. Discrepancies in prevalences between various subgroups may reflect differences in methodologies as well as status.

## V. ANEMIA

### A. Prevalence and Diagnostic Approach

To appropriately treat anemia in any age group, a correct diagnosis must be made. Hematologic problems of the elderly have been the topic of many publications and clues to greater precision in the diagnosis have been documented.[143-147] The present discussion will focus on the differentiation between anemia due to Fe, folacin, or vitamin B$_{12}$ deficiencies. As previously discussed, anemia due to dietary iron inadequacy is not a common finding in the elderly. It is important to fully characterize the anemias and to suspect causes other than Fe deficiency. Anemia due to blood loss in the elderly is commonly misdiagnosed as dietary Fe-deficiency anemia and the source of iron loss not identified.

Anemia of chronic disease is also frequently confused with that of Fe deficiency. A number of chronic diseases are accompanied by a mild anemia due to a mild to moderate shortening of the erythrocyte life span and a compensatory bone marrow response. These anemias are usually associated with chronic infection (bacterial and fungal), malignancy (carcinoma and lymphoma), collagen and inflammatory diseases (rheumatoid arthritis, systemic lupus erythematosus, polymyalgia rheumatica, vasculitis, and regional enteritis), renal disease, and tissue injury.[148]

Folacin deficiency is a more common occurrence in the elderly than Fe deficiency, as previously discussed. Dietary inadequacy is thought to be the primary cause of the condition which is easily influenced by the socioeconomic status of the elderly individual. Vitamin $B_{12}$ deficiency most frequently results from pernicious anemia in the elderly. The differentiation of the anemia resulting from a deficiency of these two nutrients is difficult because the metabolic functions of the two vitamins are interrelated and the hematologic aberrations are the same for both deficiencies. It is imperative that a correct diagnosis be made since folacin therapy in a vitamin $B_{12}$-deficient patient may mask the real deficiency and allow the neurologic damage to progress.

A logical approach to the individual with anemia is to classify the condition according to red cell indexes as hypochromic microcytic, macrocytic, or normochromic normocytic. Examination of a stained peripheral blood smear will provide information regarding abnormalities in red cells and platelets. Red cell precursors in the bone marrow must attain a specific level of hemoglobin concentration before they cease to divide and are released into the peripheral circulation as mature red cells. Fe deficiency interferes with hemoglobin synthesis. This causes cytoplasmic maturation defects, resulting in an excessive number of cellular divisions before maturation of the cell and the consequent release of small, poorly pigmented cells. Folacin and vitamin $B_{12}$ deficiency cause nuclear maturation defects that are associated with delayed cellular division (but normal cytoplasmic maturation) and release of large cells.[149]

Classification of the anemia based on size can be done on the automated blood counter as mean corpuscular volume (MCV).[150] The average MCV in normal adults is 90 $\mu m^3$, with 80 $\mu m^3$ as the lower limit of normal. A low MCV requires determination of measurements of iron status. In typical Fe deficiency, the serum Fe is low, total Fe binding capacity (TIBC) is elevated, and percent transferrin saturation reduced to 15% or less.[151] Serum ferritin levels would also be reduced in Fe deficiency anemia unless there is a coexisting liver disease or inflammatory condition.[152]

At this point in the diagnosis, the blood smear should have been examined. If the cells are small and undercolored, the diagnosis in a majority of cases is Fe deficiency anemia. The remainder will prove to have thalassemia or one of the other conditions that may result in microcytic or hypochromic anemia.

The definitive diagnosis is the response to a therapeutic trial. Laboratory indexes return to normal following 3 months of therapeutic Fe therapy; however, at least 6 months are required to assure replenishment of body iron stores.

Anemia of chronic disease and Fe deficiency anemia may yield many similar laboratory test results. However, in contrast to Fe deficiency anemia, transferrin and TIBC are usually in the low range of normal or decreased and the serum ferritin value is high-normal or elevated.[148]

The explanation that Fe deficiency anemia is due exclusively to poor diet should never be accepted until investigation for a bleeding GI lesion has been completed. Testing for occult blood in the stool is essential when Fe deficiency anemia is suspected.[148] The conditions that frequently account for bleeding in the elderly and subsequent anemia are summarized by Walsh et al.[148]

If the MCV is above 100 $\mu m^3$ and the peripheral blood smear indicates megaloblastic anemia, folacin and vitamin $B_{12}$ deficiencies should be suspected. Some patients with elevated MCVs have macrocytic (not macroovalocytic) round (rather than oval) red cells that may be caused by a variety of factors including hypothyroidism, hypoplastic anemia, hemolytic anemia, sideroblastic anemia, neoplasia, liver disease, or some other disease often associated with reticulocytosis.[153] If the MCV and the peripheral blood smear indicate megaloblastic anemia, the next step is to test serum for vitamin $B_{12}$ and folacin concentrations and the whole blood for red blood cell folacin concentrations.

When blood levels of folacin and vitamin $B_{12}$ are determined microbiologically, it is important to screen for antibiotic use which may inhibit the growth of the microbe and result in a false diagnosis of folacin deficiency. Examination of a bone marrow aspirate will confirm megaloblastosis. The deoxyuridine suppression test and the reticulocyte count may be useful in differentiating between folate and vitamin $B_{12}$ deficiencies.[153] The reticulocyte count is commonly below 1% in a $B_{12}$ deficiency but 1.5 to 8% in a folate deficiency. When a vitamin $B_{12}$ deficiency is diagnosed, the Schilling test is the simplest method of determining if $B_{12}$ malabsorption is a problem and sorting out possible causes.

Surveys designed to determine the prevalence of anemia, Fe, folate, and vitamin $B_{12}$ deficiencies may differ from the clinical situation in which diagnostic work can be performed in individual patients. However, with the advent of automated procedures, large-scale screening and characterization of anemia is possible.

## B. Hematologic Standards

The prevalence of anemia in the elderly is dependent on which criteria is used to define the condition. Numerous surveys[154,171] support the conclusion that hemoglobin levels fall slightly in the elderly and the sex difference is diminished. The suggestion that age-adjusted hematologic norms should be used when evaluating the elderly is supported by the data of Lipschitz et al.[172] However, evidence to support the use of similar standards for hematologic normality in the aged as in younger subjects is convincing.[25,173,174] Freedman and Marcus concluded from a study done in a group of 292 elderly that it would be inadvisable to establish new geriatric norms. This is based on the fact that the current "normal values" for erythrocyte parameters in use in clinical laboratories is broad enough to include more than 80% of an unselected population of elderly.[173] As pointed out by Walsh,[174] borderline anemia in the elderly is often incorrectly attributed to the aging process and may result in a failure to detect treatable conditions. It is recommended that anemia in the elderly not be approached as an inevitable consequence of aging.

Various criteria for defining anemia have been published and it is recognized that there is overlap of normal values and values indicative of anemia. When a single cut-off point is selected, a large number of false negative and false positive results may occur.[175]

Race is an important factor that may complicate the interpretation of hemoglobin levels in the elderly. Blacks have a lower hemoglobin value than whites even when matched for economic status. The difference, which ranges between 0.3 and 1 g, cannot be explained by the inclusion of subjects with inherited abnormalities of hemoglobin and persists in the elderly.[176-178]

It should be recognized that latent or chronic diseases are known to affect hematologic parameters, especially hemoglobin. This is especially pertinent in elderly population groups in which diseases are more prevalent than in the younger age groups.

Although hemoglobin concentration may not be the best criterion with which to evaluate the prevalence of anemia in the elderly, it is the one most often measured and the basis for comparison between studies. Assuming the same cut-off points, differences in prevalence of anemia may reflect differences in health status and racial composition of the population groups. To illustrate, data from several regional and one national survey can be compared using the same cut-off point for anemia (<12.0 g/dℓ for women and <14.0 g/dℓ for men). In a population group carefully screened for normal health status in Albuquerque, none of the females and 2.3% of the males were anemic. In the national HANES I survey, 4 and

10% of white elderly (65 to 74 years) women and men, respectively, were classified as anemic.[179] Anemia was identified in 8 and 33% of white elderly women and men, respectively, in Missouri.[180] Fisher et al.[181] reported a 3 and 14% prevalence of anemia in white elderly women and men in Utah. Respectively, 14 and 42.5% of predominately white elderly women and men were classified as anemic in a New York study.[182] Lipschitz et al.[172] found that 11.0 and 20% of white elderly women and men were anemic in Arkansas. In addition, in a racially mixed group in the Arkansas study, 21.4 and 34% of elderly females and males were anemic. The higher prevalence of anemia in the males vs. females may simply reflect the use of a higher cut-off point (i.e., Hb $< 14$ g/d$\ell$). In that the sex difference in hemoglobin concentration is reduced in the elderly, it would appear more reasonable to use 12 g/d$\ell$ as the cut-off for both sexes.[154-171] Bailey et al.[43] reported a 14% prevalence of anemia (Hb $<$ 12 g/d$\ell$) in a predominately black population group of males and females.

Socioeconomic factors should not be ignored when comparing prevalences of anemia between population groups. For example, the prevalences of anemia in elderly subjects from low-income households in Arkansas and Florida were higher than that found in elderly from high-income households in Albuquerque.[25,57,172] Poor socioeconomic status in elderly groups reduces the amount of resources available to ensure adequate nutrient intake as well as preventive health care costs.

It is important to separate reports of prevalences of anemia in hospitalized elderly patients from those in free-living population groups. The recognized influence of factors such as chronic disease, neoplasms, blood loss, and drug therapy on the development of anemia in the elderly prevents a true estimate of the prevalence of nutritional anemia in the hospital setting.

## VI. SUMMARY

This chapter presented current knowledge related to the nutritional anemias in the elderly. The widely held belief that Fe deficiency anemia is common in this age group is incorrect. Dietary inadequacy of iron in this age group is rare and is seldom responsible for the development of Fe deficiency in the elderly. Caution is urged when equating anemia with Fe deficiency. Other etiological factors, such as blood loss, disease, and neoplasms, must be evaluated.

Folacin status of the elderly is associated with their socioeconomic level. Unlike Fe status, folacin status may be compromised by reductions in caloric intake and by poor food selection and preparation techniques associated with low socioeconomic environments. Subclinical folacin inadequacy in this age group should be evaluated.

Vitamin $B_{12}$ deficiency in the elderly is generally associated with malabsorption and age-related changes. The diagnosis of vitamin $B_{12}$-deficiency anemia and its differentiation from anemia due to folacin-deficiency precede tests for malabsorption disorders.

Since anemia can be a very debilitating condition in the elderly, it is important to understand how to prevent and to treat this condition. It is recognized that many factors, in addition to nutrient deficiencies, contribute to the development of anemia in this age group. Accurate diagnosis is essential so that the anemia can be treated effectively and thus enhance the quality of life of the aged individual.

# REFERENCES

1. Iron-Interactions, Measurements, and Bioavailability in Foods, Symposium at the IFT Meetings, 1983, *Food Technol.*, 37, 115, 1983.
2. **Monsen, R. R. and Balintfy, J. L.,** Calculating dietary iron bioavailability refinement and computerization, *J. Am. Dietet. Assoc.,* 80, 307, 1982.
3. **Lynch, S. R., Finch, C. A., Monsen, E. R., and Cook, J. D.,** Iron status of elderly Americans, *Am. J. Clin Nutr ,* 36, 1032, 1982.
4. **Bothwell, T. H., Charlton, R. W., Cook, J. D., and Finch, C. A.,** *Iron Metabolism in Man,* Blackwell Scientific, Oxford, 1979.
5. **Cook, J. D., Layrisse, M., Martinez-Torres, C., Walker, R., Monsen, R., and Finch, C. A.,** Food iron absorption measured by an extrinsic tag, *J. Clin. Invest.,* 51, 805, 1972.
6. **Layrisse, M., Martinez-Torres, C., and Roche, M.,** Effect of interaction of various foods on iron absorption, *Am. J. Clin. Nutr.,* 21, 1175, 1968.
7. **Bennett, J. D., Hagedorn, A. B., and Owen, C. A.,** A quantitative method for measuring the gastrointestinal absorption of iron, *Blood,* 15, 36, 1960.
8. **Freiman, H. D., Tauber, S. A., and Tulsky, E. G.,** Iron absorption in the healthy aged, *Geriatrics,* 18, 716, 1963.
9. **Jacobs, A. M. and Owen, G. M.,** The effect of age on iron absorption, *J. Gerontol.,* 24, 95, 1969.
10. **Marx, J. J. M.,** Normal iron absorption and decreased red cell iron uptake in the aged, *Blood,* 53, 204, 1979.
11. **Loh, T. T. and Chang, L. L.,** Hepatic iron status in Malaysians and Singaporeans, *Southeast Asian J. Trop. Med. Public Health,* 2, 131, 1980.
12. **Charlton, R. W., Hawkins, D. M., Mavor, W. O., and Bothwell, T. H.,** Hepatic storage iron concentrations in different population groups, *Am. J. Clin. Nutr.,* 23, 358, 1970
13. **Sturgeon, P. and Shoden, A.,** Total liver storage iron in normal populations of the USA, *Am. J. Clin. Nutr.,* 24, 469, 1971.
14. **Weinfeld, A.,** Storage iron in man, *Acta. Med. Scand.,* 177(Suppl. 427), 1, 1964.
15. **Shah, B. and Belonje, B.,** Liver storage iron in Canadians, *Am. J. Clin. Nutr.,* 29, 66, 1976.
16. **Celada, A., Herreros, V., and DeCastro, S.,** Liver iron storage in Spanish aging population, *Am. J. Clin. Nutr.,* 33, 2662, 1980.
17. **Gautier de Defaix, H., Puente, R., Vidal, B., Perenz, B., and Vidal, H.,** Liver storage iron in normal population in Cuba, *Am. J. Clin. Nutr.,* 33, 133, 1980.
18. **Benzie, R. McD.,** The influence of age upon the iron content of bone marrow, *Lancet,* 1, 1074, 1963.
19. **Burkhardt, R.,** Iron overload of bone marrow and bone, in *Iron Metabolism and Its Disorders,* Kief, H., Ed., Excerpta Medica, Amsterdam, 1975, 264.
20. **Cook, J. D., Finch, C. A., and Smith, N.,** Evaluation of the iron status of a population, *Blood,* 48, 449, 1976.
21. **Valberg, L. S., Sorbie, J., Ludwig, J., and Pelletier, O.,** Serum ferritin and the iron status of Canadians, *Can. Med Assoc. J.,* 14, 417, 1976.
22. **Leyland, M. J., Harris, H., and Brown, P. J.,** Iron status in a general practice and its relationship to morbidity, *Br. J Nutr.,* 41, 291, 1970.
23. **Qvist, E., Norden, A., and Olofsson, T.,** Serum ferritin in the elderly, *Scand. J. Clin. Lab. Invest.,* 40, 609, 1980.
24 **Casale, G., Bonora, C., Migliavacca, A., Zurita, I. E., and deNicola, P.,** Serum ferritin and ageing, *Age and Ageing,* 10, 119, 1981
25. **Garry, P. J., Goodwin, J. S., and Hunt, W. C.,** Iron status and anemia in the elderly: new findings and a review of previous studies, *J. Am. Geriatr. Soc.,* 31, 389, 1983.
26. **Finch, C. A.,** Body iron exchange in man, *J. Clin. Invest.,* 38, 392, 1959.
27. **Green, R., Charlton, R. W., Steflel, H., Bothwell, T., Mayet, F., Adams, B., Finch, C., and Layrisse, M.,** Body iron excretion in man. A collaborative study, *Am. J. Med.,* 45, 336, 1968.
28. **Underwood, E. J.,** Iron, in *Trace Elements in Human and Animal Nutrition,* 4th ed., Academic Press, New York, 1977, 13.
29. Food and Nutrition Board, in *Recommended Dietary Allowances,* National Academy of Sciences, Washington, D.C., 1980.
30. Ten-State Nutrition Survey 1969—70 V Dietary, DHEW Publ. No. (HSM) 72-8133, Center for Disease Control, U.S. Department of Health, Education, and Welfare, Atlanta, Ga., 1972.
31. **O'Hanlon, P. and Kohrs, M. B.,** Dietary studies of older Americans, *Am. J. Clin. Nutr.,* 31, 1257, 1978.
32. **Kohrs, M. B., O'Neal, R., Preston, A., Eklund, D., and Abrahams, O.,** Nutritional status of elderly residents in Missouri, *Am. J. Clin. Nutr.,* 31, 2186, 1978
33. **Yearick, E. S., Wang, M.-S. L., and Pisias, S. J.,** Nutritional status of the elderly: Dietary and biochemical findings, *J. Gerontol.,* 35, 663, 1980.

34. U S. Department of Agriculture, Science and Education Administration, Food and nutrient intakes of individuals in 1 day in the United States, Spring 1977, Preliminary Rep. No. 2, U S. Department of Agriculture, Washington, D.C., 1980.

35. U.S. Department of Agriculture, Science and Education Administration, Food and nutrient intake of individuals in 1 day in Hawaii, Winter 1978, Preliminary Rep. No 5, U.S. Department of Agriculture, Washington, D.C., 1981.

36. **Garry, P. J., Goodwin, J. S., Hunt, W. C., Hooper, E. M., and Leonard, A. G.,** Nutritional status in a healthy elderly population: dietary and supplemental intakes, *Am. J. Clin. Nutr.,* 36, 319, 1982.

37. **Bailey, L. B., Wagner, P. A., Jernigan, J. A., Nickens, C., and Brazzi, G.,** Folacin and iron status of an elderly population, *Fed Proc Fed. Am Soc Exp. Biol ,* 41, 951, 1982

38. **Greger, J. L.,** Dietary intake and nutritional status in regard to zinc of institutionalized aged, *J. Gerontol.,* 32, 549, 1977.

39. **Cook, J. D.,** Clinical evaluation of iron deficiency, *Semin Hematol.,* 19, 6, 1982.

40. **Pirrie, R.,** The influence of age upon serum iron in normal subjects, *J. Clin. Pathol.,* 5, 10, 1952.

41. **Lipschitz, D. A., Cook, J. D., and Finch, C. A.,** A clinical evaluation of serum ferritin as an index of iron stores, *N. Engl. J. Med.,* 290, 1213, 1974.

42. Health and Nutrition Examination Survey United States, 1971—1974, Dietary Intake Source Data, DHEW Publ. No. (PHS) Hyattsville, Md., 1979, 79.

43. **Bailey, L. B., Wagner, P. A., Christaksis, G. J., Araujo, P. E., Appledorf, H., Davis, C. G., Masteryanni, J., and Dinning, J. S.,** Folacin and iron status and hematological findings in predominately black elderly persons from urban low-income households, *Am J. Clin. Nutr.,* 32, 2346, 1979.

44. **Fisher, S., Hendricks, D. G., and Mahoney, A. W.,** Nutritional assessment of senior rural Utahns by biochemical and physical measurements, *Am. J. Clin. Nutr.,* 31, 667, 1978.

45. **Yearick, E. S., Wang, M.-S. L., and Pisias, S. J.,** Nutritional status of the elderly: dietary and biochemical findings, *J. Gerontol.,* 35, 663, 1980.

46. **Gershoff, S. N., Brusis, O. A., Nino, H. V., and Huber, A. M.,** Studies of the elderly in Boston. I. The effects of iron fortification on moderately anemic people, *Am. J. Clin. Nutr.,* 30, 226, 1977.

47. **Harant, Z. and Goldberger, J. V.,** Treatment of anemia in the aged: a common problem and challenge, *J. Am. Geriatr. Soc.,* 28, 127, 1975.

48. **Lawson, I. R.,** Anemia in a group of elderly patients, *Gerontol. Clin.,* 2, 87, 1960.

49. **Hale, W. E., Stewart, R. B., Cerda, J. J., Marks, R. G., and May, F. E.,** Use of nutritional supplements in an ambulatory elderly population, *J. Am. Geriatr. Soc.,* 30, 401, 1982.

50. **Garry, P. J., Goodwin, J. S., Hunt, W. C., Hooper, E. M., and Leonard, A. G.,** Nutritional status in a healthy elderly population: dietary and supplemental intakes, *Am. J. Clin Nutr.,* 36, 319, 1982.

51. **Reizenstein, P., Ljunggren, G., Smedby, B., Agenas, I., and Penchansky, M.,** Overprescribing iron tablets to elderly people in Sweden, *Br. Med. J.,* 2, 962, 1979.

52. **Edwards, C. Q., Solnick, M. H., Kushner, J. P.,** Hereditary hemochromatosis: contributions of genetic analyses, *Prog. Hematol.,* 12, 43, 1981.

53. **Bothwell, T. H., Charlton, R. W., Cook, J. D., and Finch, C. A.,** *Iron Metabolism in Man,* Blackwell Scientific, Oxford, 1979.

54. **Butterworth, C. E., Santini, R., and Frommeyer, W. B.,** The pteroylglutamate components of American diets as determined by chromatographic fractionation, *J. Clin. Invest.,* 42, 1929, 1963.

55. **Perloff, B. P. and Butrum, R. R.,** Folacin in selected foods, *J. Am. Dietet. Assoc.,* 70, 161, 1977.

56. **Halsted, C. H.,** The intestinal absorption of folates, *Am. J Clin. Nutr.,* 32, 846, 1979.

57. **Huskisson, Y. J. and Retief, F. P.,** Folaatinhound van voedsel, *S. Afr. Med. J.,* 44, 362, 1970.

58. **Reisenauer, A. M., Krumdieck, C. L., and Halsted, C. H.,** Folate conjugase: two separate activities in human jejunum, *Science,* 198, 196, 1977.

59. **Strum, W. B.,** Characteristics of the transport of the pteroylglutamate and amethopterin in rat jejunum, *J. Pharm. Exp. Ther.,* 216, 329, 1981.

60. **Russell, R. M., Dhar, G. J., Dutta, S. K., and Rosenberg, I. H.,** Influence of intraluminal pH on folate absorption: studies in control subjects and in patients with pancreatic insufficiency, *J. Lab. Clin. Med.,* 93, 428, 1979

61. **MacKenzie, J. F. and Russell, R. I.,** The effect of pH on folic acid absorption in man, *Clin. Sci. Mol. Med.,* 51, 363, 1976.

62. **Rosenberg, I. H., Bowman, B. B., Cooper, B. A., Halsted, C. H., and Lindenbaum, J.,** Folate nutrition in the elderly, *Am. J. Clin. Nutr.,* 36, 1060, 1982.

63. **Schuckitt, M. A. and Pastor, P. A.,** The elderly as a unique population: alcoholism, *Alcoholism Clin. Exp. Res.,* 2, 31, 1978.

64. **Zimberg, S.,** Alcohol and the elderly, in *Drugs and the Elderly,* Peterson, D. M., Whittington, F. J., and Payne, B. P., Eds., Charles C Thomas, Springfield, Ill., 1979, 28

65. **Halsted, C. H., Griggs, R. C., and Harris, J. W.,** The effects of alcoholism on the absorption of folic acid ($^3$H-PGA) evaluated by plasma levels and urine excretion, *J. Lab. Clin. Med.,* 69, 116, 1967.

66. **Halsted, C. H., Robles, E. A., and Mezey, E.,** Decreased jejunal uptake of folic acid ($^3$H-PGA) in alcoholic patients: roles of alcohol and nutrition, *N. Engl. J Med.,* 285, 701, 1971.

67. **Halsted, C. H., Robles, E. A., and Mezey, E.,** Intestinal malabsorption in folate deficient alcoholics, *Gastroenterology,* 64, 526, 1973.

68. **Waxman, S., Corcino, J. J., and Herbert, V.,** Drugs, toxins and dietary amino acids affecting vitamin $B_{12}$ on folic acid absorption or utilization, *Am. J. Med.,* 48, 599, 1970.

69. **Franklin, J. L. and Roseberg, I. H.,** Impaired folic acid absorption in inflammatory bowel disease: effects of salicylazosulfapyridine (azulfidine), *Gastroenterology,* 64, 517, 1973.

70. **Bhanthumnavin, K., Wright, J. R., and Halsted, C. H.,** Intestinal transport of tritiated folic acid ($H^3$-PGA) in the everted gut sac of different aged rats, *Johns Hopkins Med. J.,* 135, 152, 1974.

71. **Hurdle, A. D. F. and Williams, T. C. P.,** Folic acid deficiency in elderly patients admitted to hospital, *Br. Med. J.,* 2, 202, 1966.

72. **Kasavan, V. and Noronka, J. M.,** Folate malabsorption in aged rats related to low levels of pancreatic folyl conjugase, *Am. J. Clin. Nutr ,* 37, 262, 1983.

73 **Baker, H., Jaslow, S., and Frank, O.,** Severe impairment of dietary folate utilization in the elderly, *J. Am. Geriatr. Soc.,* 26, 218, 1978.

74. **Bailey, L. B., Cerda, J. J., Bloch, B. S., Busby, J., Vargas, L., Chandler, C. J., and Halstead, C. H.,** Effect of aging on folate absorption and conjugase enzyme activities in humans, *Fed. Proc. Fed. Am Soc. Exp. Biol.,* 43, 1984

75. **Bianchi, A., Chipman, D., Dreskin, A., and Rosensweig, N. S.,** Nutritional folic acid deficiency with megaloblastic changes in the small-bowel epithelium, *N. Engl. J. Med.,* 282, 859, 1970.

76. **Veeger, W., Tan Thije, O. J., Hellemans, N., Mandema, E., and Nieweg, H. O.,** Sprue with a characteristic lesion of the small intestine associated with folic acid deficiency, *Acta Med. Scand.,* 177, 493, 1965.

77. **Berg, N. O., Dahlquist, A., Lindberg, K., and Norden, A.,** Morphology dipeptidases and dissacharidases of small intestinal mucosa in vitamin $B_{12}$ and folic acid deficiency, *Scand J. Haematol.,* 9, 167, 1972.

78. **Elsborg, L.,** Reversible malabsorption of folic acid in the elderly with nutritional folate deficiency, *Acta Haematol.,* 55, 140, 1976.

79. **Herbert, V., Colman, N., and Jacob, E.,** Folic acid and vitamin $B_{12}$, in *Modern Nutrition in Health and Disease,* 6th ed., Goodhart, R S. and Shils, M. E., Eds., Lea & Febiger, Philadelphia, 1980, 229.

80. **Hoppner, K. and Lampi, B.,** Folate levels in human liver from autopsies in Canada, *Am. J. Clin. Nutr.,* 33, 862, 1980.

81. **Chanarin, I.,** Transport of folate, in *The Megaloblastic Anemias,* Blackwell Scientific, Oxford, 1979, 163.

82. **Herbert, V.,** Nutritional requirements for vitamin $B_{12}$ and folic acid, *Am. J. Clin. Nutr.,* 21, 743, 1968.

83. Nutrition Canada, Food Consumption Patterns Report, Department of National Health and Welfare, Ottawa, Canada, 1977.

84. **Herbert, V.,** Experimental nutritional folate deficiency in man, *Trans. Assoc. Am. Phys.,* 70, 307, 1962.

85. **Wu, A., Chanarin, I., Slavin, G., and Levi, A. J.,** Folate deficiency in the alcoholic — its relationship to clinical and haematological abnormalities, liver disease and folate stores, *Br. J. Haematol.,* 29, 469, 1975.

86. **Sauberlich, H. E., Dowdy, R. P., and Skala, J. H.,** Folacin, in *Laboratory Tests for the Assessment of Nutritional Status,* CRC Press, Cleveland, 1974, 49

87. **Goodwin, J. S., Goodwin, J. M., Garry, J. M., and Garry, P. J.,** Association between nutritional status and cognitive functioning in a healthy elderly population, *JAMA,* 249, 2917, 1983.

88. **Lowenstein, F. W.,** Nutritional status of the elderly in the United States of America, 1971—1974, *J. Am. Coll. Nutr.,* 1, 165, 1982.

89. **Wagner, P. A., Bailey, L. B., Krista, M. L., Jernigan, J. A., Robinson, J. D., and Cerda, J. J.,** Comparison of zinc and folacin status in elderly women from differing socioeconomic backgrounds, *Nutr. Res.,* 1, 565, 1981.

90. **Baker, H., Frank, O., Thind, I. S., Jaslow, S. P., and Louria, D. P.,** Vitamin profiles in elderly persons living at home or in nursing homes, versus profile in healthy young subject, *J. Am. Geriatr. Soc.,* 27, 444, 1979.

91. **Read, A. E., Gough, K. R., Pardoe, J. L., and Nicholas, A.,** Nutritional studies on the entrants to an old people's home, with particular reference to folic acid deficiency, *Br. Med. J.,* 2, 843, 1965

92. **Hurdle, A. D. F. and Picton Williams, T. C.,** Folic acid deficiency in elderly patients admitted to hospital, *Br. Med. J.,* 2, 202, 1966.

93. **Batata, M., Spray, G. H., Bolton, F. G., Higgins, G., and Wollner, L.,** Blood and bone marrow changes in elderly patients, with special reference to folic acid, vitamin $B_{12}$, iron and ascorbic acid, *Br. Med. J.,* 2, 667, 1967.

94. **Elwood, P. C., Shinton, N. K., Wilson, C. I. D., Sweetnam, P., and Franzer, A. C.,** Haemoglobin, vitamin $B_{12}$ and folate levels in the elderly, *Br. J. Haematol.,* 21, 557, 1971.

95. **Morgan, A. G., Kelleher, B. E., Losowsky, M. S., Droller, H., and Middleton, R. S. W.,** A nutritional survey in the elderly: haematological aspects, *Int. J. Vitam. Nutr. Res.*, 43, 461, 1973.

96. **Magnus, E. M., Bache-Wiig, J. E., Aanderson, T. R., and Melbostad, E.,** Folate and vitamin $B_{12}$ (cobalamin) blood levels in elderly persons in geriatric homes, *Scand. J. Haematol.*, 28, 360, 1982.

97. **Meindoks, H. and Dvorsky, R.,** Serum folate and vitamin $B_{12}$ levels in the elderly, *J. Am. Geriatr Soc.*, 18, 317, 1970.

98. **Webster, S. G. P. and Leeming, J. T.,** Erythrocyte folate levels in young and old, *J. Am. Geriatr. Soc.*, 27(Suppl. 10), 451, 1979.

99. **Toskes, P. P.,** Current concepts of cobalamin (vitamin $B_{12}$) absorption and malabsorption, *J. Clin. Gastroenterol.*, 2, 287, 1980.

100. **Chow, B. F., Rosen, D. A., and Lang, C. A.,** Vitamin $B_{12}$ serum levels and diabetic retinopathy, *Proc Soc. Exp. Biol. Med.*, 87, 38, 1954.

101. **Chow, B. F., Wood, R., Horonick, A., and Okuda, K.,** Agewise variation of vitamin $B_{12}'$ serum levels, *J. Gerontol.*, 11, 142, 1956.

102. **Glass, G. J. B., Goldbloom, A., Boyd, L. J., Laughton, R., Rosen, S., and Rich, M.,** Intestinal absorption and hepatic uptake of radioactive vitamin $B_{12}$ in various age groups and the effect of intrinsic factor preparations, *Am. J. Clin. Nutr.*, 4, 125, 1956.

103. **Chernish, S. M., Helmer, O. M., Fouts, P. J., and Kohlstaedt, K. G.,** The effect of intrinsic factor on the absorption of vitamin $B_{12}$ in older people, *Am. J. Clin. Nutr.*, 5, 651, 1957

104. **Tauber, S. A., Goodhart, R. S., Hsu, J. M., Blumberg, N., Kassab, J., and Chow, B. F.,** Vitamin $B_{12}$ deficiency in the aged, *Geriatrics*, 12, 368, 1957.

105. **Hyams, D. E.,** The absorption of vitamin $B_{12}$ in the elderly, *Gerontol. Clin.*, 6, 193, 1964.

106. **McEvoy, A. W., Fenwick, J. D., Boddy, K., and James, O. F. W.,** Vitamin $B_{12}$ absorption from the gut does not decline with age in normal elderly humans, *Age Ageing*, 11, 180, 1982.

107. **King, C. E., Leibach, J., and Toskes, P. P.,** Clinically significant vitamin $B_{12}$ deficiency secondary to malabsorption of protein-bound vitamin $B_{12}$, *Dig. Dis. Sci.*, 24, 397, 1979.

108. **Gaffney, G. W., Watkin, D. M., and Chow, B. F.,** Vitamin $B_{12}$ absorption: relationship between oral administration and urinary excretion of cobalt-60 labeled cyanocobalamin following a parenteral dose, *J. Lab. Clin. Med.*, 53, 525, 1959.

109. **Davis, R. L., Lawton, A. H., Prouty, R., and Chow, B. F.,** The absorption of oral vitamin $B_{12}$ in an aged population, *J. Gerontol.*, 20, 169, 1965.

110. **Hyams, D. E.,** The blood, in *Textbook of Geriatric Medicine and Gerontology*, 2nd ed., Brocklehurst, J. C., Ed , Churchill Livingstone, New York, 1978, 560.

111. **Toskes, P. P., Hansell, J., Cerda, J., and Deren, J. J.,** Vitamin $B_{12}$ malabsorption in chronic pancreatic insufficiency: studies suggesting the presence of a pancreatic "intrinsic factor", *N. Engl. J. Med.*, 284, 627, 1971.

112. **Giannella, R. A., Broitman, S. A., and Zamcheck, N.,** Competition between bacteria and intrinsic factor for vitamin $B_{12}$: implications for vitamin $B_{12}$ malabsorption in intestinal bacterial overgrowth, *Gastroenterology*, 62, 255, 1972.

113. **Filipsson, S., Hu Hen, L., and Lindstedt, G.,** Malabsorption of fat and vitamin $B_{12}$ before and after intestinal resection for Crohn's disease, *Scand J. Gastroenterol.*, 13, 529, 1978.

114. **Heinivaara, O. and Palva, I. P.,** Malabsorption of vitamin $B_{12}$ during treatment with *para*-aminosalicylic acid. A preliminary report, *Acta Med. Scand.*, 175, 469, 1964.

115. **Tomkin, G. H., Hadden, D. R., Weaver, J. A., and Montgomery, D. A. D.,** Vitamin $B_{12}$ status of patients on long-term metformin therapy, *Br. Med. J.*, 2, 685, 1971.

116. **Chanarin, I.,** Vitamin $B_{12}$ binding proteins, in *The Megaloblastic Anemias*, 2nd ed., Blackwell Scientific, Oxford, 1979, 59

117. **Linnell, J. C., Hofferand, A. V., Hussein, A., Wise, I. J., and Matthews, D. M.,** Tissue distribution of coenzyme and other forms of vitamin $B_{12}$ in control subjects and patients with pernicious anemia, *Clin. Sci. Mol. Med.*, 46, 163, 1974.

118. **Hsu, J. M., Kawin, B., Minor, P., and Mitchell, J. A.,** Vitamin $B_{12}$ concentrations in human tissues, *Nature (London)*, 210, 1264, 1966.

119. **Halsted, J. A., Carroll, J., and Rubert, S.,** Serum and tissue concentration of vitamin $B_{12}$ in certain pathologic states, *N. Engl. J. Med.*, 260, 575, 1959.

120. **Nelp, W. B., Wagner, H. N., and Reba, R. C.,** Renal excretion of vitamin $B_{12}$ and its use in the measurement of glomerular filtration rate in man, *J. Lab. Clin. Med.*, 63, 480, 1964.

121. **Chung, A. S. M., Pearson, W. N., Darby, W. J., Miller, O. N., and Goldsmith, G. A.,** Folic acid, vitamin $B_6$, pantothenic acid, and vitamin $B_{12}$ in human dietaries, *Am. J. Clin. Nutr.*, 9, 573, 1961.

122. **Heyssel, R. M., Bozian, R. C., Darby, W. J., and Bell, M. D.,** Vitamin $B_{12}$ turnover in man: the assimilation of vitamin $B_{12}$ from natural foodstuff by man and estimates of minimal daily dietary requirements, *Am. J. Clin. Nutr.*, 18, 176, 1966.

123. **Schilling, R. F.**, Intrinsic factor studies II The effect of gastric juice on the urinary excretion of radioactivity after the oral administration of radioactive vitamin B₁₂, *J. Lab Clin. Med*, 42, 860, 1953

124. **Chanarin, I.**, *The Megaloblastic Anemias*, 2nd ed., Blackwell Scientific, Oxford, 1979, 126.

125. **Nexo, E.**, Variation with age of reference values for P-cobalamins, *Scand. J. Haematol.*, 30, 430, 1983.

126. **Mollin, D. L. and Ross, G. I. M.**, The vitamin B₁₂ concentrations of serum and urine of normals and patients with megaloblastic anemias and other disease, *J. Clin. Pathol*, 5, 129, 1952.

127. **Gaffney, G. W., Horonick, A., Okuda, K., Meier, P., Chow, B. F., and Shock, N. W.**, Vitamin B₁₂ serum concentrations in 528 apparently healthy subjects of ages 12—94, *J Gerontol*, 12, 32, 1957.

128. **Boger, W. P., Wright, L. D., Strickland, S. C., Gylfe, J. S., and Ciminera, J. L.**, Vitamin B₁₂: correlation of serum concentrations and age, *Proc. Soc. Exp. Biol. Med.*, 89, 375, 1955.

129. **Chow, B. F., Wood, R., Horonick, A., and Okuda, K.**, Agewise variation of vitamin B₁₂ serum levels, *J. Gerontol.*, 11, 142, 1956.

130. **Swendseid, M. E., Hvolboll, E., Schick, G., and Halsted, J. A.**, The vitamin B₁₂ content of human liver tissue and its nutritional significance, *Blood*, 12, 24, 1957

131 **Batata, M., Spray, G. H., Bolton, F. G., Higgins, G., and Wollner, L.**, Blood and bone marrow changes in elderly patients with special reference to folic acid, vitamin B₁₂, iron and ascorbic acid, *Br Med. J.*, 2, 667, 1967

132. **Meindok, H. and Dvorsky, R.**, Serum folate and vitamin B₁₂ levels in the elderly, *J. Am. Geriatr. Soc*, 18, 317, 1970.

133. **Watkin, D. M., Lang, C. A., Shock, N. W., and Chow, B. F.**, Agewise differences in the urinary excretion of vitamin B₁₂ following intramuscular administration, *J. Nutr.*, 50, 341, 1953.

134. **Adams, J. F. and Boddy, K.**, Studies in cobalamin metabolism, in *The Cobalamins: A Glaxo Symposium*, Arnstein, H. R. V. and Wrighton, R. J., Eds., Churchill Livingstone, Edinburgh, 1971, 153.

135 **Munasinghe, D. R. and Pritchard, J. G.**, The relationship between mean corpuscular volume, serum B₁₂ and serum folate status in aged persons admitted to a geriatric unit, *Br. J. Clin Pract.*, 32, 16, 1978.

136. **Elsborg, L., Lund, V., and Bastrup-Madsen, P.**, Serum vitamin B₁₂ levels in the aged, *Acta Med. Scand.*, 200, 309, 1976

137. **Hughes, D., Elwood, P. C., Shinton, N. K., and Wrighton, R. J.**, Clinical trial of the effect of vitamin B₁₂ in elderly subjects with low serum B₁₂ levels, *Br. Med J*, 2, 458, 1970.

138. **Elwood, P. D., Shinton, N. K., Wilson, C. I. D., Sweetnam, P., and Frazer, A. C.**, Haemoglobin, vitamin B₁₂ and folate levels in the elderly, *Br J Haematol.*, 21, 557, 1971.

139. **Killander, A.**, The serum vitamin B₁₂ levels at various ages, *Acta Paediatr. Scand.*, 46, 585, 1957.

140. **Waters, W. E., Withey, J. L., Kilpatrick, G. S., and Wood, P. H. N.**, Serum vitamin B₁₂ concentrations in the general population: a ten-year follow-up, *Br. J. Haematol.*, 20, 521, 171

141. **Magnus, E. M., Bache-Wiig, J. E., Aanderson, T. R., and Melbostad, E.**, Folate and vitamin B₁₂ (cobalamin) blood levels in elderly persons in geriatric homes, *Scand J Haematol*, 28, 360, 1982.

142. **Bailey, L. B., Wagner, P. A., Christakis, G. J., Araujo, P. E., Appledorf, H., Davis, C. G., Dorsey, E., and Dinning, J. S.**, Vitamin B₁₂ status of elderly persons from urban low-income households, *J. Am. Geriatr Soc*, 28, 276, 1980

143. **Howe, R.**, Tips on diagnosing and treating anemia in the aging, *Geriatr Med*, December 29, 1979

144. **Kravitz, S. C.**, Anemia in the elderly, in *Clinical Aspects of Aging*, Reichel, W., Ed., Williams & Wilkins, New York, 1978, chap 30.

145. **Baisden, R.**, Hematologic problems of the elderly, *Hosp Pract*, January, 93, 1978.

146. **Howe, R. B.**, Anemia in the elderly. Common causes and suggested diagnostic approach, *Postgrad. Med.*, 73, 153, 1983.

147. **Christensen, D. J.**, Diagnosis of anemia, clues to greater precision, *Postgrad. Med.*, 73, 293, 1983.

148. **Walsh, J. R., Cassel, C. K., and Madler, J. J.**, Iron deficiency in the elderly: it's often nondietary, *Geriatrics*, 36, 121, 1981.

149. **Howe, R. B.**, Anemia in the elderly, *Postgrad. Med.*, 73, 153, 1983.

150. **Wintrobe, M. M.**, *Clinical Hematology*, Lea & Febiger, Philadelphia, 1981, 529.

151. **Cook, J. D., Finch, C. A., and Smith, N.**, Evaluation of the iron status of a population, *Blood*, 48, 449, 1976.

152. **Christensen, D. J.**, Diagnosis of anemia, *Postgrad. Med.*, 73, 293, 1983.

153. **Herbert, V.**, The nutritional anemias, *Hosp. Pract.*, March, 65, 1980.

154. **Cooper, W. M., Hieber, R. D., and Chapman, W. L.**, Anemia in the aged, *J. Am. Geriatr. Soc.*, 15, 568, 1967.

155. **Garby, L., Irnell, L., and Werner, I.**, Iron deficiency in women of fertile age in a Swedish community. Estimation of prevalence based on response to iron supplementation, *Acta Med. Scand.*, 185, 113, 1969.

156. **Cook, J. D., Alvarado, J., Gutnisky, J. M., Labardini, J., Layrisse, M., Linares, J., Lova, A., Maspes, V., Restrepo, A., Reynafarje, C., Sanchez-Medal, L., Velez, H., and Viteri, F.**, Nutritional deficiency and anemia in Latin America: a collaborative study, *Blood*, 38, 591, 1971.

157 **Jefferson, D. M., Hawkins, W. W., and Blanchaer, M. C.,** Haematological values in elderly people, *Can. Med. Assoc. J.,* 68, 347, 1953.

158. **Hobson, W. and Blackburn, E. K.,** Haemoglobin levels in a group of elderly persons living at home alone or with spouse, *Br. Med. J.,* 1, 647, 1953.

159. **Hawkins, W. W., Speck, E., and Leonard, V. G.,** Variation of the hemoglobin level with age and sex, *Blood,* 9, 999, 1954.

160. **Gillum, H. L. and Morgan, A. F.,** Nutritional status of the aging. I. Hemoglobin levels, packed cell volumes and sedimentation rates of 577 normal men and women over 50 years of age, *J. Nutr.,* 52, 265, 1955.

161. **Semmence, A.,** Anemia in the elderly, *Br. Med. J.,* 2, 1153, 1959.

162 **Natvig, H.,** Studies on hemoglobin values in Norway. I Hemoglobin levels in adults, *Acta Med. Scand.,* 173, 423, 1963.

163. **Vellar, O. D.,** Studies on hemoglobin values in Norway. IX. Hemoglobin, hematocrit and MCHC values in old men and women, *Acta Med. Scand.,* 182, 681, 1967.

164. **Smith, J. S. and Whitelaw, D. M.,** Hemoglobin values in aged men, *Can. Med. Assoc. J.,* 105, 816, 1971.

165. **Elwood, P. C.,** Epidemiological aspects of iron deficiency in the elderly, *Gerontol. Clin.,* 13, 2, 1971.

166. **Milne, J. S. and Williamson, J.,** Hemoglobin, hematocrit, leukocyte count, and blood grouping in older people, *Geriatrics,* 27, 118, 1972.

167 **Earney, W. W. and Earney, A. J.,** Geriatric hematology, *J. Am. Geriatr. Soc.,* 20, 174, 1972.

168. **Helman, N. and Rubenstein, L. S.,** The effect of age, sex, and smoking on erythrocytes and leukocytes, *Am. J. Clin. Pathol.,* 63, 35, 1975.

169. **Kelly, A. and Munan, L.,** Haematologic profile of natural populations: red cell parameters, *Br. J. Haematol ,* 35, 153, 1977.

170 **Dvilansky, A., Bar-Am, J., Nathan, I., Kaplan, H., and Galinsky, D.,** Hematologic values for healthy older people in the Negev area, *Isr. J. Med. Sci.,* 15, 821, 1979.

171. **Htoo, M. S. H., Kofkoff, R. L., and Freedman, M. L.,** Erythrocyte parameters in the elderly: An argument against new geriatric normal values, *J. Am. Geriatr. Soc.,* 27, 547, 1979.

172. **Lipschitz, D. A., Mitchell, C. O., and Thompson, C.,** The anemia of senescence, *Am. J. Hematol ,* 11, 47, 1981.

173. **Freedman, M. L. and Marcus, D. L.,** Anemia and the elderly: is it physiology or pathology?, *Am. J. Med. Sci.,* 280, 81, 1980.

174. **Walsh, J. R.,** Borderline anemia in old age, *Geriat. Med. Today,* 2(Suppl. 3), 25, 1983.

175. WHO Technical Report Series No. 503, *Nutritional Anemias,* Geneva, 1972.

176. **Garn, S. M., Smith, N. J., and Clark, D. C.,** Lifelong differences in hemoglobin levels between blacks and whites, *J. Natl. Med. Assoc.,* 67, 91, 1975.

177. **Garn, S. M., Smith, N. J., and Clark, D. C.,** The magnitude and implications of apparent race differences in hemoglobin values, *Am. J Clin. Nutr.,* 28, 563, 1975.

178. **Meyers, L. D., Habicht, J. P., and Johnson, C. L.,** Components of the difference in hemoglobin concentrations in blood between black and white women in the United States, *Am. J. Epidemiol.,* 109, 539, 1979.

179 Hemoglobin and Selected Iron-Related Findings of Persons 1—74 Years of Age: United States 1971—1974, PHS Publ. No. 46, Department of Health, Education, and Welfare, Hyattsville, Md., 1979.

180. **O'Neal, R. M., Abrahams, O. G., Kohrs, M. B., and Eklund, D. L.,** The incidence of anemia in residents of Missouri, *Am. J. Clin. Nutr.,* 29, 1158, 1976.

181. **Fisher, S., Hendricks, D. G., and Mahoney, A. W.,** Nutritional assessment of senior rural Utahns by biochemical and physical measurements, *Am. J. Clin. Nutr.,* 31, 667, 1978.

182. **Htoo, M. S., Kotkoff, R. L., and Freedman, M. L.,** Erythrocyte parameters in the elderly: an argument against new geriatric normal values, *J. Am. Geriatr. Soc.,* 27, 547, 1979.

183. **Bailey, L. B.,** Unpublished data.

*Index*

# INDEX

**Q**

**R**

**S**

**T**

Printed and bound by CPI Group (UK) Ltd, Croydon, CR0 4YY

22/10/2024

01777632-0006